Waste-to-Resources 2019

Matthias Kuehle-Weidemeier (ed.)

WASTE-TO-RESOURCES 2019

VIII INTERNATIONAL SYMPOSIUM MBT, MRF AND RECYCLING

Resources and Energy from Waste

Proceedings

14th -16th of May 2019

Organisers

www.icp-ing.de

wasteconsult
INTERNATIONAL

www.wasteconsult.de

Gold Sponsor

www.ume-ag.de

Silver Sponsor

www.sutco.de

Cuvillier Verlag Göttingen
Internationaler wissenschaftlicher Fachverlag

Bibliografische Information der Deutschen Nationalbibliothek
Die Deutsche Nationalbibliothek verzeichnet diese Publikation in der
Deutschen Nationalbibliografie; detaillierte bibliographische Daten sind im Internet
über http://dnb.d-nb.de abrufbar.
1. Aufl. - Göttingen: Cuvillier, 2019

© CUVILLIER VERLAG, Göttingen 2019
Nonnenstieg 8, 37075 Göttingen
Telefon: 0551-54724-0
Telefax: 0551-54724-21
www.cuvillier.de

ISBN 978-3-7369-7009-0
eISBN 978-3-7369-6009-1

Gold Sponsor:

www.ume-ag.de

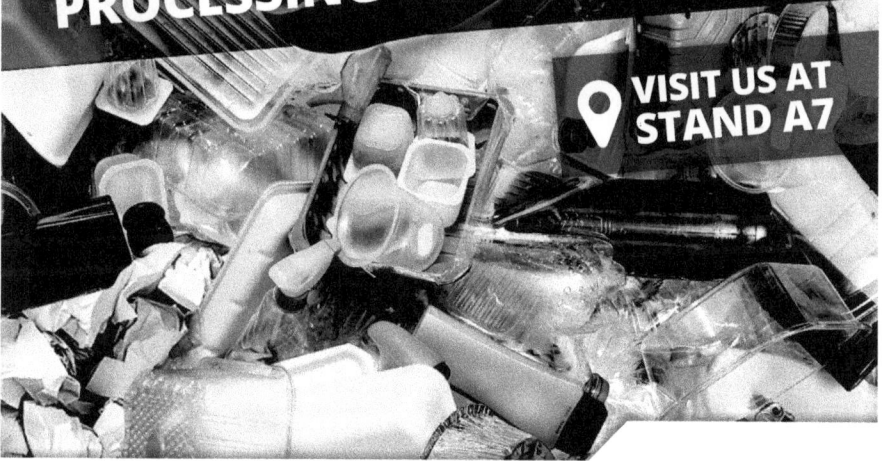

ICP Ingenieurgesellschaft
Prof. Czurda und Partner mbH

ICP - Ingenieurgesellschaft Prof. Czurda and Partner mbH

Ingenieurgesellschaft Prof. Czurda and Partner mbH is an independent and internationally recognized Engineering Consultancy based in Karlsruhe, Germany. ICP was founded in 1990 by Prof. Dr. Dr. Kurt Czurda and the managing directors Dr. Thomas Egloffstein and Gerd Burkhardt. Since then has grown in size with a staff of 25 at the head office in Karlsruhe and 50 based throughout Germany. Before 1995, ICP projects were predominately based in Germany, later ICP became increasingly engaged in international projects.

The ICP expertise covers all areas of design and engineering activities in the fields of waste treatment, (including landfill technologies, waste treatment and handling plant, recycling facilities, collection and storage and transfer stations etc.), contaminated ground and buildings, renovation and extension of existing buildings, environmental design, public works, construction quality assurance, geodesy, geo-technics, geology and hydro-geology.

A further core activity is to support public authorities world-wide in evaluating technical studies, technical offers and financial offers prepared by other suppliers in the framework of solid waste management projects. ICP has used its technical expertise to help Contracting Authorities from various countries (such as Algeria, Northern Cyprus, Qatar and Tunisia) to correctly evaluate and compare different technical solutions presented by bidders in international solid waste management tenders.

ICP has been contracted by private organisations as well as public authorities, international development banks, or development aid institutions (KfW, EIB, World Bank, GIZ etc.). We understand the requirements and needs of our private clients as well as the committee based budget systems of the public authorities and their corresponding decision making processes. We are familiar with the working methods of development agencies and investment institutions.

Our membership in the *"Verband Beratender Ingenieure"* (the German Association of Consulting Engineers) as well as numerous specialist organisations and working groups in Germany ensure the continuous professional development of our staff. Due to the involvement of the directors in several expert committees and working groups, ICP staff is permanently updated with the latest developments in their specialist fields.

The application of the most modern software also contributes to maintaining the highest quality standards with state of the art technology for ICP clients. An example: all design and modelling in ground works or for landfills is untaken with three dimensional modelling of ground profiles. This guarantees the highest degree of accuracy for the calculation of earth volume quantities.

ICP has implemented ISO 9001 standards to guarantee quality assurance of our services, through preliminary design to supervision of construction. The ISO 9001 registration aids the company in meeting the ever changing needs of you, our customer, with consistent quality.

Fields of activity and services provided

Solid Waste Management

- Preparation of solid waste plans, strategies, concepts, feasibility studies and economic analysis;
- Landfill technology (design and construction supervision of new landfills or landfill phased extensions, including sludge landfills);

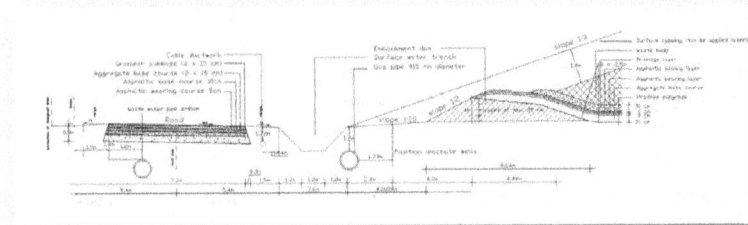

Figure 1: Example of a vertical section for a landfill in Muscat, Oman

- Landfill closure (landfill gas collection technology, leachate treatment, refurbishment of landfill drainage systems, site investigations and feasibility studies, monitoring etc.);
- Design and construction supervision of waste treatment plant (mechanical-biological pre-treatment, composting, fermentation, sorting and handling etc.); and
- Tender preparation and evaluation – technical and financial evaluation of offers submitted for solid waste management services, works and/ or supplies.

Contaminated site investigation and rehabilitation

- Contaminated site investigation (from historical records to technical sampling and ground water modelling);
- Clean up of contaminated sites (design and supervision of treatment methods such as Pump and Treat, Soil Vapor Extraction, Permeable Reactive Barrier, etc.);

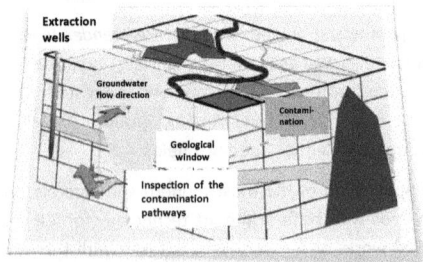

- Securing of contaminated sites / stabilization of existing hazards (design and supervision of treatment measures such as surface sealing, slurry wall containment, etc.); and
- Monitoring systems.

Renewable Energies

- Waste to energy (Biogas plants and RDF production plants);
- Landfill gas utilization; and
- Solar plants.

Planning of building decommissioning and demolition

- Site investigation / Groundwater management;
- Stability of deep excavations also in densely built areas;
- Controlled demolition;
- Investigation of hazardous materials;
- Measurement of emissions and imissions; and
- Surveys of current condition / Conservation of evidence.

Plant and working safety

- Health and safety plans;
- Health and safety coordination;
- Work place health and safety; and
- Plant safety.

Environmental planning

- Environmental and social impact assessment;
- Risk assessment;
- Environmental expert advice; and
- Environmental damage assessment.

Geotechnical, soil mechanics and physical investigations

- Building assessment and foundation recommendations;
- Stability calculations;
- Settlement calculations;
- Execution of field tests and probing;
- Specification and supervision of boreholes;
- Disturbed and undisturbed sampling;
- Soil mechanical tests according to DIN ISO und ASTM;
- Soil physical tests according to DIN ISO und ASTM;
- Development of sealing materials from residues; and
- Testing of construction materials before being used as landfill sealing materials.

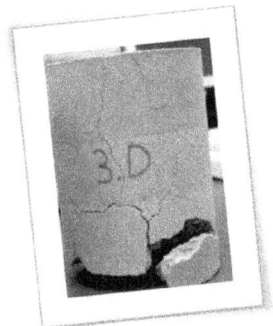

Geology und Hydrogeology

- Expert opinions;
- Ground water measurements;
- Drinking water protection areas;
- Execution of field tests and probing;
- Specification and supervision of boreholes; and
- Borehole probe, analysis and review.

Geodesy

- Terrestrial surveys;
- Settlement monitoring; and
- Inclinometer measurement.

Knowledge Transfer

- Organisation of symposiums and conferences;
- Training for professionals and local authorities; and
- Capacity building.

International Experience

ICP has been implementing international projects **since 1991**. The company has been providing waste management expertise to beneficiaries from **48 countries** on **4 continents**.

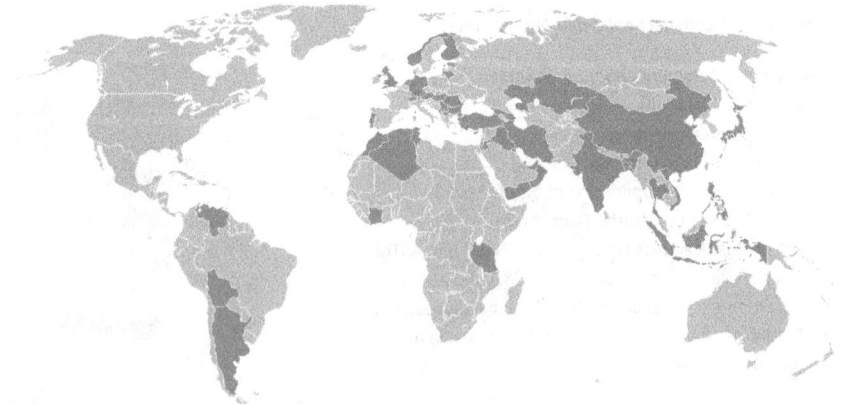

Figure 2: ICP's experience in other countries

Content

I   MBT and Recycling

II   Climate protection by circular economy

III   Current plants and system concepts

IV   Concepts for circular economy

V   Liquefaction and carbonation of waste

VI   Waste analysis

VII   Production, optimization and utilization of alternative fuels

VIII   Fire safety and occupational safety in waste treatment and recycling plants

IX   Organic waste fractions and anaerobic waste treatment

X   Biogas

XI   Lechate from waste treatment

XII   Plastics and other waste fractions

XIII   Processing and recycling of mineral waste

Note

The proceedings were arranged by ICP Ingenieurgesellschaft Prof. Czurda und Partner mbH with high accuracy. Nevertheless, errors cannot be fully excluded. ICP and the editors are not liable or responsible for the correctness of the information in this book. The authors take responsibility for their content.

If a brand or trade name was used, there might be trademark rights valid, even if they were not explicitly stated.

The proceedings contain complicated technical terms. ICP do not take any liability for the correctness of the translation. Please check the plausibility of the content you are reading.

Potential of waste management to achieve the recycling rates of the EU Waste Framework Directive

Bertram Zwisele, Carsten Böhm

ARGUS – Statistik und Informationssysteme in Umwelt und Gesundheit GmbH
(Statistics and Information Systems in Environment and Public Health)

Karl-Heinrich-Ulrichs-Straße 20a, 10785 Berlin, Germany

Abstract

Future waste streams and corresponding recycling rates are projected by use of a fore-cast model for the context of Germany. The model produces a forecast for the municipal waste amount based on the past waste generation, material compositions of the waste streams, economic (growth rates) and demographic (population) developments. Each waste stream is defined by a respective material composition that allows stream model-ling on the level of material categories (e.g. paper, glass, plastics, organic) in the waste stream (e.g. residual household waste) by defining growth rates for the amount of each material category. The contribution of each municipal waste stream to the new recycling targets is estimated by respective factors taken from DESTATIS data and own calcula-tions. In the chosen baseline scenario, the new WFD recycling targets will not be achieved. For this reason, further scenarios are defined with higher collection rates of certain waste material categories in the main municipal waste streams. This follows the approach to "divert" recyclable material from such municipal waste streams that contrib-ute low amounts to recycling according to the new WFD (e.g. residual household waste), towards those municipal waste streams that are related to high recycling contri-butions (e.g. separately collected paper, glass, organic). It is the objective to explicitly point out the great challenge to meet future recycling rates, to express the need of ap-propriate and validated data for reliable scenario definitions and to demonstrate the functionality of the model. This results into the question how to implement appropriate waste management strategies that will realize the necessary higher collection rates to reach the future recycling targets.

Keywords

WFD recycling targets, municipal waste streams, forecast model, collection rates, recy-cling rates, waste management strategy

1   Introduction

The European Commission has adopted changes to the Waste Framework Directive (WFD). This affects the existing reporting system in Germany and the methodology for determining recycling rates for municipal waste (EU, 2018).

At present, the determination of recycling rates in Germany is based on the measure-ment of material input into waste treatment plants. As a result of the revision of the Directive the measurement points for recycling will have to be changed in future. The Commission proposes to determine the recycled amount primarily on the basis of the input into final recycling plants or, under certain conditions, on the basis of the output of sorting plants. In addition, it is proposed to estimate recycling quantities by means of average waste specific loss rates.

The amendments to the WFD may have multiple and far-reaching consequences for the collection, processing and management of the statistical data and the legal framework. The municipal waste recycling target for 2020 remains unchanged at 50%, but will be gradually raised in the following years to 55% (2025), 60% (2030) and finally 65% (2035).

2   Modelling approach

2.1   Forecast of waste streams

2.1.1   Waste streams

For the forecast of the German recycling rates a projection of the amounts of the relevant municipal waste streams is necessary. For this purpose, the material-specific municipal waste streams are defined according to the municipal waste definition (EU, 2018) on the basis of the List-of-Waste (LoW) codes. The displayed amounts for the municipal waste streams correspond to the annual input into German waste treatment plants, reported under the respective LoW codes. This study focuses on five major municipal waste streams as material specific aggregates of the respective LoW codes:

- Residual Household & Business Waste (LoW: 20 03 01 00 - 02)
- Organic Waste (20 03 01 04, 20 02 01, 20 01 08, 20 03 02, 20 01 25)
- Paper & Cardboard Waste (15 01 01, 20 01 01)
- Mixed Packaging & Recyclables (15 01 06 00 - 02, 20 01 99 01, 15 01 05)
- Glass (20 01 02, 15 01 07)

It is important to note that the above mentioned waste streams are usually collected separately at source, directly. And so, the waste separation habits of the population have a major impact on the collection rates.

Further existing municipal waste streams are defined and modelled separately. But for reasons of readability and their quantitative importance (approx. 14% of the total municipal waste amount) they are aggregated here to one waste stream (Rest):

- Rest: bulky waste (20 30 07), wood (15 01 03, 20 01 38), plastic (15 01 02, 20 01 39), street cleaning (20 03 03), electronic equipment (20 01 23, 20 01 35, 20 01 36, 20 01 21), metal (15 01 04, 20 01 40, 15 01 11), textile (15 01 09, 20 01 10, 20 01 11), others (20 01 13-19, 20 01 26-34, 20 01 99 00, 20 03 99, 20 02 03, 15 01 10, 20 01 37, 20 01 41)

In comparison to the above mentioned major waste streams, the specific collection systems for these waste streams are usually not close to private households (except for bulky waste maybe). These collection systems are located closer to recycling centres, waste disposal companies or business actors. Waste separation habits of private consumers have less impact. It is assumed that collection rates are already high due to efficient waste separation practices in the non-private sector.

The Table 1 shows the corresponding amounts of municipal waste measured as input into German waste treatment plants with domestic origin (DESTATIS, 2011-2018). These inputs are considered as the past municipal waste potential. The future projection of these waste streams is the basis for the calculation of the respective recycling rates. Here, the latest available eight reporting years are used as period for trend analysis, and 2016 is considered as starting point for the forecast, so that 2017 will be the first forecast year.

Table 1   Reported municipal waste inputs into German waste treatment plants [Mio. t]

Waste stream / year	2009	2010	2011	2012	2013	2014	2015	2016
Res. Household & Business	17.9	18.0	18.1	17.7	17.7	17.6	17.5	17.8
Organic	9.3	9.6	9.9	10.0	9.8	10.8	11.0	11.5
Paper & Cardboard	8.1	8.0	8.1	8.1	7.6	8.0	8.1	7.8
Mix. Packaging & Recyclables	4.3	4.4	4.5	4.6	4.6	4.7	4.9	4.8
Glass	2.4	2.5	2.6	2.4	2.5	2.4	2.6	2.6
Rest	5.9	6.3	6.4	6.5	6.8	7.0	6.8	7.2
TOTAL	47.9	48.8	49.6	49.2	49.0	50.6	50.9	51.6

[DESTATIS, 2011-2018]

2.1.2 Composition of waste streams

Each projected municipal waste stream is defined by an average material composition. As validated average compositions for the domestic context are not yet available, the compositions are assumed, based on extensive experience from related projects and studies. However, the average domestic material composition will be determined as result of an on-going waste analysis of the domestic residual household waste stream. The following Table 2 shows the assumed average material compositions for the major waste streams (only on the main material level for better readability).

Table 2   Average waste stream compositions [mass %]

waste stream material category	Res. Household & Business Waste (stream)	Organic Waste (stream)	Paper Waste (stream)	Mix. Packag- ing/Rec. Waste (stream)
Paper/cardboard	10.7%	0.0%	95.2%	7.4%
Glass	6.4%	0.0%	0.0%	3.2%
Plastic	7.2%	0.0%	1.9%	42.3%
Metal	2.0%	0.0%	2.9%	10.2%
Organic	47.0%	94.6%	0.0%	10.0%
Wood	1.0%	0.0%	0.0%	1.2%
Textile	3.6%	0.0%	0.0%	1.3%
Inert	2.0%	0.0%	0.0%	0.0%
Compound	6.9%	0.0%	0.0%	16.8%
Hazardous	0.5%	0.0%	0.0%	0.0%
Others	8.6%	5.4%	0.0%	0.3%
< 10 mm	4.2%	0.0%	0.0%	7.3%

[Example data from a municipality]

2.1.3 Growth rates of waste stream material categories

A further requirement for the forecast of municipal waste streams is the definition of the future development of the waste stream material categories. This is realized by definition of (constant or graded) growth rates that describe the expected change of mass per capita with reference to the previous year. In this case, constant annual growth rates are used.

Growth rates are defined on various levels. For a certain number of waste material categories, the future trends are projected based on an analysis of the net-domestic consumption, e.g. the annual net-domestic consumption of print paper per capita is associated to the generation of paper waste per capita (with certain time delay determined through average product use time). As further requirement of the forecast, the generated waste per capita at the beginning of the forecast period (2016) has to be defined. This data is obtained from the reported municipal waste amounts (Chapter 2.1.1), the waste stream compositions (Chapter 2.1.2) and the past population figures of the respective years (Chapter 2.1.4).

The Table 3 summarizes the starting values of generated waste per capita and the respective expected annual growth rates (constant) obtained from net-domestic consumption trends. These growth definitions are applied to all waste streams.

Table 3   Growth rates of waste stream material categories by analysis of net-domestic consumption

material category	Generated waste (2016) [kg per cap.]	Expected future annual growth [+/-% p.a.]
Paper packaging	28.3	2.3%
Print paper & others	91.3	-1.1%
Glass	45.2	-2.0%
Plastic packaging	27.4	2.5%
Other plastics	29.0	-1.1%
Ferrous metal packaging	6.1	-0.8%
Compound packaging	12.4	-1.0%
Electronic equipment	12.2	-1.0% *
		*) modified value, based on net-domestic consumption trend

[DESTATIS, UBA, VDP, BKV, EUROSTAT, MULTIPLE YEARS, modified]

In a similar way, per capita starting values (2016) and growth rates are defined for all the other material categories that form part of a waste stream (Chapter 2.1.2). As an example, the following growth rates are considered in the forecast:

- Ferrous metals, not packaging (+0.7% p.a.)
- Wood (+1.8% p.a.)
- Textiles (+6% p.a.)
- ...

The respective trend analysis is not based on net-domestic consumption but on the reported municipal waste amounts, the waste stream compositions and the population figures. These growth rates are applied to all municipal waste streams, apart from the

Residual Household & Business Waste. In the case of the Residual Household & Business Waste, the material category growth rates result from specific waste analyses from several years.

2.1.4 Demography

Population figures from the past and the expected population size within the forecast period are necessary in order to normalize the municipal waste amounts, and to calculate the absolute waste amounts, respectively.

The official population data from the past and a certain population forecast scenario are integrated into the model. The Table 4 shows population figures for the starting year 2016 (DESTATIS, 2018a) and for selected years from the chosen future demographic scenario ("BEV-VARIANTE-05", DESTATIS, 2018b).

Table 4   Population scenario (selected years)

Year	2016 (start)	2017	2018	2020	2025	2030	2035
Population [Mio.]	82.5	82.8	81.7	81.6	81.1	80.2	79.0

As a remark, it should be noted that there is a major gap of approx. 1 million persons due to the shift from the past to the future time series between the years 2017 (past) and 2018 (scenario). This circumstance will have an effect on the absolute waste amounts in form of a discontinuity around these years.

For this study it is assumed that half of the population lives in (rural) conditions that lead to typically "rural" waste compositions, while the other half lives in (urban) conditions generating typical "urban" waste. This is especially relevant in terms of organic garden and kitchen waste generation.

2.2   Waste stream specific recycling factors

In future, the overall recycling rate has to be determined based on the output of waste treatment plants, or based on the input into recycling plants (Chapter 1). The recycling targets mentioned in Chapter 1 refer to this modification of the measurement points.

In a previous (not yet published) study, waste stream specific "recycling factors" were determined for each municipal waste stream that express the relation between the reported municipal waste input into the waste treatment plants and the expected output that can be considered recycled after the complete waste treatment process. Table 5 shows first rough approximations for the average recycling factors, based on the previous study results (for reference year 2015) and used in this forecast.

Table 5   Waste stream specific recycling factors (reference year 2015)

Waste stream	Recycling factors (2015)
Res. Household & Business	5%
Organic	70%
Paper & Cardboard	95%
Mix. Packaging & Recyclables	40%
Glass	90%
Rest	35%

Such factors are used to determine the future recycling amounts of each municipal waste stream by multiplication of the forecasted municipal waste amounts with the respective factors, i.e.

Municipal Waste Stream [t] x Recycling Factor [%] ≈ Recycled Amount [t]

This means for example, that only 5% of the residual household waste is recycled, while other municipal waste streams collected separately close to households show far higher recycling factors.

The recycling factors are also employed under the assumption that they will increase in future (see Chapter 2.3.3). This can be understood in the way that the existing waste treatment plant technology and the corresponding recycling efficiency will improve during the forecast period.

2.3   Scenario definition

2.3.1 Baseline

The baseline scenario describes the situation that the municipal waste streams are projected under the constraints mentioned in the previous chapters. Upon this base, the (constant) recycling factors (Table 5) are applied in order to calculate the respective overall recycling rate, i.e. without consideration of measures that aim to raise collection rates.

2.3.2 Scenario A: Higher collection rates in collection systems close to private households

This scenario is defined by far higher collection rates of certain material categories in the main municipal waste streams than in the baseline scenario. This follows the approach to "divert" recyclable material from such municipal waste streams that contribute low amounts to recycling according to the new WFD (e.g. residual household waste), towards those municipal waste streams that are related to high recycling contributions (e.g. separately collected paper, glass, organic). Recycling factors in this scenario are the same as mentioned in Table 5 (constant over time).

For this scenario the collection rates defined in Table 6 are used. For example, the actual collection rate of paper packaging in the Paper & Cardboard Waste Stream is about 64%, i.e. this share of the total paper packaging potential in the municipal waste can be expected in this waste stream at the beginning of the forecast period (2016). It is targeted to raise the collection rates linearly to 90% in 2025 and then to 95% in 2035.

Table 6   Collection rates of Scenario A

Collection rate		actual	target	target
OF material category:	IN waste stream:	**2016**	2025	2035
Paper packaging	Paper & Cardboard	**64%**	90%	95%
Print paper & others	Paper & Cardboard	**80%**	90%	95%
Glass	Glass	**65%**	90%	93%
Plastic packaging	Mix. Packaging & Rec.	**53%**	70%	85%
Compound packaging	Mix. Packaging & Rec.	**49%**	50%	55%
Ferrous metal pack.	Mix. Packaging & Rec.	**62%**	80%	85%
Other ferrous metals	Mix. Packaging & Rec.	**25%**	40%	40%
Other plastics	Mix. Packaging & Rec.	**32%**	40%	45%
Garden Waste	Organic (urban collection)	**91%**	95%	95%
Kitchen Waste	Organic (urban collection)	**34%**	65%	70%
Garden Waste	Organic (rural collection)	**86%**	95%	95%
Kitchen Waste	Organic (rural collection)	**9%**	65%	70%

All collection rates are determined with the aim to divert the necessary mass for reaching the targeted collection rates from the Residual Household & Business Waste. The limited presence of the necessary material mass to be shifted is a limitation for the definition of the collection rates. In principle, the masses could be taken from other waste streams, too.

2.3.3 Scenario B: Higher recycling factors of selected waste streams and higher collection rates in collection systems close to private households

This scenario setting is similar to the one in Scenario A, with the difference that the recycling factors of selected waste streams are raised linearly over time until reaching +10% in 2035 (with reference to the starting values in 2015 mentioned in Table 5). As recycling factors of the waste streams Glass and Paper & Cardboard are already on a high level, the increase in recycling efficiency is only expected for the waste streams Residual Household & Business, Organic, Mixed Packaging & Recyclables and Rest (see Table 7).

Table 7   Expected recycling factors (2020 to 2035)

Waste stream	Recycling factors (2015)	2020	2025	2030	2035
Res. Household & Business	5%	5%	5%	5%	6%
Organic	70%	72%	74%	75%	77%
Paper & Cardboard	95%	95%	95%	95%	95%
Mix. Packaging & Recyclables	40%	41%	42%	43%	44%
Glass	90%	90%	90%	90%	90%
Rest	35%	36%	37%	38%	39%

The recycling factors are increased in order to observe the impact on the overall recycling rate if these waste streams would be recycled with higher efficiency. If recycling targets are met or even surpassed then certain collection rates could be decreased perhaps (compared to Scenario A).

3   Forecast results

3.1   Waste stream potential

As a next step, the total expected municipal waste amount is projected on the level of material categories. The forecast results for the municipal waste streams of Scenarios Baseline, A and B are given in Table 8.

Table 8   Expected municipal waste amount for Scenarios Baseline, A and B [Mio. t]

waste stream	2020 Base	2020 A/B	2025 Base	2025 A/B	2030 Base	2030 A/B	2035 Base	2035 A/B
Res. Household & Business	17.3	13.9	17.0	9.2	16.7	8.1	16.4	7.0
Organic	11.4	13.5	11.3	16.0	11.2	16.1	11.1	16.2
Paper & Cardboard	7.8	8.4	7.6	9.0	7.5	9.1	7.4	9.3
Mix. Packaging & Recyclables	4.7	5.1	4.8	5.7	5.0	6.2	5.1	6.7
Glass	2.3	2.7	2.1	2.9	1.9	2.6	1.7	2.4
Rest	7.1	7.1	7.4	7.4	7.8	7.8	8.2	8.2
TOTAL	50.6	50.6	50.3	50.3	50.0	50.0	49.8	49.8

For the baseline scenario in general, it can be expected that the major municipal waste streams will stay on a constant level, however with slightly falling tendency.

The huge impact of the specified collection rates (Chapter 2.3.2) on the waste stream amounts (and thus on the composition of the municipal waste stream) can be observed (Scenario A and B). For example, the total amount of Residual Household & Business Waste is strongly decreased during the whole forecast period. By 2025, approx. 8 Million t of this waste stream would be shifted to other waste streams (mainly towards Organic, i.e. +5 Million t), as a consequence of the defined challenging collection rates and because the residual household waste stream acts as the "source" for other waste streams. All other waste streams are increasing compared to the baseline (except waste stream Rest that is excluded from modifications by collection rates, see Chapter 2.1.1).

3.2   Overall recycling rate

Which overall recycling rate can be expected based on the municipal waste stream forecast of Chapter 3.1? The specific recycling factors for each projected municipal waste stream (Chapter 2.2) reveal that the new recycling targets will not be met in the baseline scenario (see Table 9). The expected overall recycling rate is always below the targets defined in the WFD.

Table 9   Expected contributions to recycling from each municipal waste stream, Scenarios Baseline, A and B [Mio. t]

waste stream	year 2020 Base	2020 A B	2025 Base	2025 A B	2030 Base	2030 A B	2035 Base	2035 A B
Res. Household & Business	0.9	0.7 / 0.7	0.9	0.5 / 0.5	0.8	0.4 / 0.4	0.8	0.4 / 0.4
Organic	7.9	9.4 / 9.7	7.9	11.2 / 11.8	7.8	11.3 / 12.1	7.7	11.3 / 12.4
Paper & Cardboard	7.4	8.0 / 8.0	7.2	8.6 / 8.6	7.1	8.7 / 8.7	7.0	8.8 / 8.8
Mix. Packaging & Recyclables	1.9	2.0 / 2.1	1.9	2.3 / 2.4	2.0	2.5 / 2.6	2.0	2.7 / 3.0
Glass	2.1	2.4 / 2.4	1.9	2.6 / 2.6	1.7	2.3 / 2.3	1.5	2.1 / 2.1
Rest	2.5	2.5 / 2.6	2.6	2.6 / 2.7	2.7	2.7 / 2.9	2.9	2.9 / 3.2
TOTAL RECYCLING	22.6	25.0 / 25.4	22.4	27.7 / 28.5	22.2	27.9 / 29.2	22.0	28.2 / 29.9
Expected municipal waste potential	50.6	50.6	50.3	50.3	50.0	50.0	49.8	49.8
Expected overall recycling rate	44.7%	49.4% / 50.2%	44.5%	55.1% / 56.7%	44.3%	55.9% / 58.4%	44.1%	56.6% / 60.1%
Targeted overall recycling rate	50%	50%	55%	55%	60%	60%	65%	65%

For 2020 and 2025, the collection rates of scenario A would lead to overall recycling rates (49.4% and 55.1%) that are in the range of the defined targets (50% and 55%). However, they would only be reached very tightly, without any "reserve". In more distant years, recycling targets would not be met, while the gap seems to become even wider in the future.

For scenario B, it can be observed that the increase of the recycling factors might lead to sufficient overall recycling rates (50.2% and 56.7%) in 2020 and 2025. However, again there is not much reserve to expect, so that there is no reason to decrease the defined collection rates. For 2030 and following years, even more effort in improvement of recycling and collection seems necessary.

4   Discussion

The forecast of waste streams for municipal solid waste was based on data from 2009 to 2016 published by DESTATIS according to LoW categorization. These waste streams have been broken down to material categories using the composition of each waste stream. The currently used estimation of the waste composition for each waste stream can be significantly improved as soon as the nationwide analysis results (material composition of Household and Bulky Waste) for Germany will be available (expected by end of 2019).

Further assumptions for the forecast are validated data for population published by DESTATIS, and net domestic consumption data from Federal Environment Agency, industry associations and others. Current recycling rates were calculated from the waste streams according to an output (from waste treatment plants) based calculation model developed by ARGUS. Future recycling rates were simulated in order to meet the defined recycling rates of the WFD.

The results show that huge efforts will be necessary to increase the separate collection systems and to improve the current residual waste treatment system towards material recycling instead of incineration or energy recovery. Furthermore, the model does not consider potential waste streams which are so far not collected and fed into waste treatment plants such as garden waste treated by home composting.

Assuming the recycling rates for currently separate collected material categories (ref. to table 5) will stay unchanged in the future and Residual Household & Business Waste is the main source to increase recycling rates by separate collection, it will be impossible to meet the European targets for 2030 and 2035 by increasing the collection rates only. This means, that additional efforts are necessary to improve the whole recycling process at each stage from separate collection until the final recycling facility.

5 Literature

BKV et al. | multiple years | „Produktion, Verarbeitung und Verwertung von Kunststoffen in Deutschland (Kurzfassung)", BKV GmbH, Frankfurt am Main

Destatis | 2011-2018 | "Fachserie 19, Reihe 1, Umwelt, Abfallentsorgung", reference years 2009-2016 (annual report), Statistisches Bundesamt, Wiesbaden, annually published from 2011-2018

Destatis | 2018a | Bevölkerung: Deutschland, Stichtag. Fortschreibung des Bevölkerungsstandes. Tabelle 12411-0001. Statistisches Bundesamt (Destatis), 2018, Stand: 21.03.2019 / 23:55:02, available at https://www-genesis.destatis.de/genesis/online

Destatis | 2018b | Vorausberechneter Bevölkerungsstand: Deutschland, Stichtag, Varianten der Bevölkerungsvorausberechnung. Tabelle 12421-0001. Statistisches Bundesamt (Destatis), 2018, Stand: 21.03.2019 / 23:52:19, available at https://www-genesis.destatis.de/genesis/online/link/tabelleErgebnis/12421-0001

EU | 2018 | Directive (EU) 2018/851 of the European Parliament and of the Council of 30 May 2018 (OJ L 150, 14.06.2018, p. 109)

Eurostat | multiple years | Elektro- und Elektronikgeräte-Abfall (WEEE) nach Abfallbewirtschaftungsmaßnahmen [env_waselee], Database Eurostat, http://ec.europa.eu/eurostat/web/waste/data/database

UBA | multiple years | „Aufkommen und Verwertung von Verpackungsabfällen in Deutschland", (annual report), Umweltbundesamt (UBA), Dessau-Roßlau

VDP | multiple years | „Rechnerischer Verbrauch von Papier, Karton und Pappe." (annual report), Verband Deutscher Papierfabriken e.V. (German Pulp and Paper Association), Bonn

Author's addresses

Dr.-Ing. Bertram Zwisele, Dipl.-Ing. Carsten Böhm
ARGUS - Statistik und Informationssysteme in Umwelt und Gesundheit GmbH
Karl-Heinrich-Ulrichs-Straße 20a
D - 10785 Berlin
Tel +49 (0)30 / 39 80 60-11, -16
E-Mail bertram.zwisele@argus-statistik.de, carsten.boehm@argus-statistik.de
Website: http://www.argus-statistik.de

Recycling Region Harz – The Education Offensive

Petra Hauschild and Jürgen Poerschke

University of Applied Sciences in Nordhausen

Abstract
The Harz region has a long and diverse tradition when it comes to primary raw materials and is therefore severely affected by structural change. Under the leadership of the University of Applied Sciences in Nordhausen, four university partners from the federal states of Thuringia, Saxony-Anhalt and Lower Saxony are working on issues related to recycling recyclable materials as well as their quality assurance and appropriate treatment technologies. The results are intended to advance companies based in the Harz region and further boost the use of secondary raw materials. The main focus is on consumers taking back small electrical appliances. This group of substances is often incorrectly disposed of by the public and lost from the recycling loop. An educational campaign is to be carried out to make children and young people, in particular, aware of these problems in an interactive way.

1 Introduction

The collection and take-back of waste and the extraction of the recyclable materials it contains are essential to future raw material supply and environmental protection. Just like the green energy revolution already under way in Germany, a recycling material revolution is an indispensable foundation for growth and employment in society.

Nowadays, raising awareness and informing people about how to deal with increasingly scarce raw materials are especially important in the light of these primarily anthropogenic waste streams. The launch of a recyclable material revolution can only be successful if public understanding of the problem of 'waste' is created and fostered. Our task is to inform people and encourage citizens to become active participants. An educational campaign aims to make children and young people, in particular, aware of these issues in an interactive way, thereby laying the foundation for a sustainable and resource-efficient recyclable material revolution as early as school age.

Relevant questions can be answered by providing and sharing information on all issues relating to terms such as:

- Recyclable materials
- Take-back
- Processing
- Recovery.

Everybody can play their part in the recyclable material revolution. People can save raw materials, for instance, by sorting waste properly at home or only buying something new if they really need it. Public outreach work must identify risks to the environment caused by everyday behaviour. We are moving one step closer to the recyclable material revolution by offering practical suggestions for change and opportunities to get actively involved.

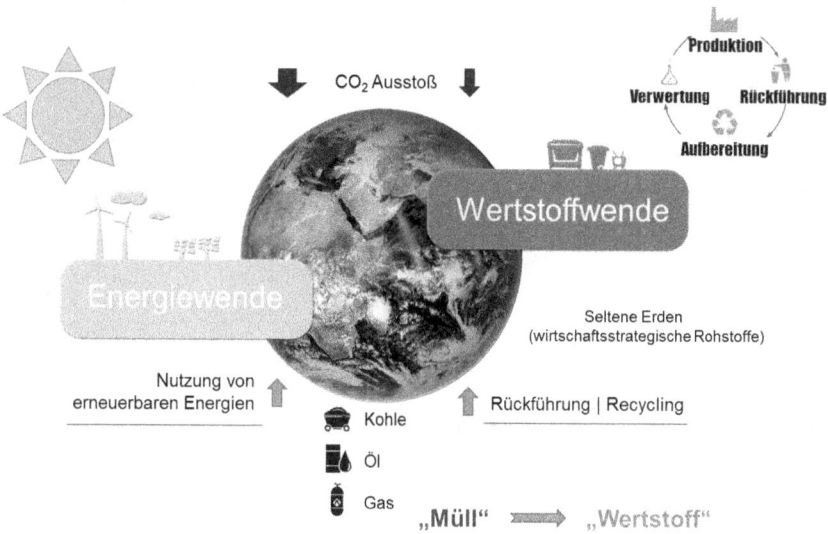

Figure 1: Interplay between the energy revolution and recyclable material revolution

2   Harz recycling region

Four universities from three German federal states have joined forces and created the 'Harz recycling region' as part of the German Federal Ministry for Education and Research's Twenty20 – Partnership for Innovation. The University of Applied Sciences in Nordhausen (HSN), the Technical University of Clausthal, the Magdeburg-Stendal University of Applied Sciences and Otto-von-Guericke University in Magdeburg have set themselves the goal of harnessing the previously untapped raw material potential contained in waste and landfills in a project under the leadership of Professor Jürgen Poerschke from HSN. Using waste electrical and electronic equipment (WEEE) as an example, the project aims to identify possible ways to optimise recovery operations, starting with take-back, and develop and test new potential solutions. An interactive educational campaign should be designed to raise awareness of this issue among children and young people, in particular. This project's main focus is putting in place the conditions for a sustainable and resource-efficient recyclable material revolution.

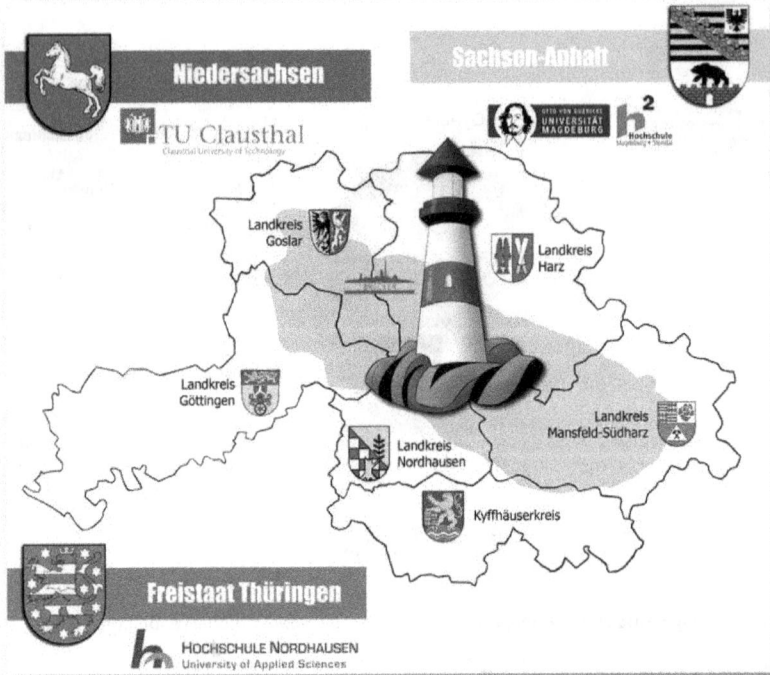

Figure 2: Project network for establishing and expanding the Harz recycling region

Simple take-back systems in touch with the people are the first pre-requisite for recycling WEEE. Short pathways and convenient take-back opportunities for all citizens may mean that WEEE does not end up in the bin or being littered. 'Simple' and 'in touch with the people' also means that the needs of shift workers, people who work far away from where they live, older residents and people with different levels of education are taken into consideration.

3   Education campaign

Many citizens are not interested in or motivated to discard their 'waste' in a way that makes it available as a source of secondary raw materials. A lack of information about the importance of taking back waste and/or about the benefits of 'waste' as a recyclable material is often behind this wrongdoing. Consequently, it is important to bring about a shift in public behaviour and a change in a material's image from waste to recyclable. Children are a vital interface to create the necessary social pressure. Integrating this issue into school curricula is an important tool here. This information has to be easy for children and young people to experience and grasp. For instance, children can experience the concept of 'returning resources' in a playful way in primary schools. Children often share experiences that they are excited about immediately with their social circles. In the process, they become a 'supervisory body' for their parents, and parents want to be role models for their children, which gets the ball rolling. One way to get young people – whose lives are closely connected to communication using mobile phones or tablets – on board would be to develop an app showing how to sort recyclable materials.

Moreover, converting Germany's waste management sector into a circular and resource economy – a task that will require new skilled workers and their continuing education in the long term - poses challenges not only for managing complex material streams and for technical development of efficient waste treatment technologies. It is also conceivable that new career opportunities will emerge in a circular economy model based on resource recovery.

3.1   Target groups

The target groups comprise primary and secondary school pupils, university students, unemployed people undergoing vocational retraining, apprentices and employees in the recycling industry, people working in businesses and citizens in their role as consumers.

Teaching materials, exhibitions and exhibits should facilitate age-appropriate knowledge transfer. All activities should focus on how interested parties interact with one another so that knowledge is absorbed in society and people can share what they have learned with one another. This promotes collective learning and strengthens a sense of community.

Events should be designed with the target group in mind and announced through a website, which has yet to be created. Additional parameters for target groups include:

- Use of these materials in the classroom, tailored to a variety of different schools, e.g. primary school – inclusion in science and local studies classes – adaptation of the visual aids to be used.
- School trips to WEEE manufacturers or processors, for instance (e.g. Nordthüringer Werkstätten – preliminary dismantling of WEEE).
- Drawing up flyers for WEEE or creating a brochure about the recyclable material revolution for all citizens.
- For apprentices – training on how to become a resource scout – learning to go through premises with open eyes, identify and dispose of waste. By doing so, they will encourage other employees to use resources more carefully (Source: IHK Erfurt, 2014, "Apprentices as Energy Scouts project").

3.2   Education on sustainable development

Sustainability in everyday teaching means creating ways of giving more space to this topic during regular lessons. Standard subjects could thus be incorporated into curricula, allowing sustainable thinking to be shared.

Curricula in Thuringia

Primary school – e.g. inclusion in science and local studies classes

Figure 3: Recyclable materials licence for children

Secondary schools – e.g. inclusion in chemistry and geography classes

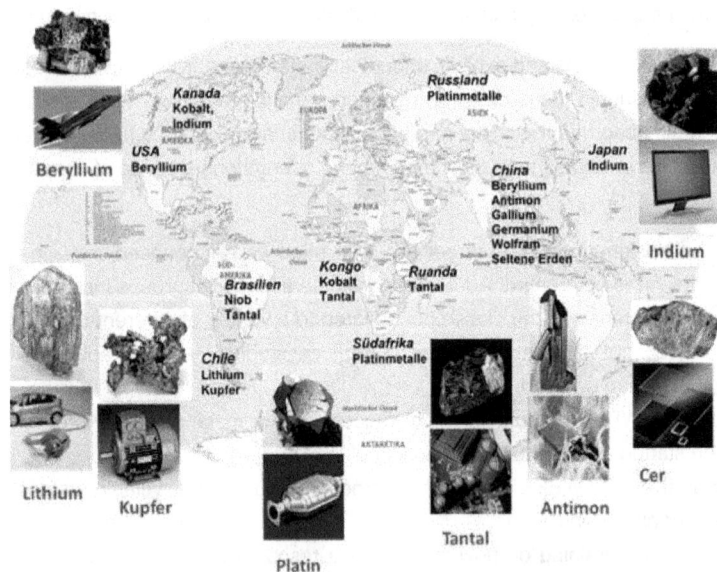

Figure 4: Key elements and their deposits (DERA, Nov. 2017, Britta Bookhagen)

This creates ways for students to use the knowledge they have gained directly in schools and/or their personal surroundings. Taking part in competitions may be a helpful tool to deal with the issue of the environment and sustainability and to understand the complexity of this issue.

However, a severe staff shortage means that educators are not in a position to integrate the issue of the environment and recycling into their teaching plans in the desired form without giving them detailed prepared information. In other words, school materials must be customised so that they are available and can be used by educators without them spending a lot of time. They must also be easy for teachers in the classroom and for educators teaching other subjects to handle. Teaching materials have already been developed for primary schools, which include both student workbooks and accompanying booklets for teachers.

Teaching materials should be made accessible to educators with the help of bodies including the Thuringia Institute for Teacher Training, Curriculum Development and Media (ThILLM).

4   The 'education cabinet'

The goal is to raise public awareness of how to deal properly with all kinds of recyclable materials. Besides projects on the ground in nurseries, schools and other educational institutions, local and supra-regional institutions are also addressing this issue in their own buildings with the help of educational theory elements. The 'education cabinet' is to use technology-based learning aids.

Figure 5: Modern digital learning strategies and the 'raw material suitcase' as visual aids

Content is also shared using displays, exhibitions, and practical and visual tests.

The education cabinet will go beyond the strategies already employed by other educators and allow a direct link to be forged with the manufacturing industry. Potential examples here include dismantling and sorting lines for recyclable materials. One of the primary objectives of the educational area is to create a link to the real recycling process, raising awareness of processing technology at an early stage. If prepared properly from an educational perspective, this can promote acceptance of training or studies in technical fields.

The education cabinet is being integrated into the new Thuringia Innovation Centre for Recyclable Materials (ThIWert), which was named as a priority by the European Fund for Regional Development (EFRD). The ERFD is a European Commission funding instrument for all regions of Europe, which works to even out differences in levels of development. The ThIWert views itself as a network bringing together industry, service providers, researchers and developers in the recyclable materials and circular economy.

5   Looking to the future

Environmental awareness measures focus on educating children and young adults. The education cabinet should help to create opportunities to encourage students, to sharpen perception of recycling and sustainability and to bring them on board as the next generation of researchers.

Activities are limited not only to those on offer in the education cabinet. There is also an opportunity to carry out projects and events related to sustainability issues directly at schools and to encourage sustainable thinking throughout the entire school community. This teaching material lays the foundation for including sustainability in the classroom in a simple manner. This material has to be continued, expanded and be usable for students of all ages.

The goal is to integrate resources and recycling more into curricula for educational facilities, thereby creating an opportunity for children and young adults to play an active role in bringing about the recyclable material revolution.

6   Literature

IHK Erfurt	2014	Apprentices as energy scouts project
Bookhagen, Britta	Nov.	"Key elements and their deposits"
DERA	2017	

Zero Waste: Material recovery instead of incineration and landfilling of MBT output fractions.

Matthias Kuehle-Weidemeier

ICP Ingenieurgesellschaft Prof. Czurda und Partner mbH, Karlsruhe, Germany

Abstract

By implementing modern thermal and non-thermal waste treatment technologies, waste management hast made a big leap forward. In the last two decades, Central Europe has been the motor of the development. The first decade of this century has been a period of development and installation of new processes in numerous locations. This was followed by nearly a decade of minor progress. This article will analyse the current situation and show how higher recycling rates even from mixed and residual waste can be achieved.

Keywords

Incineration, mechanical biological treatment, material recovery, energy efficiency, waste hierarchy, wet mechanical separation, optical sorting, urban mining.

1 History and current situation

1.1 Introduction and historical development

By implementing modern thermal and non-thermal waste treatment technologies, waste management hast made a big leap forward in central Europe. The first decade of this century has been a period of development and installation of new processes in numerous locations. A significant progress in landfill diversion was achieved. This period was followed by nearly a decade of stagnation or minor progress. This was a period of recovering from the huge investments that have been done and establishing the waste management system based on this new technologies and the changes they have brought.

1.2 Benefits of MBTs with most common techniques

Figure 1 shows the average mass-balance of all German MBTs (different types) based on a study in 2007. About 25% of the MBT Inputs finally ends on a landfill (ashes from RDF incineration not included), which means, that an average landfill diversion of 75% is reached. This is a huge progress in saving landfill volume and emissions.

Methane emission potential of MBT output (fine fraction going to landfill) is reduced by about 90-95% compared to untreated waste, if boundary values are kept, that are common in European countries (AT_4 = 5-10mg O_2/g). This is indicated by various investigations (Kühle-Weidemeier, Bogon, 2008). Considering, that additionally the tonnage of landfilled material is reduced by MBT by an average of 75%, MBT brought a giant leap forward in reducing greenhouse gas emissions from waste management.

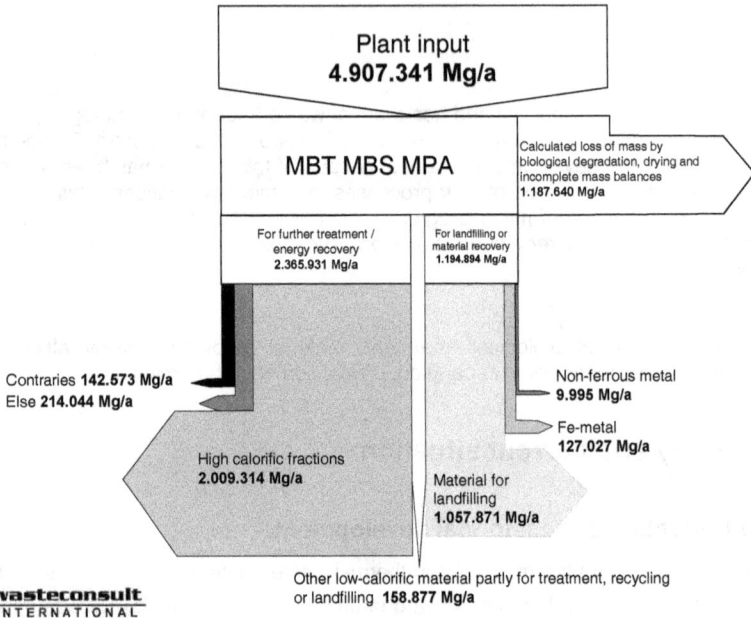

Plant input
4.907.341 Mg/a

MBT MBS MPA

Calculated loss of mass by biological degradation, drying and incomplete mass balances
1.187.640 Mg/a

For further treatment / energy recovery
2.365.931 Mg/a

For landfilling or material recovery
1.194.894 Mg/a

Contraries **142.573 Mg/a**
Else **214.044 Mg/a**

Non-ferrous metal
9.995 Mg/a

Fe-metal
127.027 Mg/a

High calorific fractions
2.009.314 Mg/a

Material for landfilling
1.057.871 Mg/a

wasteconsult
INTERNATIONAL

Other low-calorific material partly for treatment, recycling or landfilling **158.877 Mg/a**

Figure 1: Mass-balance of the German MBTs (Kühle-Weidemeier et al., 2007)

Figure 1 shows the average mass-balance of the German MBTs based on a study in 2007.

Compared to direct combustion of the waste, using MBT only an average of 40% of the waste ends in incineration. These 40% have a high calorific value and become a useful solid fuel that changes disposal oriented combustion to energy recovery by incineration in RDF power plants. Less incineration / combustion means less highly toxic residues from flue gas cleaning and less resources degraded or lost in incinerator bottom ash.

1.3   Potential improvements of MBT processes

This is the moment to look at potential improvements in MBT. These improvements are based on the fact that still 25% of the MBT input mass ends on landfill and up to 40% in RDF power plants. **A higher material recovery rate would be favourable.**

1.4   Why material recovery (recycling) of the coarse / high caloric fraction is better than application in RDF power plants

Depending on its quality (leaching test) incinerator bottom ash is used as construction material (mainly for roads) or landfilled. The long term behaviour of incinerator bottom ash is a subject of controversial discussion. The main concern is, that a real long term stability (immobilisation of heavy metals) is possibly not given.

A part of the exhaust gas cleaning residues is highly toxic and gets stored in subsurface hazardous waste landfills.

Ferrous metals are removed from incinerator residues by magnetic separation. These metals are heavily oxidised. Non-ferrous metals are often irrecoverably lost in the bottom ash.

Concerning the conservation of resources, waste incinerators are energy and resource destruction plants. Table 1 reveals how much energy is lost if only the energy represented by the calorific value is recovered.

Table 1: Calorific value and energy equivalent (cal. value + energy effort for production) of some plastic materials (Reimann 1988)

Material	Calorific value [kJ/kg]	Energy equivalent [kJ/kg]
Polyethylen (PE)	43,000	70,000
Polypropylen (PP)	44,000	73,000
Polystrol (PS)	40,000	80,000
PVC hard	18,000	53,000

2   General reasons for more material recovery

2.1   Legal requirements

The EU has specified a five step waste hierarchy, shown in Figure 2. EU policy and authorities push waste management regulations forward to ensure that superior grades in
the hierarchy are achieved for all kinds of waste. Disposal (e.g. landfilling) and recovery
(e.g. incineration) only reach a low grade in the hierarchy, while recycling (e.g. with output from sorting plants) has a superior grade in the hierarchy.

Additionally, the EU Waste Framework Directive requires increasing recycling rates.

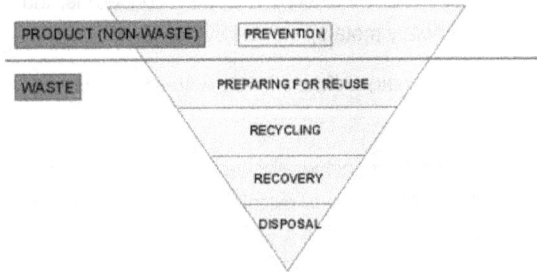

Figure 2: European Waste Hierarchy (European Commission, 2008)

2.2   Ecological requirements

2.2.1 Population growth and material consumption

The approaching exhaustion of many raw materials and expanding demand for resources due to fast growth of word population and increasing prosperity in many developing countries are a challenge for the world economy and will become a driving factor
for enhanced waste treatment / material recovery technology. Quantity and quality of
recovered resources from residual waste depend on the kind of waste treatment.

The world population will grow to round about 9.1 billion in 2050 (UN, 2009). That corresponds to an average annual growth of 56 million.

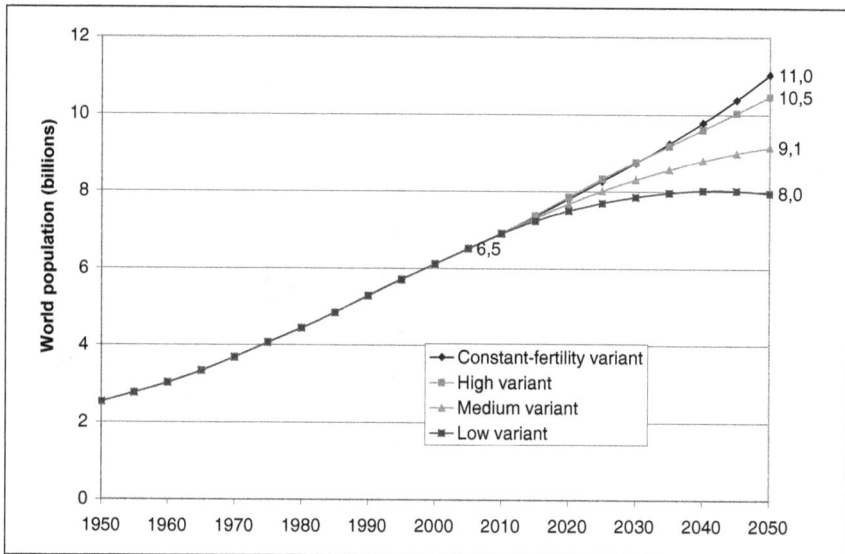

Figure 3: Different scenarios of world population growth (data source: UN, 2009)

The German Foundation for World Population (DSW) reports 2017 on their web site a current world population growth of about 81 million people per year. This is nearly as much as the total number of Germany's inhabitants.

2.2.2 Reduction of energy consumption and CO_2- emissions by recycling

Recycling is important for climate protection too. Figure 4 shows that recycling saves an enormous amount of CO_2 emissions and thus energy. For example, copper recycling saves 36%, steel recycling 56%, PE recycling 70%, PET recycling 85% and aluminium recycling even 95% compared to primary material production.

The calculated emissions of the recycling process consider collection, transport and the recycling process itself. Considered transport distances to the recycling facilities are based on the true situation. In case of PET this is the transport to south east Asia. It has to be mentioned, that plastics, paper and wood are only feasible for a small number of recycling cycles. Paper fibres can be re-used 5 - 7 times.

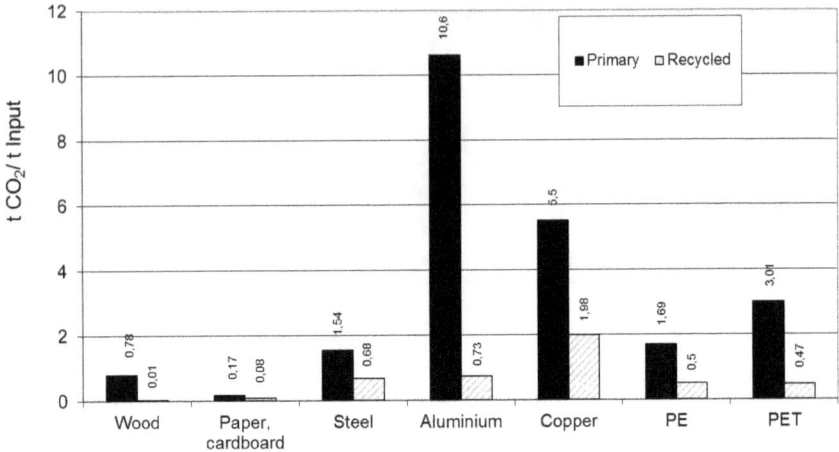

Figure 4: CO_2-emissions by primary and secondary material production and avoided emissions by recycling (data: Interseroh, Umsicht, 2008)

3   Necessary improvements for higher material recovery rates and getting closer to Zero Waste solutions

3.1   Coarse, high calorific fraction

Sensor based sorting is applicable to the coarse high caloric fraction. Better sensors, higher computing power, continuous improvements of the software as well as mechanical improvements resulted in huge progresses in sensor based sorting. This allows a higher separation rate as well as improved quality of the separated fractions. Integration of more sensor based sorting machines in waste treatment plants, especially MBTs will increase the potential for higher recycling rates of waste components

A continuing limitation are impurities on the waste component surfaces, especially caused by humid (or dried) organic materials, that result in smelling and often dirty looking material from sorting processes of mixed or residual waste. Directly recycled plastic from such fraction smells and won't be accepted by consumers. Due to this limitations, most of the fractions (except metals and a few plastics) sorted from the coarse or a biologically dried full fraction do not have a positive market value. Hence, there is often no motivation to run an optical sorter for other purposes than removing PVC.

For a higher market value, at least washing of such materials would be required. Additional costs that have to be compared to the potentially higher value of the output materials.

3.2   Fine fraction

Except of metals, the fine fraction consists mainly of materials that often have no positive market value and are more difficult to separate with classic dry sorting methods: A
mixture of humid organics and minerals. Wet separation methods are the appropriate
way to separate this materials. In combination with biological drying, also dry sorting
and separation techniques can be applied.

3.3   Including wet separation steps

Much higher recycling rates are possible with waste treatment plants that combine sensor based sorting (usually optical NIR) with wet separation techniques. Water can be
given a dual function: Separating and cleaning. This is much more efficient than using
the water just for washing.

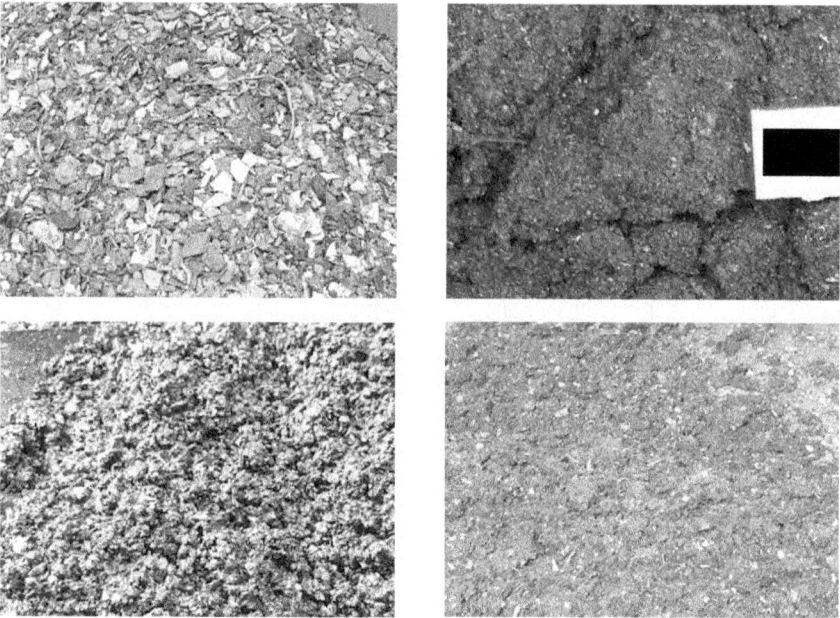

Figure 5: Various fractions from a biological and wet mechanical treatment step of an MBT

An early form of wet separation techniques is found in MBTs with wet anaerobic digestion (AD). Pipes, pumps and reactors of wet AD plant are very sensitive to blocking by
minerals like sand or by fibrous substances. Sedimentation of minerals in the reactors
can also be a major problem. To avoid or minimize these issues, MBTs with wet AD
have a frontend that includes wet separation techniques that separate organic waste

components from mineral fractions of selected particle sizes as well as fibrous materials.

Figure 5 shows the organic and 3 mineral fraction from an initial wet separation of a wet AD MBT. This separation was just done to protect the plant equipment but not with the target of recycling additional fractions. Refining this technology is the way to more recycling. It is obvious that improvements and intelligent changes in the plant concept can be the key to convert MBTs from pre-treatment plants before disposal or incineration to real material separation plants with an output that consists of clean recyclables and not of disposal fractions and RDF.

4   Already available technical solutions

4.1   OWS wet process SORDISEP

OWS provides a process called SORDISEP (SORting, DIgestion & SEParation). It was developed to treat contaminated digestate from mixed (or residual) waste organics in order to assure the production of a high quality compost and recyclable fractions from mixed waste digestate. It removes / reduces visual and other contamination in compost derived from mixed waste household organics. The process is described by De Baere and Mattheeuws (2017) as follows:

During the DRANCO digestion, about 60 to 65% of the volatile solids (representing the easily degradable and often wet and sticky/odorous organic components of the waste) are converted to biogas in the digester. The remaining volatile solids are the more fibrous materials that do not decompose easily as well as microbial biomass. The resulting digestate can be easily separated using screens and other wet separation equipment, as developed in the SORDISEP process. Sand, short fibers and inerts can be recovered and cleaned in order to produce marketable end products. This increases landfill diversion up to 85% and recovery of materials out of mixed waste to 50%.

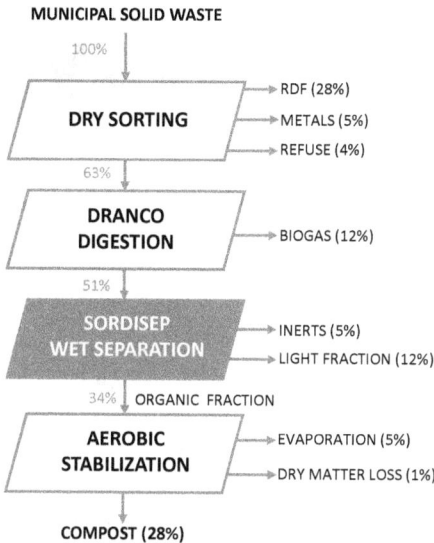

MUNICIPAL SOLID WASTE

100%

DRY SORTING
→ RDF (28%)
→ METALS (5%)
→ REFUSE (4%)

63%

DRANCO DIGESTION
→ BIOGAS (12%)

51%

SORDISEP WET SEPARATION
→ INERTS (5%)
→ LIGHT FRACTION (12%)

34% ORGANIC FRACTION

AEROBIC STABILIZATION
→ EVAPORATION (5%)
→ DRY MATTER LOSS (1%)

COMPOST (28%)

Figure 6   Schematic overview of the SORDISEP process and integration in a MWPF

SORDISEP's wet separation technology primarily makes use of density differences (like air classifiers do) but using water rather than air as the floatation medium. SORDISEP is ideal to separate particles of glass, sand, light and heavy particles from organic materials prior to composting.

And of particular importance, SORDISEP continuously recycles the water used for separation, avoiding the need for added water and producing minimal liquid effluent.

A first full-scale plant including a SORDISEP-unit was integrated in the Mixed Waste Processing Facility (MWPF) of Bourg-en-Bresse, in France, with operations starting in 2016.

The final compost produced complies with the French NFU 44-051 norm. PCB's, petroleum based compounds (plastics) and pesticides as well as small inerts were all way below the limits. In Table 2, the values for heavy metals are shown.

Table 2   *Compost values & French and USA norms for heavy metals*

Metals mg/kg TS	Compost Bourg-en-Bresse mg/kg TS	Norm France mg/kg TS	Norm Ontario Class A mg/kg TS	Norm US EPA mg/kg TS
Arsenic	2.4	18	13	75
Cadmium	0.8	3	3	85
Chromium	67	120	210	3 000
Copper	126	300	400	4 300
Lead	66	180	150	840
Mercury	0.2	2	0.8	57
Nickel	57	60	62	420
Zinc	402	600	700	7 500

In Table 3, *results for PCB's, pesticides, herbicides & petroleum ae shown.*

Table 3   *Compost values & factor lower than standard*

Product	Factor lower than standard*
PCB's	10-100
Pesticides	10-1000
Herbicides	100-1000
Petroleum based mineral oils	Absent

* Hawaii standard

For more details, please refer to De Baere and Mattheeuws (2017).

4.2   Urban Mining Engineers AG wet process SchuBio®

Based on experience since about 1991 with the WABIO / DBA WABIO process, an advanced MBT process with wet mechanical separation was developed with support of the German Environmental Fund (Deutsche Bundesstiftung Umwelt, DBU). To verify the applicability of the process for various materials and waste compositions in larger scale, a mobile demonstration plant in a container was built in 2004. This technical scale plant was tested in various locations. The process got the name SchuBio®. It separates the waste components in clean fractions, which become commodities and can be flexibly utilised in adoption to local and changing market conditions.

Figure 7: Different locations of the SCHUBIO®-Demonstration Plant (Schu, 2009)

An important fact that was discovered during the development of the process chain until its current level is that biological treatment handicaps the separability of the waste components and also prevents the removal of heavy metals from the organic fraction. Heavy metal depletion is the latest development of the SchuBio® process.

Several wet separation steps are combined. *"The separation of inert matter and the separation into fractions with different particle size are preconditions for the thermo-mechanical celllysis. The celllysis is causing the organic fibres to fray and separate, thus breaking down the cell walls so that cell water is released.*

The inert fractions are rinsed, first with circulated water then with fresh water, and can be recycled as building material. If required, further treatment in a demolition waste recycling plant yield even better quality. The following products are obtained from the waste:

- *stones -   gravel -   sand -   fine sand   -   silt.*

The organic fractions are dewatered by screw presses after sieving. Dewatering includes also the cell water due to the thermo-mechanical celllysis as described above." (Schu, 2009).

The current process is visualized in Figure 8. A closer look shows, that this is a clever combination of well-known and approved process steps in an improved sequence combined with inventions that result in ground-breaking abilities of material recycling. This is reached by the combination of wet mechanical separation, celllysis and sensor based sorting devices.

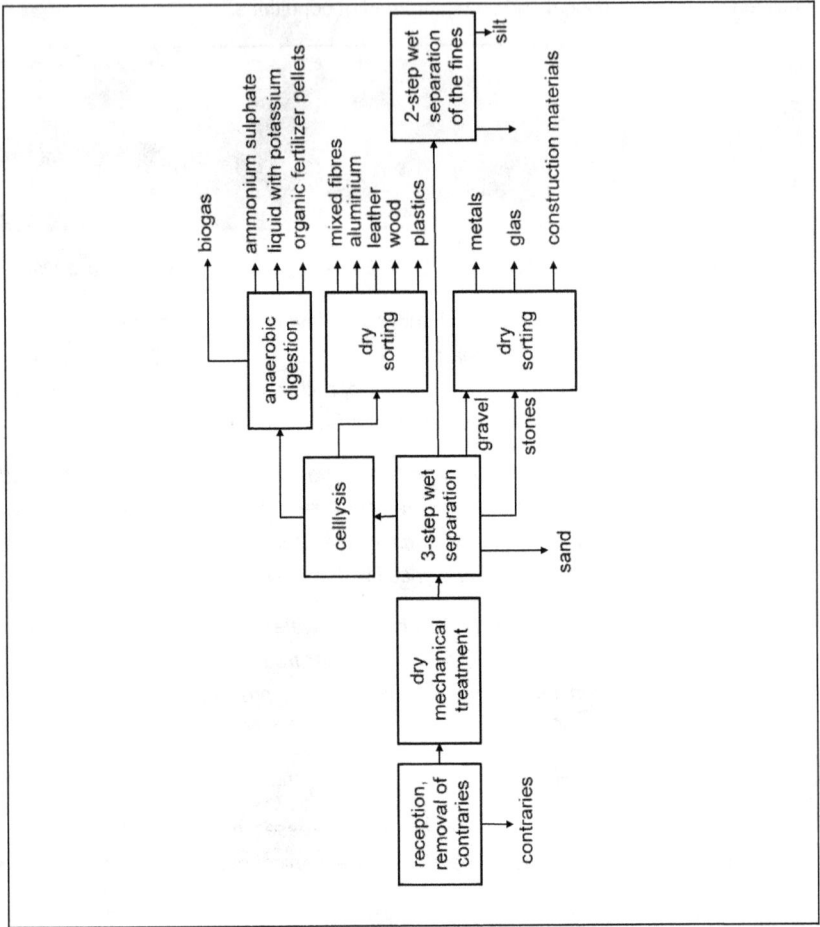

Figure 8: Simplified chart of the SCHUBIO® process

Figure 9 - 11 show various output fractions of the SchuBio® demonstration plant. The materials leave the plant washed and hygenized.

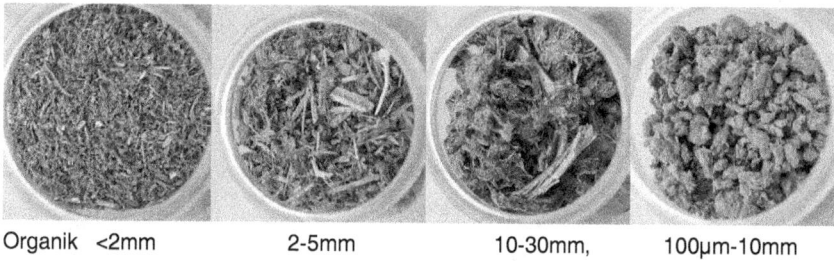

Organik <2mm   2-5mm   10-30mm,   100µm-10mm

Figure 9: Organic output fractions of the SchuBio® demonstration plant

Minerals <100µm   100µm-2mm   2-15mm   15-80mm

Figure 10: Mineral / inert output fractions of the SchuBio® demonstration plant

RDF 30-80mm   Organic pellets

Figure 11: RDF and externally made pellets from the organic output fractions of the SchuBio® demonstration plant

After successful operation of the container size demonstration plant and further process optimization, a full scale plant was built in Switzerland (Figure 12). Commissioning and start of regular operation were successful and biogas production of the integrated AD step quickly reached the expected level. Technical acceptance of the SchuBio-Process took place at the End of 2012. Unfortunately, the AD of household waste and biowaste was rendered impossible in 2013. "The digestate from MSW and biowaste was supposed to be dried together with sewage sludge and recycled materially as phosphate fertilizer by a subsequent thermochemical phosphate recycling process" that was a separate plant not in the responsibility of SchuBio / Schu AG and a concept of other companies.

"However, this thermochemical phosphate recycling process turned out not to be working. Responsible for the development of this failed process for phosphate recycling were the Swiss-Government, Lonza AG, Envirotherm GmbH, Outotec, Paul Scherrer Institut and BAM (Bundesanstalt für Materialforschung und –prüfung).

In consequence, the SchuBio plant was changed from dry fermentation to wet fermentation without any conversions and is now processing leftovers and other biomass and is operated till today at designed biogas capacity." (Schu 2017)

Figure 12: Full scale MBT with wet separation, celllysis and AD (photo EcoEnergy)

Afterwards, the world wide patented SchuBio® process was further improved and new inventions were added, thus rendering the process completely independent from subsequent thermochemical processes, incineration or landfill.

Figure 13 gives a simplified example of a possible mass flow.

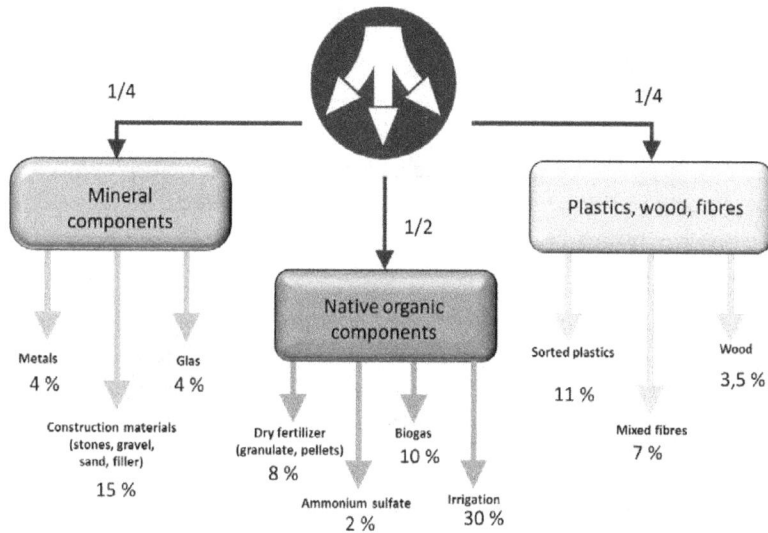

Figure 13: Example for the mass flow of a plant following the current level of development (Schu, 2017)

4.3   Achievements of the new plant type

- Waste is separated and converted to marketable raw materials

- Hygenized output

- Recovery of washed minerals, reducing landscape damage caused by mining

- Biomass is separated and depleated in pollutants

- Separation of heavy metals

- Separation and sorting of washed plastics

- Source segregated collection of putrescible organics and plastics might become technically avoidable, lower collection costs

- Production of fertilzer and irrigation water (only where permitted)

- Recycling of nearly total of the input, minimum disposal and groundbreaking landfill diversion rate

4.4   Economical viability

Costs for a SchuBio® plant are on a similar level like a conventional MBT with wet AD, but the output are clean commodities that have a higher market value. Hence, operation costs can be lower than those of conventional MBTs. The heavy metal removal and destruction of possible organic pollutants by long lasting, thermophilic dry AD even produce an optically and chemically clean organic output fraction that can be used as pelletised fuel or fertilizer, depending on local legal regulations and market requirements.

Large scale plants (>350 000 Mg/a) open the possibility to make new products from mixed plastics and processed fibrous materials on site. These high quality rugged products for outdoor use have a significant value and can turn the whole complex in a profitable enterprise under appropriate conditions.

5   Summary and conclusion

MBT was able to significantly improve waste management by minimising rates of landfill diversion and incineration and reducing greenhouse emissions. Current widespread MBT plants recover energy, but achieve only low recycling rates. These plants can be vastly improved by integration of up to date wet mechanical separation steps and sensor based sorting devices. The progress in MBT technology technically facilitates previously unimagined recycling levels.

6   Literature

Biebeler, H., Mahammadzadeh, M. und Selke, J.-W.	2008	Globaler Wandel aus Sicht der Wirtschaft. Chancen und Risiken, Forschungsbedarf und Innovationshemmnisse. Deutscher Instituts-Verlag, Köln, ISBN 978-3-602-14791-5
De Baere, L., Mattheeuws, B.	2017	Production of clean compost out of mixed MSW: a giant leap towards Zero Waste. In: Kuehle-Weidemeier, M., Buescher, K. (2017): Waste-to-Resources 2017, proceedings.
Deutsche Stiftung Weltbevölkerung	2017	https://www.dsw.org/infografiken/
European Commission	2008	http://ec.europa.eu/environment/waste/frame work/
Faulstich, M.	2008	Abfallwirtschaft und Ressourcenschutz. Welchen Beitrag leistet Recycling zur Nachhaltigkeit? Präsentation zum Rohstoffkongress 2008, Berlin.

Fraunhofer Institut UMSICHT, IN-   2008   Recycling für den Klimaschutz. Ergebnisse
TERSEROH AG (Hrsg.)   der Studie von Fraunhofer UMSICHT und
                                              INTERSEROH zur CO$_2$-Einsparung durch
                                              den Einsatz von Sekundärrohstoffen, Bro-
                                              schüre.

Kühle-Weidemeier, M.; Langer, U.;   2007   Anlagen zur mechanisch-biologischen
Hohmann, F.   Restabfallbehandlung. Schlussbericht. By
                                              order of the German Environment Agency
                                              (Umweltbundesamt) UFOPLAN 206 33 301

Kühle-Weidemeier, M., Bogon, H.   2008   Methanemissionen aus passiv entgasten
                                              Deponien und der Ablagerung von mecha-
                                              nisch-biologisch behandelten Abfällen -
                                              Emissionsprognose und Wirksamkeit der
                                              biologischen Methanoxidation -. Schlussbe-
                                              richt. Im Auftrag des Umweltbundesamtes.
                                              FKZ: 360 16 015

Schu, R.; Schu, K.   2009   IModernization of a Swiss MBT-plant with the
                                              SCHUBIO®-Process. In: Kühle-Weidemeier,
                                              M. (Hrsg.): International Symposium MBT
                                              2009. Proceedings.

Schu, R.   2017   Personal notices

SERI (Sustainable Europe Research   2009   http://www.materialflows.net/mfa/
Institute)   Visited 10.03.2009

UN (United Nations)   2009   World Population Prospects: The 2008 Revi-
                                              sion. Population Division of the Department
                                              of Economic and Social Affairs of the United
                                              Nations Secretariat, http://esa.un.org/unpp

Visvanathan, C.; Norbu, T.;   2007   Applying Mechanical Pre-Treatment and
Chiemchaisri, C.; Charnnok, B.   Landfill Mining. Approach in Recovering Re-
                                              fuse Derived Fuel (RDF) from Dumpsite
                                              Waste: Thailand Case Study. In: Kühle-
                                              Weidemeier, M. (Hrsg.): International Sym-
                                              posium MBT 2009. Proceedings.

Author's address

Dr.-Ing. Matthias Kuehle-Weidemeier
ICP Ingenieurgesellschaft Prof. Czurda und Partner mbH
Gartenstr. 4
D-30851 Langenhagen
Phone +49 511 6558 1775; ICP main office: +49 721 944 77-0
Main office: +49 721 944 77-0

kuehle@icp-ing.de
kuehle@wasteconsult.de

www.icp-ing.de
www.wasteconsult.de

Recycling Instead of Landfilling –
a Contribution to Climate Protection

Werner P. Bauer

WtERT Germany GmbH, Munich

Abstract

The paper starts with discussing the huge problems regarding RRR-Communication especially due to global warming effects because no one seems to consider closing landfills. The paper concentrates on asking for awareness and new essential aims in MSW management (closing landfills and its tools) and concludes in showing how the WtERT Network and its' Decision Support System could and should help in international waste management communication.

Keywords

Circular Economy, Climate Change, Landfill, Waste to Energy, Recycling, Incineration

1 About Priorities

1.1 Perception

Frequently we ask our international contacts what their top motivation is in optimizing waste management in their respective country.

They often respond that they want to keep the cities clean or would like to create a modern recycling system. When landfills need to be closed, this mainly relates to old dumping grounds and is never a matter of negotiating landfilling and the associated contribution to climate protection.

Of course, stakeholders in the EU try to do what they are told.

1.2 Public Concern

In the press release IP/05/1673 "Recycling: Europe's New Waste Strategy" dated December 21st, 2005 the European Commission reports on its new strategy: Europe is to transform into a recycling society "which tries to prevent waste and utilizes waste as a resource".

Years later, on June 14th, 2018 the new EU guidelines regarding the "Circular Economy Package" were published in the Official Journal of the European Union (L150) stating clear goals with respect to waste prevention and recycling. In addition to the distinct plastic strategy, the new recycling quotas of 65% for household waste and 75% for packaging waste (each by 2030) have, in particular, become the center of attention of the public.

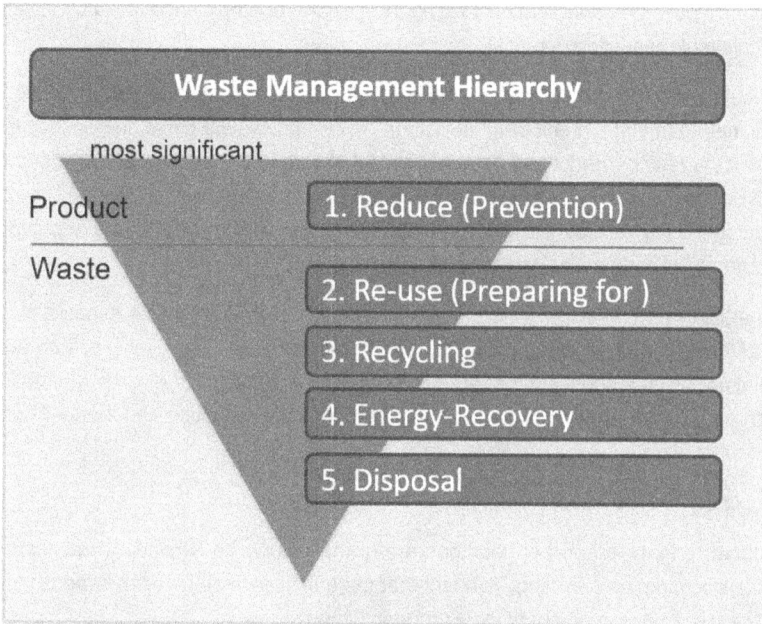

Diagram 1   European Waste Management Hierarchy

The fact that the EU has limited (or is it more a permission?) the depositing of household waste on landfills to a maximum of 10% by 2030 without having binding intermediate goals and, in addition, has incorporated exception clauses in the law of up to 5 years, should convey even to the last attentive observer that the closing of landfills, when compared to the significance of recycling, is a rather subordinate goal.

1.3   Different Perspective

This is so distant from the state of knowledge in terms of the real and possible contributions of waste management to climate protection as can be abstracted from the German Federal Ministry of Environment, Nature Conservation and Nuclear Safety´s status report from August 2005 regarding the "contribution of waste management to climate protection and possible potentialities".

In the report´s summary there are three remarkable sentences which, in my opinion, are distinct enough to reevaluate the significance of recycling from a global viewpoint.

1) "In total, the disposal procedures of the waste incineration plants and the co-incineration contain the highest potential for reducing greenhouse gases…"

2) "All energetic procedures, considering the general conditions, have a 90% share of the achievable reduction potential."

3) "Instead of producing a carbon dioxide equivalent of presently (2005) 87 m with waste management, a credit of 47 m carbon dioxide equivalent could be depicted in the future. As a result, a reduction potential of 134 m t carbon dioxide equivalent could be achieved from 2000 to 2020 for the municipal waste management of EU-15. The major portion stems from the almost 100 m t of carbon dioxide equivalents from the prevented methane emissions as a result of the role of landfilling."

In an article from 1968 the general secretary of one of the first waste associations, H. Straub, having in mind the situation at the time in Germany, intuitively recognized the correlation: "In essence it's a matter of reducing waste in terms of volume … if possible by carrying out recycling". Of foremost importance is the reduction and then – "if possible" the recycling.

And now back to the question referred to at the beginning:

Of course, it is excellent that countries are giving thought to secondary raw materials and are encouraging recycling. The fact that the anti-litter-strategies are seriously being discussed is an achievement of the European Commission.

Imagine how much faster the worldwide carbon footprint could be improved if the European Commission, in addition to the well-known waste hierarchy, would clearly identify the goal of "overcoming the landfills" and would evaluate all of the existing "tools" such as incineration (with energy recovery), recycling, and prevention based on their impact on a climate-relevant volume reduction. In this chronological order – incineration, recycling and prevention – waste management has developed in Germany.

But it seems that it has still not found its way into the minds of the law-making body that the idea of "incineration" is transforming into a concept of "powerplants" driven with waste to produce valuable steam energy and other sorts of energy recovery.

Step by step incineration plants in West Europe are being turned into waste fired powerplants, which focus more and more on energy recovery.

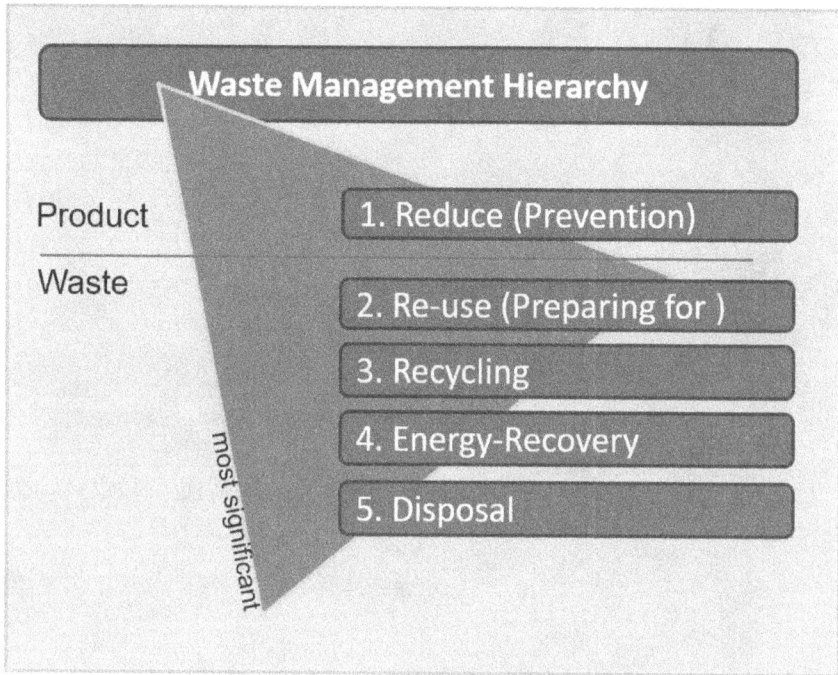

Waste Management Hierarchy

Product

Waste

1. Reduce (Prevention)

2. Re-use (Preparing for)

3. Recycling

4. Energy-Recovery

5. Disposal

most significant

Diagram 2   European Waste Management Hierarchy from a Different Angle

Due to the prevailing landfill crisis incineration plants had to be built. The associated rise in costs of waste management enabled recycling to establish itself much faster.

2   Awareness

Often recycling is scaled down to material recycling when perceived in an abridged form.

2.1   Climate Change

In an integrated waste management, energy recovery must receive a higher evaluation when viewing the time factor related to closing landfills as rapidly as possible.

When considering the ever so important reduction of climate gases, the aim should lie on reducing landfilling and reassessing any tools available to achieve this goal.

Waste Management Hierarchy

Product

Waste

Priority of Tools

1. Reduce (Prevention)

2. Re-use (Preparing for)

3. Recycling

4. Energy-Recovery

Basic Aim!

5. Reduce Disposal!!!!

most significant

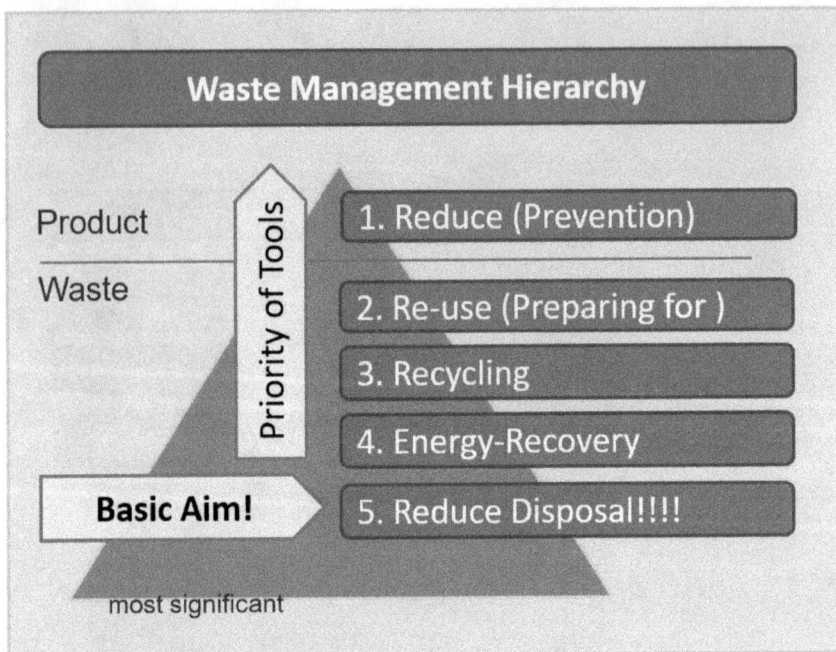

Diagram 3   European Waste Management Hierarchy for Action

In addition to the waste management objective of reducing disposal more attention should be given to the structures on waste disposal sites and the technical goal to close landfills. A significant reduction in climate gases can be achieved even in existing waste disposal sites by implementing technical measures.

Here one can differentiate between the closing of landfills resulting in a reduction of gas emissions and the additional benefit of conducting energy recovery by utilizing gas.

It is unlikely that this will occur with the measures of the CDM Joint Implementation Initiatives through the purchase of Certified Emission Reductions (CERs) when viewing the still low prices of CO_2 certificates.

2.2 Wording

Extending the term "Reduce Disposal" to "Close Landfills" is an important new wording, which, in addition to the greater weight on the technical components, can be fixed in the minds of those concerned.

Those who read the following wording in law texts

a. Recycling and Reduce Disposal

or

b. Recycling and Close Landfills

and who want to implement these will set other priorities and act differently.

Using the right words is very important, because it is strongly connected to your inner motivation for doing things.

You can feel the difference by using the two harmless words "Climate Warming" against "Climate Heating". "Heating" has nothing to do with cozy feelings.

And even more, let us start to avoid the term "landfill", which sounds as if there is a hole in the earth waiting to be filled.

Some specified wording is anchored to a vague labeling of our childhood. So, it might feel that a term like "Incineration" is deeply connected with the uncertainty of technical discussion longing to the past.

Therefore, to use the Word "Power plant, run by scrap" instead of using "Incineration" makes a great difference.

3 Necessity of a New Communication on a Global Scale

Those who are following the dimension of the climate change with interest, but without apprehension are likely to conclude that all measures taken to counteract this phenomenon must be conducted on a global level.

3.1 The 80/20 Principle

The "80/20 Rule" (Vilfredo Pareto) says that 80% of the results are achieved with 20% of the total effort. The remaining 20% of the results with 80% total effort require quantitatively the most work.

Taking the time component of climate change into consideration, it is understandable that we must begin where we can achieve the most with the least amount of effort (20%). Here, too, reduction in mass and the closing of landfills have priority.

3.2 Dissemination of Knowledge without Preconceptions

If one follows the statements of this speech so far, then it is evident that a global dissemination of knowledge without preconceptions is essential.

The following diagram is intended for public discussion to help develop an unbiased side-by-side of activities in place of the often ideologically struck counterparts of material and thermal recycling – and that apart from EU hierarchy.

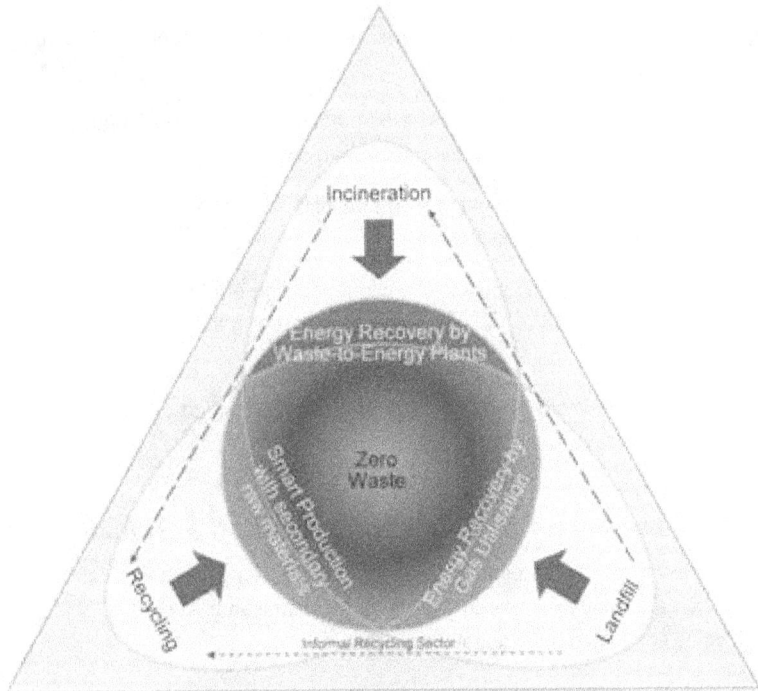

Diagram 4   Recycling, Landfilling and Incineration to Achieve Zero Waste
Copyright WtERT Germany GmbH

The collection of materials from waste, their energy recovery, and their recycling in smart production are processes that go hand in hand. Especially in modern countries with a high rate of energy recovery, recycling is growing in importance and becoming a significant economic factor.

3.3   Integrity of Creation

The waste community must be developed into a sharing community with the aim to find out its part for the integrity of creation.

3.3.1 Free Your Mind

If we free our mind and

- talk about recycling with additionally emphasizing the issue of landfill problems while looking for better solutions for the entire system,

- encourage people to be interested in new ways of solution making and if

- we have an open-minded approach towards a diversity of thinking

then we need a new slogan where we can embrace all the methods and activities in an integral waste management.

Till we find something better – we shall continue to use the slogan *Waste to Energy (WtE)*.

3.3.2 Way of Acting

We pursue our keen interest to identify and provide outstanding practical examples of the Waste to Energy concept and to consolidate practical and expert knowledge and experience. We do this on a very broad level because we want to demonstrate that the concept of transforming waste to energy offers valuable solutions in all areas of the waste and resource industry. Furthermore, unlike in the hierarchies of other common waste management programs the sectors of landfill, incineration, and recycling are given equal status and presented free of bias.

3.3.3 Sharing Community

In 2002 the Waste-to-Energy Research and Technology Council (WtERT) was founded by the Earth Engineering Center of Columbia University, New York, and the U.S. Energy Recovery Council. The goal was to identify and help develop the most suitable means for managing various solid wastes research, and to disseminate this information by means of publications, the web, and technical meetings. At the end of 2011, the Global WtERT Council (GWC) was created as a U.S. non-profit organization. Today GWC is the umbrella organization for 19 WtERT organizations in 18 countries, which consist of Brazil, Chile, China, Colombia, Cuba, Czech Republic, France, Germany, Greece, India, Italy, Korea, Pakistan, the Philippines, Serbia, Singapore, UK and the USA.

With all the above-mentioned arguments in mind, in January 2017 WtERT Germany was assigned by GWC to collect and showcase knowledge from all WtERT organizations.

Through many discussions on the topic of WtE we developed the vision to succeed in not only compiling the knowledge of our WtERT Partner Organizations, but also in uniting this with the experience of multiple partners in research organizations, industry and municipalities. To reach a high credibility of the whole WtE System, we believe it to be important to show existing solutions (case studies) and share personal recommendations from international WtE experts.

In August 2017 we relaunched www.wtert.net and thereby offer the opportunity to join the WtERT Decision Support System (WtERT-DSS).

The WtERT DSS provides a platform for stakeholders from all over the world to

•   get informed and inform about state-of-the-art methods and technologies for sustainable waste management,

•   inform about the status of waste management in their country and learn about solutions in neighboring and other countries toward approaching sustainable waste management,

•   to provide a database of realized solutions by means of case studies from all over the world,

•   to get in contact with scientists, local decision makers, associations and companies who may assist with the implementation of the required technology.

Global WtERT-Organisations

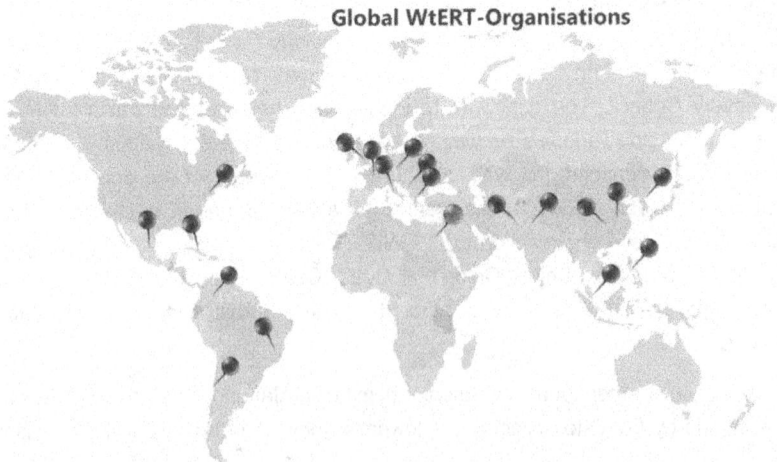

Diagram 5   Locations of Global WtERT Organisations

3.4   EU Including Candidate Countries

Looking at the chart below depicting the quotas achieved for the treatment of municipal waste up until 2017, one can identify the vast potential of reducing landfills merely in the European Union including the current candidate countries.

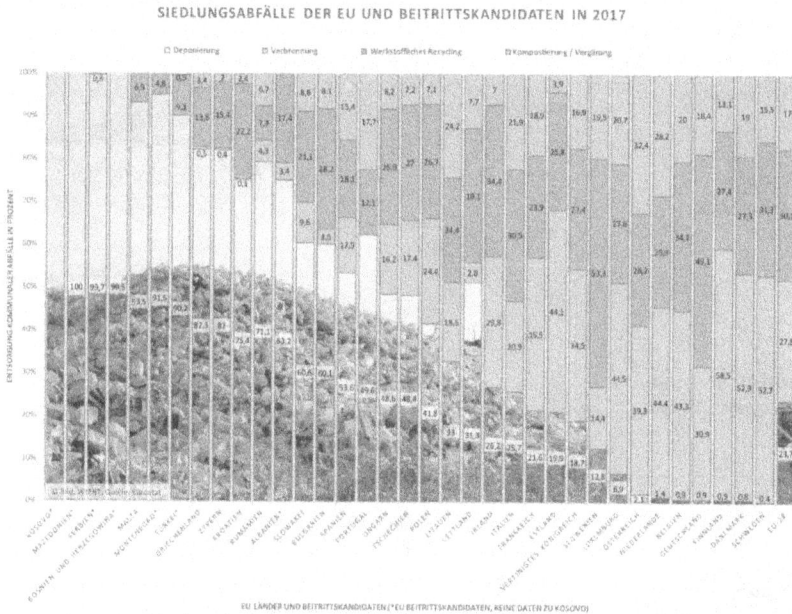

SIEDLUNGSABFÄLLE DER EU UND BEITRITTSKANDIDATEN IN 2017

Diagram 6   Current State of Waste Management in the EU

With the *Waste to Energy – Support for Decisions* as an offer for stakeholders, we will continually bundle a wide range of possible waste management solutions and show them with the goal of doing our contribution of reducing CO_2 emissions.

4   References

Dehoust, G., **Wieg-**   2005   *Umweltforschungsplan des Bundesministe-*
mann, K., **Fritsche**, U.,   *riums für Umwelt; Naturschutz und Reak-*
Stahl, H., **Jenseit**, W.,   *torsicherheit. Statusbericht zum Beitrag der*
Herold, A., **Cames**, M.,   *Abfallwirtschaft zum Klimaschutz und mög-*
Gebhardt, P., **Vogt**, R.,   *liche Potentiale.* Forschungsbericht 205 33
Giegrich, J.   314. On behalf of the Federal Environment
Agency. Retrieved from:
https://www.umweltbundesamt.de/sites/def
ault/files/medien/publikation/long/3006.pdf

European   2005,   *New waste strategy: Making Europe a recy-*
Commission   *cling society.* [Press release]. Retrieved
         Dec. 21st   from: http://europa.eu/rapid/press-
release_IP-05-1673_en.htm?locale=en

European   2018,   *Circular economy: More recycling of house-*
Parliament   April   *hold waste, less landfilling.* [Press release].
         18th   Retrieved from:

                       http://www.europarl.europa.eu/news/en/pr
essroom/20180411IPR01518/circular-
economy-more-recycling-of-household-
waste-less-landfilling

Official Journal of the   2018,   *DIRECTIVE (EU) 2018/852 OF THE EURO-*
European Union   June   *PEAN PARLIAMENT AND OF THE COUNCIL of*
         14th   *30 May 2018 amending Directive 94/62/EC*
on packaging and packaging waste
L150/2018. Retrieved from:

                       https://eur-
lex.europa.eu/eli/dir/2018/852/oj

Straub, Hans   2018   *Arbeitsgemeinschaft für Abfallbeseitigung*
(AFA). Reissued in: Müll & Abfall, Fachzeit-
schrift für Abfall- und Ressourcenwirtschaft,
12/2018. P. 640-643. Berlin: Erich Schmidt
Verlag

Address of author:

Dipl.–Ing (TU) Werner P. Bauer

WtERT Germany GmbH

Lipowskystraße 8

81373 Munich

Germany

Phone +49 (0)160 531 3624

E-mail: bauer@wtert.net

Waste management and climate protection internationally – Cooperation for sustainable waste management in Brazil

Cora Buchenberger, Christiane Pereira, Klaus Fricke

Technische Universität Braunschweig, Leichtweiß-Institut für Wasserbau,
Abteilung Abfall- und Ressourcenwirtschaft, Braunschweig, Germany

Abstract

Since 2010, Brazil has a very ambitious waste management policy, however waste management has not improved substantially and potentials for e.g. reduction of GHG emissions remain high. The German-Brazilian cooperation project ProtegeEr aims to improve the climate-friendly waste management on several levels, one of them is cooperation with Brazilian universities, in order to tackle the lack of skilled personnel, the lack of applied research projects and the relationship of waste management and climate protection. The project works on the federal, the municipal and the academic level and is supposed to contribute to the implementation of a sustainable waste management for Brazil.

Keywords

Climate protection; sustainable development; emerging countries; Brazil; Capacity development; sustainable waste management; sustainability.

1 Introduction

1.1 Status quo in Brazil

According to official data from the Brazilian Ministry of Science and Technology, in 2016 the waste sector in Brazil produced around 32 million tons CO_2eq, with a rising trend. In contrast, estimations from Technische Universität Braunschweig show a reduction potential of at least 58 million tons CO_2eq through a transition towards more sustainable measures in waste management.

In order to exploit this potential, i.e. to support the integration of waste management and climate protection, the German institutions *Deutsche Gesellschaft für Internationale Zusammenarbeit* (GIZ) Gmbh and *Technische Universität* Braunschweig elaborated the project "Support to the implementation of the National Policy for Solid Waste considering climate protection". This project emerged within the scope of the Internationale Klimaschutzinitiative (International Climate Ínitiative, IKI) promoted by the *"Federal Ministry for the Environment, Nature Conservation and Nuclear Safety (BMU)"* and was approved with Brazilian political partners such as the Brazilian Ministries of Environment and the former Ministry of Cities, which was merged in the beginning of this year into the new Ministry of Regional Development.

Regarding waste and pollution, the current situation in Brazil is in a critical state. Many municipalities count innumerous contaminated sites and high rates of water pollution due to illegally deposited waste. Until 2014, only a share of 58% of municipalities managed to rehabilitate their wild dumps into sanitary landfills or constructed new landfills without any previous waste treatment. This means that a share of 42% of total Municipal Solid Waste (MSW) is still disposed in an inappropriate way. According to ABRELPE, the Brazilian Association of Public Cleaning and Special Waste, about 80 million t of MSW are produced per year[1], and this amount is expected to grow by more than 3 % per year, whereas the population grows at a rate of about 0.8 %.[2]

Currently, the extension of sanitary landfill sites is the main effort on municipal and state level. In order to reach the aims and utilize the potential for climate protection in the Brazilian waste sector, aspects like waste treatment before final disposal, increase of recycling rates, adaptation and improvement of financial frameworks for municipalities, as well as capacity development at ministries, public and private banks and especially for municipalities must be addressed. The Brazilian government issued the National Solid Waste Policy (PNRS) in 2010, promoting the introduction of selective waste collection and reverse logistics, recycling of household waste, composting of organic waste, and generating renewable energy through biomass and biogas before final disposal throughout the whole country. Despite this ambitious policy, the situation regarding waste management has not improved much since its introduction, recycling rate for instance remain at a very low level of about 3 %, including composting.

In the past years, the Brazilian government rather promoted the construction of new landfills by a subsidy of 100% in economically weak municipalities and, by doing so, simultaneously impeded measures with greater climate protection impacts in other municipalities. In few states, environmental agencies approve new landfill sites only with a corresponding treatment plan. In the past, even simple sorting or composting plants were not operated efficiently and, therefore, closed. Today, because of these facts, many municipalities hesitate to introduce new technologies for the treatment of MSW.

1.2   Political situation

President Bolsonaro's election at the end of 2018 brought some political insecurity and big changes in the political departments. One of the partner ministries of the project, the Ministry of the Cities, was merged with the Ministry for Regional Integration into the new

---

[1] ABRELPE (2018) Panorama dos Residuos Sólidos no Brasil, em 2017

Ministry of Regional Development. Several responsible persons for the project were substituted. Despite the declaration not to host 2019's climate conference COP, the climate-related activities of the project have not been impeded or stopped and the project has a good standing within the new ministry.

2   Approach

2.1   Project design

The project ProteGEEr (derived from the Portuguese verb "proteger" which means *to protect*, and highlighting *GEE*, the Portuguese equivalent of GHG) started its activities in 2017 and is designed to support the implementation of a sustainable waste management in Brazil on three levels. First, on the federal level, working directly with the Ministries of Environment and Regional Development; second, on the municipal level, along with eight selected cities of different sizes all over Brazil; third, on the academic level, along with seven of the best ranked (according to CAPES) public universities in the areas of environmental and civil engineering from all over Brazil. The Technische Universität Braunschweig is one of three execution partners in this project, jointly with the consulting company GOPA-Adelphi and GIZ.

In the following, some major obstacles for the implementation of a sustainable waste management in Brazil are listed:

- The lack of skilled personnel to operate processes and plants, whereas the demand for workforce in the waste management sector is expected to grow.

- The lack of skilled personnel on municipal or judiciary stages: in municipalities, skilled workers able to evaluate plant projects and tenders, are missing; on the judiciary level, skilled workforce for law enforcement is missing.

- The lack of applied research, meaning cooperation between universities and municipalities, in order to support planning, construction and/or operation of plants.

- The lack of cooperation between universities and companies, in order to capacitate workers and/or students according to their specific needs, to develop specific, technological solutions and to promote innovations.

- The lack of awareness for the connection between adequate waste management and climate protection, as well as for the impact on air, water and soil and eventually for human health, within the population.

---

[2] http://www.brasil.gov.br/noticias/cidadania-e-inclusao/2018/08/populacao-brasileira-ultrapassa-208-milhoes-de-pessoas-revela-ibge

The TU Braunschweig as coordinator of the Brazilian partner universities aims to tackle the above mentioned obstacles by several activities, as shown in the next section.

The project aims to improve framework conditions for the utilization of climate protection potentials of the waste sector. Therefore, climate relevant criteria must be integrated in the regulations of ministries. For a successful introduction of sustainable waste management, municipality and company employees will be trained and municipalities will receive decision support (*capacity development*). During the project course, all significant players like representatives of municipalities, associations, cooperatives as well as public and private companies will be integrated, in order to achieve an intensive collaboration. Moreover, through the integration of climate-relevant aspects into the guidelines of the National Waste Policy, the quantification of specific reduction goals for the waste sector will be enabled. Thus, it is important for a national waste policy to not only including short-term, but also medium and long-term measures which may strengthen Brazil´s (so far) leading role in climate-friendly waste management in Latin America. From a social and economic point of view, the project provides an impulse for the development of an integrated circular economy within the scope of modern waste treatment technologies and strategies. Moreover, this initiative supports the further development of the Brazilian economy for a Green Economy and Green Jobs. The expansion of waste collection systems and the partly automated separation and treatment of municipal waste enable a better quality and larger quantity of different available waste fractions, hereby indirectly involving 500,000 informal waste collectors, who are integrated into the collecting and recycling systems by the municipalities.

Regarding ecological aspects, it is obvious that recycling and energy recovery of waste enable a significant reduction of fossil energy consumption. The introduction of separate collection and stabilization (composting and fermentation), and utilization of the organic fractions as a fertilizer and humus generator is encouraged. This will lead to an improvement of the quality of Brazil´s nutrient and humus poor soils, as well as groundwater. Negative impacts through insufficiently controlled landfills are minimized, and land consumption for new landfills is reduced, due to the increase of volume and lifespan of existing landfills.

2.2   Activities

Since the project's start, several technical visits have taken place, in order to demonstrate the functioning of processes and machines to e.g. political decision makers. Collaborators of the project have participated at important expositions and fairs of the sector, like IFAT 2018 in Munich or WasteExpo 2017 and 2018 in São Paulo and presented the approach to the public. In addition, delegations from Brazil participated in the IFAT and several site visits in Germany.

Within the academic line of the project, workshops in Brazil have taken place, in order to discuss research projects, possibilities of adapting the academic education and funding opportunities in Germany, Brazil and the EU. As a result, several smaller project proposals have been handed in in Brazil and one bigger research project was handed in in Germany. If the project proposals are successful, some initial financing for applied research will be available. Since some of the projects involve private companies, the cooperation between universities and the private sector is expected to grow, with the positive impact of more practical research and more practically skilled students.

In addition, partner municipalities and the universities were brought together to discuss possible common research projects, according to the cities' needs. First ideas have been collected and some initial analysis will be carried out. During the next months, a solid growth in the project number can be expected.

The universities have created a network to improve the internal communication and to increase their visibility as excellence institutions for waste management and sustainability in Brazil. The network chose a speaker who will represent the network when necessary. Since the beginning of the project, the number of participating universities has grown, from initially five to now eight universities and one research institution, located all over Brazil: Federal University of Ceará, research centre of the state of Ceará, Federal University of Pernambuco, University of Pernambuco, University of Brasilia, Federal University of Rio de Janeiro, State University of Rio de Janeiro, State University of Londrina and Federal University of Santa Catarina. The growth in number of participating institutions came from the institutions themselves, because they are interested in the constant exchange and knowledge transfer between Germany and Brazil. Due to the long-term experience of TU Braunschweig in Brazil, the project's network encompasses a lot of crucial players for the implementation of a climate-friendly waste management, as can be seen in the following image.

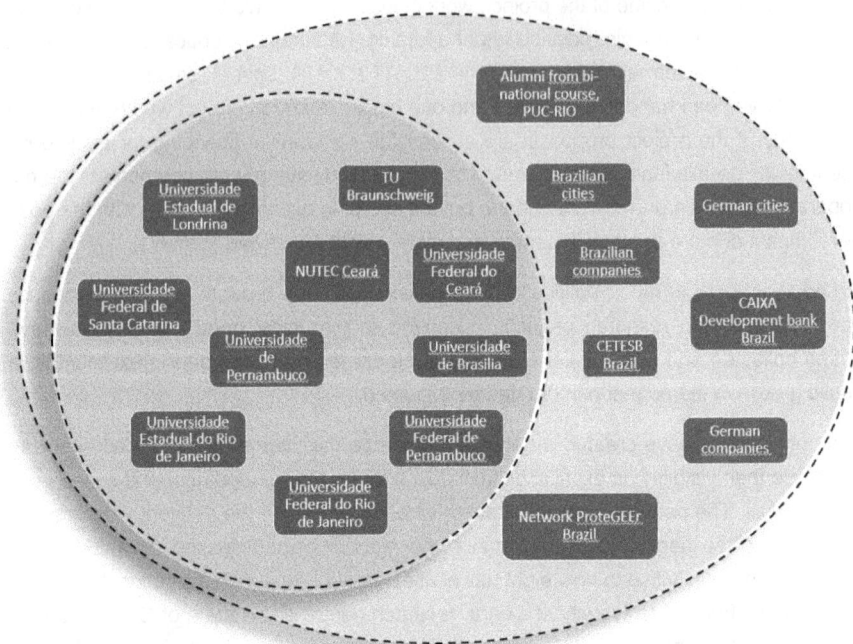

Figure 1: Project network

2.3   Teach4waste

Teach4waste is an internet based multimedia platform which is being developed by TU Braunschweig, with support of the Deutsche Bundesstiftung Umwelt and is one of the pillars for (sustainable) success of the project. The platform provides e-learning classes on all essential topics of sustainable waste management and is intended for all kind of persons interested in learning about sustainable waste management. It is designed to be self-explaining, thereby easy to navigate for self-learners, but also provides a variety of material for experienced staff, e.g. professors.

The material encompasses ready-to-use sets of presentation slides, videos of certain processes, animations, experiments with the necessary forms and instructions, films showing examples of creative and innovative waste management solutions worldwide, extensive scripts and quiz for learners to assess their (newly acquired) knowledge. The access to the platform is free of charge.

Within the scope of ProteGEEr, one activity is to discuss the teaching material with the partner universities and to adapt it to Brazilian legal requirements and other framework conditions. After this adaption, the material (or parts of it) shall be introduced into existing courses, so the climate protection potential of waste management becomes more visible. At the same time, teaching is expected to become more practical.

3   Outlook

Despite the initially described current situation and challenges, the Brazilian market offers wide opportunities. The cement industry for instance pushes for waste recovery which opens opportunities for the production of RDF and a reduction of waste to be landfilled.

The development of the market for Secondary resources is expected to bring more momentum to the development of the whole sector. The following table shows specific numbers for Secondary Resources.

Secondary Resources	Market situation	Comments
Recyclables	Average price 200 Euro/ton	Based on over 370 cooperatives and large informal sector – small scale plants – 150 tons/m
Compost	30-50 Euro/ton for selective waste Actual production 6 million tons/a. Stable market with growth potential	Currently no limitation for mixed waste but a new law is under preparation that will prohibit mixed waste compost
RDF	Average price 20 Euro/ton	Cement industry being business partner, long term contracts Strategy: 30% substitution of energy source in 5 years
Biogas	Average price 45 Euro/MW	Volatile tendency, currently high potential

4   Conclusion

The National Policy for Solid Waste, issued 2010 in Brazil, moved the topic *solid waste* to another level, extrapolating discussions focused exclusively on forms of final disposal on landfills. The new legal framework incorporates the consciousness of wealth and potential possibilities in waste management, and also reveals the errors and omissions that accumulated over the past 30 years. Waste management has changed significantly in recent years, towards a pillar of sustainable development, greatly contributing to environmental protection and, through recycling of waste, also ensuring better environmental conditions.

Until 2021, the project ProteGEEr will continue to provide comprehensive knowledge about the new market in Brazil and to build an inter-relationship with the waste sectors of Brazil and Germany. Thereby, an exchange with iconic German institutions on best practices to ensure climate protection will be established and natural resources will be preserved, thus providing a continuous exchange of experiences through vocational and technological education. The support and the dissemination of theory and practical knowledge of German management will bring to Brazil an innovative vision and inspiration to transform the current system in an efficient and continuous reality that meets the premises of the National Solid Waste Policy and the global trends for climate protection. These efforts should cause a cultural change that will protect natural resources and the climate, ensuring a better future for next generations.

5   Literature

ABRELPE (2018): Panorama dos Residuos Sólidos no Brasil 2017, Available online http://abrelpe.org.br/panorama/

Giersdorf, J.; Fricke, K.; Pereira, C; (2017): Climate Protection through sustainable waste management, presentation held during Regional Conference of DAAD in Recife, 2017.

ProteGEEr project information, available online http://www.protegeer.gov.br/

Author's address(es):

Dipl.-Reg. Wiss. Latam. Cora Buchenberger, Dipl-Ing. Christiane Pereira, Prof. Dr.-Ing. Klaus Fricke
Technische Universität Braunschweig
Leichtweiß-Institut für Wasserbau, Abt. Abfall- und Ressourcenwirtschaft
Beethovenstraße 51a
D-38106 Braunschweig
Telefon +49 531 391 3990
E-Mail c.buchenberger@tu-braunschweig.de

GHG reduction in waste management in Brazil: Life cycle and IPCC approaches and its preliminary results

Hélinah Cardoso Moreira[1], Jürgen Giegrich[2], Tathiana A. Seraval[3], Flávio Ravier Viana de Assis[3]

1 Deutsche Gesellschaft für die Internationale Zusammenarbeit – GIZ Gmbh, Brasília, Brasil
2 Institute for Energy and Environmental Research – Ifeu-Institut, Heidelberg, Germany
3 Methanum Energy and Waste - Belo Horizonte, MG, Brasil

Abstract

Brazil proposed the instrument of ratification of the Paris Agreement in September 2016, which contains a commitment to reduce GHG emissions by 37 % in 2025 and 43 % in 2030, with reference to 2005, but didn't mentioned specific mitigation measures for the municipal solid waste sector.

In order to realize how the waste sector can contribute with the NDC goals, GHG-mitigation scenarios were proposed considering different technologies implementation levels and considered in LCA and IPCC approaches. The first results will be presented an idea about how an optimized waste management can contribute to the mitigation of GHG in Brazil, considering technological diversification through the valorization of the dry and organic fractions of waste.

Keywords

Climate protection; municipal solid waste (MSW); sustainable waste management; circular economy; Brazil; landfill; quantification of greenhouse gases emissions.

1   Introduction

1.1   Climate and waste management

Brazil published the Nationally Determined Contribution (NDC) of the Paris Agreement in September 2016. In this Agreement, the country committed itself to adopt measures to reduce GHG emissions by 37 % in 2025 and 43 % in 2030, with reference to 2005, equivalent to an emission ceiling of 1,300 and 1,200 $MtCO_2e$ in 2025 and 2030 respectively.

As presented in the "3rd National Communication of Brazil to the United Nations Framework Convention on Climate Change", the main anthropogenic sources of GHG in the country are: fossil fuel combustion (coal, oil and gas) for the generation of electricity, transportation, industry and urban agglomerates (CO_2); farming (CH_4) and changes in land use such as deforestation (CO_2); Industrial processes (CO_2); and final disposal of solid waste (CH_4). The treatment of municipal solid waste contributed just with 2.5 % of the whole emissions (MCTIC, Annual Estimates, 2017), mainly because the final disposal on soil.

The final disposal of solid waste is the main responsible for the emissions of methane in the sector due to the fact that its management in Brazil is based on the grounding of in-natura residues, occurring the anaerobic decomposition of the organic materials and, consequently, generation of methane. In this context, an integrated municipal solid waste (MSW) management, based on the principle of circular economy, can be an important ally in the reduction of GHG emissions. However, once evaluated by national inventories, these benefits would be accounted for in other sectors of the economy by the emissions measurement approach of the International Panel on Climate Change (IPCC), such as emissions in transport due to waste collection (accounted for in the transport sector), recycling (industry sector) or energy recovery (energy sector).

Another approach for quantifying the GHG-Emissions is based on Life Cycle Thinking, which presents an integrated view and is focused on the flow of materials, from the extraction of the raw material to the final disposal but including all loops of materials and energy recycling. From this point of view, all the emissions resulting from complete waste management are considered, starting from material production up to transportation and final disposal. The broader view of the waste sector from the perspective of the Life Cycle Thinking demands a new approach also in relation to the adopted strategy of quantification of emissions, and that has been gradually adopted all over the world.

According to the National Basic Sanitation and Municipal Waste Policy and the recently updated National Sanitation and Municipal Waste Plan (PLANSAB), which is in the public consultation phase, progressive targets for eliminating dumps and increasing the coverage of services for the rural and urban areas were established. Added to this, Brazil, in its National Policy on Climate Change (PNMC), Law No. 12.187/2009, established the national commitment to adopt mitigation actions to reduce GHG emissions, implying the need to implement new technological solutions of low emissions, and consequently, the reduction of the final disposal of waste in landfills and dumps, major focuses of methane generation (CH_4) and carbon dioxide (CO_2).

The National Solid Waste Policy (PNRS) - Law No. 12.305, of August 2, 2010 - presents the substantial guidelines on the integrated management of municipal solid waste (MSW) in Brazil and institutes the obligation of adequate disposal of MSW and consequently the replacement of "dumps" by landfills and emphasizes that only treated waste should be sent to landfills.

Even with a consolidated legal framework on waste, considering the 5570 municipalities, just 36.9 % of the municipalities send the MSW to sanitary landfills and the rest for environmental inappropriate disposal sites, such as open dumps or uncontrolled landfills or did not provide information (SNIS, 2017).

Under the International Climate Initiative of the German Federal Ministry for the Environment, Nature Conservation and Nuclear Safety, Brazil and Germany are implementing a technical project for climate protection in municipal solid waste management.

The project is enforcing the understanding of the contribution of the municipal solid waste sector to climate change mitigation, by promoting environmental friendly technologies for waste treatment, by understanding the contribution of the sector to other sectors of the economy and strengthening the circular approach.

In this context, the first studies were carried out on how the solid waste management sector in Brazil can contribute to the goals of the Paris Agreement, and also how an optimized waste management can contribute to the mitigation of GHG, considering technological diversification through the valorization of the dry and organic fractions of waste.

In order to comply with the National Solid Waste Policy of Brazil and with the National and Municipal Waste Plan, GHG-mitigation scenarios were proposed considering different technologies implementation levels and evaluated in Life Cycle and IPCC approaches. The first results will be presented and an idea about how the developing countries can contribute to a low carbon economy through waste measures will be showed.

1.2   IPCC and Life Cycle methodologies

The two chosen methodological approaches to account the GHG emissions of the actual situation and potential mitigation measures in waste management are the following:

- GHG reporting of a national GHG inventory according to the IPCC Guideline of 2006 (IPCC approach)

- GHG assessment of an entire waste management system for a country with a life cycle perspective – (Life Cycle approach)

It is of utmost importance to clearly distinguish the two approaches for setting GHG reduction goals and related reduction measures.

The IPCC approach is directly related to the reporting scheme of IPCC and must follow the rules of this scheme. A mitigation goal in the waste management sector for instance is restricted to the GHG reduction of direct emissions of

- the reporting time of one specific year

- the reporting territory

- the management activities of the waste sector according to IPCC (landfill, etc.)

In contrast to the IPCC approach the accounting of a life cycle leads to mitigation goals that encompass

- the entire time span of GHG emissions caused by the waste of one year

- the GHG emission without territorial limit caused by the waste of a territory

- all waste management activities including substitution effects regardless of IPCC

For the political setting of GHG mitigation objectives in waste management it is important to define which approach is applied. Normally the IPCC approach is used for setting goals since the philosophy of IPCC relates to a reporting year, a reporting territory and a clear-cut set of emission sources. Furthermore, it is recommended by the international community, since the IPCC is recognized as the most authoritative scientific and technical voice on the issue of climate change, and its assessments has been supported by the United Nations (UN) and the United Nations Framework Convention (UNFCCC). It also offer a possibility of using a single approach globally, allowing comparison and monitoring of national efforts on the issue.

But the reporting of GHG emissions following the life cycle approach is significantly useful for public communication when used in parallel because it appreciates the entire waste management system including recycling of dry and organic fractions and its contribution in terms of materials and energy substitution. In that manner, it shows the real contribution of environmental friendly technologies for waste treatment for climate protection. It also provides in a transparent way an immediate impression of the consequences for GHG accounting from decisions in waste management independent of the time and the location.

Hence the life cycle approach complements the IPCC derived GHG mitigation objectives and facilitates a helpful decision support to achieve these goals.

2   Approach

2.1   Guidance and assumptions

The scenarios were defined in both methodologies based on the following strategies:

- Increase in organic recycling

- Increased recycling of dry waste, with specific targets by type of dry waste: glass, paper, plastic, aluminum, etc.

- Closing of dumps and migration to sanitary landfills with different efficiency collection rates

- The final target was the reduction of 43 % of GHG in the waste sector by 2030.The estimated basic figures for the GHG calculations for 2030 are:

- Inhabitants of Brazil: 225.523.235

- Waste amount per capita and year: 0,978 kg/cap/year

- Total waste generated in 2030: 80,5 Mio tonnes

For estimation reasons, the composition of this solid waste for the year 2030 was assumed to be identical to the waste of the year 2005:

Table 1 Average composition of solid waste in Brazil

Components	in % (wet waste)
Food waste	48,4 %
Garden and park waste	3,0 %
Paper, cardboard	13,1 %
Plastics	13,5 %
Glass	2,4 %
Ferrous Metals	2,3 %
Aluminium	0,6 %
Textiles	2,6 %
Rubber, leather	0,7 %
Nappies (disposable diapers)	4,0 %
Wood	4,7 %
Mineral waste	0,0 %
Others	4,7 %
Total (must be 100 %)	100,00 %

Additional basic information is the distribution of types of landfilling as final disposal. The objective of PNRS is the closure of 90 % of dump sites. So the amount of dump sites from 2005 was used to calculate the reduction of dump sites until 2030 as basis for the GHG calculations. The distribution of types of landfilling in Brazil in 2005 is as follows:

Table 2 Types of landfilling in Brazil in 2005

Open dumps/unmanaged disposal site	50,3 %
Uncontrolled landfill without gas collection	22,3 %
Sanitary landfill with gas collection	27,4 %

The estimated rate of landfill gas collection of the sanitary landfill was determined to be 7,2 %. An input for the basic setting of waste management options is also the existing recycling rates achieved. Unfortunately recycling rates could only be found for 2017 with 26 % of dry recyclables in total. In lack of other data from 2005 it was assumed that these recycling rates are the breakdown of the 26 % to materials as reference:

Baseline scenario used for the life cycle approach:

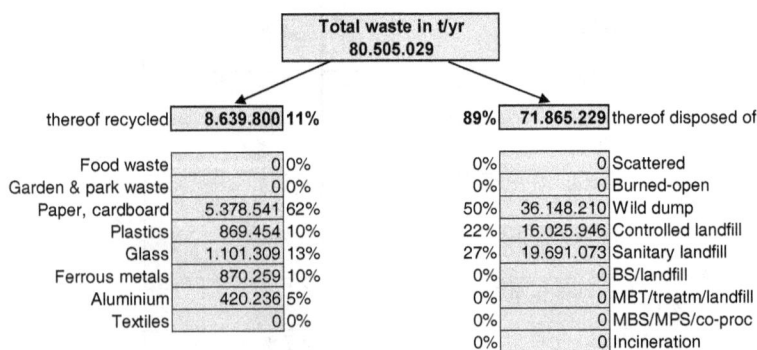

```
                              Total waste in t/yr
                                 80.505.029

thereof recycled   8.639.800 11%          89%  71.865.229 thereof disposed of

Food waste              0 0%          0%          0 Scattered
Garden & park waste     0 0%          0%          0 Burned-open
Paper, cardboard  5.378.541 62%       50% 36.148.210 Wild dump
Plastics          869.454 10%         22% 16.025.946 Controlled landfill
Glass           1.101.309 13%         27% 19.691.073 Sanitary landfill
Ferrous metals    870.259 10%          0%          0 BS/landfill
Aluminium         420.236 5%           0%          0 MBT/treatm/landfill
Textiles                0 0%           0%          0 MBS/MPS/co-proc
                                       0%          0 Incineration
```

Figure 1   Baseline scenario used for the life cycle approach

Each methodology demands specific input data and some techniques were adopted to overcome the lack of some (i.e. interpolation, expert recommendation), the quantification considered some details not mentioned on this article.

The IPCC model demand a historical series, because it is based on the first-order decay model. For this estimation was considered the period from 1990 until 2030. It's important to mention that the baseline scenario will be the basis of comparison for the effects of adopting the strategies considered for organic, dry and methane recovery in the mitigation scenarios analyzed, but not represent the exactly the reality in Brazil.

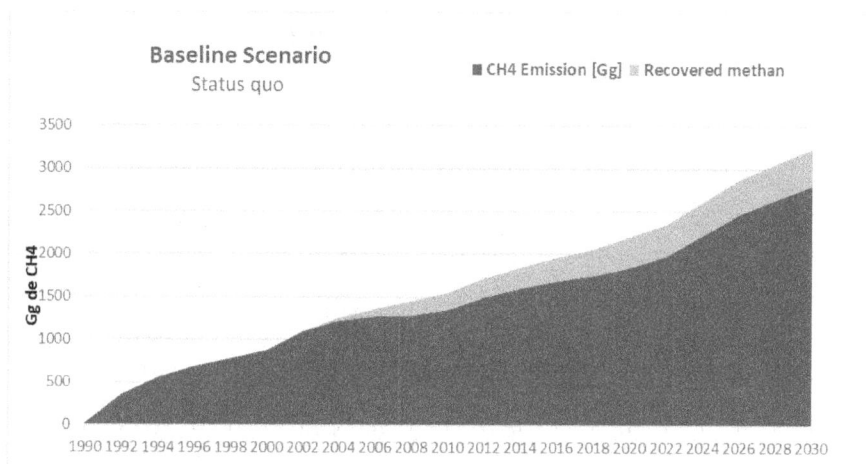

Figure 2   Baseline scenario with IPCC with accumulated methane emissions and recovery of from 1990 to 2030

2.2   Relevant scenarios for quantification

Besides the equal baseline scenario for all groups of scenarios a stepwise improvement for the sequence of scenario groups is applied. So recovery scenarios for dry recyclables are modelled with the best option of rehabilitation of dump sites and the recovery options for organic material are based on high quality landfilling and increased recovery rates of dry material.

The waste management objectives of PNRS are fulfilled subsequently:

- 90 % closure of dump sites until 2031

- 45 % reduction of dry recyclables from landfilling for 2031

- 53 % reduction of wet recyclables from landfilling for 2031

This strategy helps to identify the importance of waste management options for GHG reductions in the view of the life cycle approach.

3   Results

3.1   Life cycle approach

The results using the life cycle approach are presented in two steps:

- Closing dump sites and equipping sanitary landfills

- Overall improvement of MSW management fulfilling the objectives of PNRS

The life cycle approach calculates the GHG emissions caused by the waste generated in the target year 2030, integrated over all future years and related to the target year. This is especially important for the methane emissions from dumps and landfills governed by collection rates and treatment measures of landfill gas.

In first step the GHG emissions are analysed assuming that only the objective of converting 90 % of the wild dumps to sanitary landfills is implemented in 2030 and some forms of landfill gas capture and treatment is established for sanitary landfills. The remaining dump sites and controlled landfills – by definition – have no installations for gas collection. No further measures for recycling besides existing activities are taken into account (see figure 1). The following scenarios are chosen for step 1 calculations and results:

- Baseline scenario: The situation of today (see table 2) is extrapolated to the year 2030 without any changes. Sanitary landfills have a gas collection rate of 7 % but released without treatment.

- Landfill scenario 1: 90 % of the dump sites are converted to sanitary landfills still with a gas collection rate of 7 % but treating 50 % by flare and 50 % by electricity generation. (5 % dump sites, 20 % controlled landfills, 75 % sanitary landfills)

- Landfill scenario 2: Conversion of dump sites as in landfill scenario 1 but with a gas collection rate of 25 % treating 50 % by flare and 50 % by electricity generation.

- Landfill scenario 3: Conversion of dump sites as in landfill scenario 1 but with a gas collection rate of 40 % treating 50 % by flare and 50 % by electricity generation.

The business-as-usual baseline scenario results in a GHG emission of about 54 Mio. tonnes CO_2 eq. integrated over time. A benefit of 14 Mio t CO_2 eq. from substituting primary material production by recycling of about 10 % of the total MSW has already been taken into account. Further results can be summarized as:

- The conversion of 90 % dump sites to sanitary landfills but without changing the efficiency of gas collection (7 %) fulfills the objective of the national waste management policy for landfilling (PNRS) but increases the GHG emissions from 54 Mio t CO_2 eq. to 76 Mio t CO_2 eq. This is due to the higher methane forming potential in a

higher packed sanitary landfill compared to a more loosely packed dump site (calculated with default values of IPCC).

- The conversion of 90 % dump sites to sanitary landfills with a gas collection rate of
25 % still leads to an increase of GHG emissions by 15 %. Only a gas collection
rate of 40 % compensates the shift from dump sites to sanitary landfills reaching 52
Mio t CO_2 eq. that corresponds to a slight decrease compared to the actual situation. The consequence is to combine the conversion goal for dump sites to sanitary
landfills with efficient gas collection and treatment systems.

The second step of analysing GHG emissions from waste management in Brazil with the
help of the life cycle approach incorporates all objectives set by the national waste management policy (PNRS). Besides closing 90% of the dump sites additionally 45 % of dry
recyclables and 53 % of wet recyclables shall be diverted from landfills in 2030. Again,
calculation scenarios are used to investigate the consequences for GHG emissions of
these overall waste management improvements for the target year 2030.

- Baseline scenario: For comparison the baseline scenario is kept the same as
above. The situation of today is extrapolated to 2030 with about 50 % waste going
to dump sites and a recycling rate of 26 % for dry recyclables.

- Total Scenario 1: 90 % of the dump sites are converted to sanitary landfills with a
gas collection and treatment rate of 40 % (50 % flare, 50 % electricity generation)
and an increase of recycling for dry material from 26 % to 47 % are chosen for this
scenario. Recycling rates are set to 80 % for paper, 30 % for plastics, 80 % for
glass, 80 % for ferrous metals and 87 % for aluminium. No recycling of wet recyclables is assumed.

- Total Scenario 2: The same assumptions as for the Total Scenario 1 are combined
with a recycling rate of 40 % of organic material which is about 16.5 Mio tonnes out
of 80 Mio tonnes of waste in 2030. Separate collection, treatment with anaerobic digestion and use of the biogas and the digestion residue as compost are further
specifications.

- Total Scenario 3: Now the figures for Total Scenario 1 are calculated with 60 %
identical recycling of organic waste that is about 25 Mio tonnes separately collected
biogenic material. The span of 40 % to 60 % recycling of wet recyclables is chosen
to cover the objective of PNRS and evaluate its consequence for the GHG emissions.

Results for the GHG emissions of these total scenarios are presented in figure 3. Light
grey columns with positive numbers are related to the GHG emissions caused by waste
management technologies from diffusive methane emissions from landfills to recycling ac-

tivities of dry and wet recyclables. Striped columns with negative numbers represent the GHG emissions that are avoided by substituting primary material by recycling. The dark grey columns show the net GHG results caused by waste management activities minus avoided GHG emissions. The difference of the net GHG results between the scenarios and the baseline determines the GHG savings of waste management in Brazil for the waste handling of the year 2030.

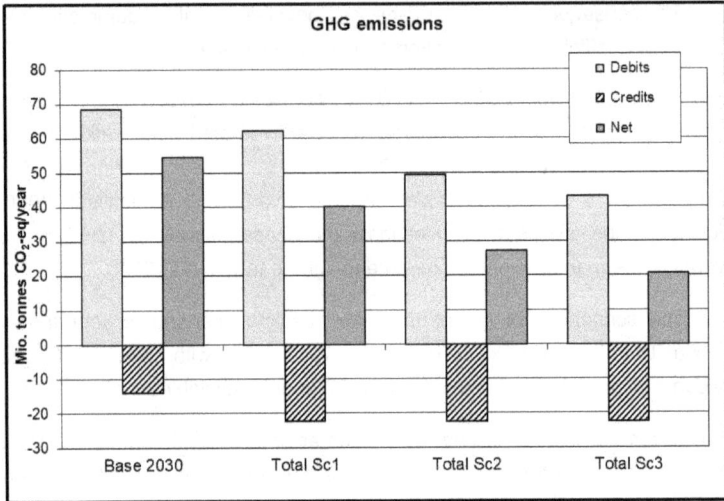

Figure 3   GHG balance for waste management scenarios in 2030 with LCA approach

The results can be summarized as follows:

- The conversion of 90 % of the dump sites to sanitary landfills with a gas collection and treatment rate of 40 % and additionally the diversion of 47 % of dry recyclables from landfill and their recycling will lead to a decrease of GHG emission of about 26 % for the waste generated in 2030. Mostly the increase of recycling of dry material in the MSW contributes to reduction of GHG emissions from about 54 Mio t CO_2 eq. to 40 Mio t CO_2 eq.

- The reduction of GHG emissions will be considerably higher if organic material is recycled. Separating, digesting and recycling additionally 40 % of wet recyclables leads to the GHG emission of only 27 Mio t CO_2 eq. and 60 % recycling to 21 Mio t CO_2 eq. for the waste generated in 2030. These are reductions to the baseline scenario representing the waste management of today by 50 % and 62 %. The effects of organic waste recycling is mainly due to avoiding methane emissions in the landfills.

3.2   IPCC approach

The scenarios established with the IPCC approach was based on the impact of each strategy and the impact of integrated policies, considering concomitant measures like increase of dry and organic recycling rates.

Table 3   Scenarios considered in IPPC approach

Scenarios	Policies (PNSB, PNRS)
	Isolated impact of policies
C1	P1: 53 % reduction in organic waste sent to landfill
C2	P2: 45 % reduction of dry waste sent to landfill
C3	P3: Expansion of methane recovery in landfills
	Impact of integrated policies
C4	P1: 53 % reduction in organic waste sent to landfill *plus* P2: 45 % reduction of dry waste sent to landfill
C5	P1: 53 % reduction in organic waste sent to landfill *plus* P2: 45 % reduction of dry waste sent to landfill *plus* P3: Expansion of methane recovery in landfills

In scenario 5, a 53 % reduction in organic matter and 45 % reduction in dry matter sent to landfill, associated with a 22 % recovery rate of methane from landfills in 2019 to 2025 and 2.75 % from 2026 to 2030, resulting in the only scenario that achieve the NDC goal.

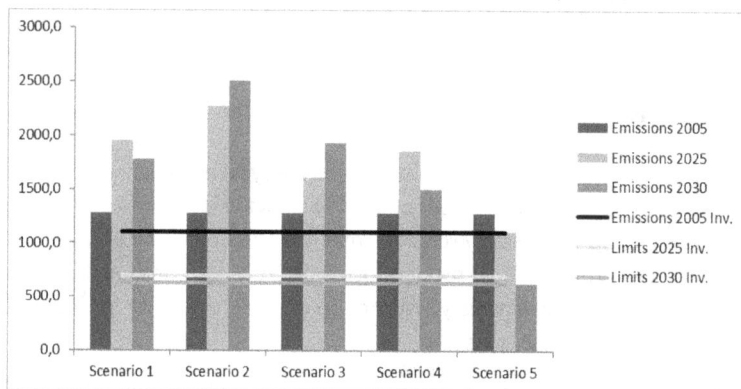

Figure 4   Scenario Emissions x Inventory Limits [Gg CH4]

According to the IPCC premises, methane generation has a gradual and cumulative be-havior over time, intrinsic to the emissions recorded and reported via IPCC. This implies that in addition to the net reduction of emissions between the base and inventory years, it is necessary to reduce the incremental emissions accumulated over the years, since a landfilled waste today generates emissions in the current year and future years, due to its fraction that has not yet completely degraded. And so on and so forth. Thus, by maintain-ing a grounding standard, even if it is significantly reduced, we add the emissions from previous years of waste already grounded, and which continue to emit methane for at least two decades.

For this reason, methane recovery is an important action with a short-term effect, since it mitigates emissions immediately (after the year in which the removal was reported), pro-vided that the process is properly monitored and reported. This situation justifies efforts on the part of the Federal Government to establish instruments for regulation, financing of actions, transparency and monitoring of progress in the sector, making them subject to accounting in inventories and national communication. However, it is important to empha-size that, in isolation, methane recovery is not capable of leading the sector to meet the goals of the Paris Agreement. It is important to emphasize that for IPCC, methane flaring is considered a useful measure to avoid it emission. Energy production in sanitary landfills will be accounted for in the energy sector.

On the other hand, methane recovery measures should be accompanied by medium-term actions to mitigate methane generation. Otherwise, the recovery efforts will have an effect in the short term but will rise again due to the incremental and cumulative nature of the emissions. In this sense, the effects of the reduction of organic grounding proved to be more significant, due to the conditions and degradability and potential emissions intrinsic to this type of material (food waste in general), and which represent 50 % of all MSW gen-erated in the country.

4   Final Conclusions

This conclusion is based on an assessment with the life cycle approach which assumes all GHG emissions to be related to the year of generation of the given amount of waste. This is different to the IPCC approach which calculates the GHG emission of a year including the waste from years before. Both methodologies can't be compared and have different scopes, which allow a complementary vision of how certain strategies in waste manage-ment can impact on GHG emissions.

A long time average of a defined waste management will lead to the same GHG emissions for IPCC reporting as determined by the life cycle approach (without debits and credits of recycling) since generation of waste and GHG emissions from landfills are in a steady state.

The GHG reduction related to recycling is only determined if the substituted primary raw material and their avoided GHG emissions is included. Following the IPCC logic for GHG accounting these activities are assigned to the reporting sectors energy generation and industry.

GHG reduction of 50 % does not result automatically if the PNRS waste management objectives are fulfilled. So sanitary landfills must be necessarily equipped with gas collection and treatment systems. A gas collection efficiency of at least 40 % should be guaranteed over the lifetime of a sanitary landfill.

If dump sites are converted to sanitary landfills without efficient gas collection systems GHG emissions may even increase considerably.

The GHG emissions of recycling activities must be reduced as much as possible. Especially biogenic treatment process must avoid the emission of fugitive methane emissions.

The reduction of GHG emissions in waste management shall contribute to the NDC (Nationally Determined Contribution) of Brazil to the Paris Agreement which is based on annual GHG accounting values. This is necessary to be in line with the other sectors and the methodologies of IPCC.

The accounting of the life cycle approach of GHG in waste management is not meant to achieve a specific mitigation goal but to give indications which measures are important and contribute to the reduction to which extent. It can be analysed if temporal and territorial effects occur outside the IPCC reporting scheme.

More pragmatically, when we evaluate waste policies from a cross-cutting perspective, we recognize their role not only in the context of climate change, but in the pursuit of global goals of sanitation, accessible and clean energy, innovation and infrastructure, sustainable cities, responsible production and consumption - all United Nations (UN) Sustainable Development Goals (SDGs). The perspective of broader and more integrated policies opens new paths towards a circular and low-carbon economy.

5   Literature

Brazil. Ministry of the Environment (MMA).	2009	Law n° 12.187, December 29, 2009. Establishes the National Policy on Climate Change (PNMC) and makes other provisions. Brasília.
Brazil. Ministry of the Environment (MMA).	2010	National Solid Waste Plan (PNRS). Brasília, 108p., 2010 (Preliminary Version).
Brazil. Ministry of the Environment (MMA).		Law n° 12.305, August 2, 2010. Establishes the Na-

2010   tional Solid Waste Policy, amends Law no. 9.605, of
       February 12, 1998; and makes other provisions.
       Brasília, 20p. 2010.

Brazil. Ministry of the Envi-   2015   Paris Agreement (2015). Available at:
ronment (MMA).   http://www.mma.gov.br/clima/convencao-das-
                                        nacoes-unidas/acordo-de-paris.html. Accessed on 15
Brazil. Ministry of Regional   April 2019
Development. National   2019   National Sanitation Information System (NHIS). Di-
Sanitation Secretariat.   agnosis of municipal solid waste management 2017.
(SNS/MDR)   Brasília, 2018.

Brazil. Ministry of Science,   2017   Annual Emissions Estimates for Brazil
Technology, Innovation and
Communication.

Giegrich, Jürgen. IFEU –   2019   Guidelines for the reduction of Greenhouse Gases
Institut für Energie- und   for Waste Management in Brazil. Preliminary phase.
Umweltforschung GmbH   Heidelberg, 21 March 2019.

                                        Report on the quantification of GHG emissions in the
Seraval; Tathiana. De   municipal solid waste sector and perspectives of
Assis, Flavio. METHANUM.   2018   alignment of climate and waste management policies
                                        in Brazil. Belo Horizonte, 2018.

Author's address(es):   MSc. Tathiana Almeida Seraval
                                                  Methanum Energy and Waste
MSc. -Ing. Hélinah Cardoso   Rua Iraí, 560, Vila Paris
Deutsche Gesellschaft für   30.380-725 Belo Horizonte-Minas Gerais,
Internationale Zusammenarbeit GmbH   Brasil
SCN Quadra 01, Bloco C, Sala 1501   Telefon +55 31 98167 1446
70.711-902 Brasília-DF, Brasil   E-Mail tathiana@methanum.com
Telefon +55 61 999855563
E-Mail helinah.cardoso@giz.de
                                                  Ing. Flávio Ravier Viana de Assis
                                                  Methanum Energy and Waste
Dipl.-Phys. Jürgen Giegrich   Rua Iraí, 560, Vila Paris
ifeu - Institut für Energie- und Umweltfor-   30.380-725 Belo Horizonte-Minas Gerais,
schung Heidelberg gGmbH   Brasil
Wilckensstr. 3   Telefon +55 31 99579 9852
69120 Heidelberg   E-Mail flavio.assis@methanum.com
Telefon: +49 6221 4767-21
E-Mail: juergen.giegrich@ifeu.de

Taking Waste Management in Hamburg to the Next Level – Construction of a Centre for Resources and Energy

Heinz-Gerd Aschhoff, Ronja Grumbrecht and Verena Höck

Stadtreinigung Hamburg, Germany

Abstract

The public waste management authority Stadtreinigung Hamburg (SRH) is building a centre for resources and energy (CRE) on the site of the old Stellinger Moor waste-to-energy plant in Hamburg. The CRE will combine mechanical, biological and thermal waste treatment processes, meaning that the site can process all of the household waste and a significant proportion of the biodegradable and green waste generated in sections of north-western Hamburg. Household waste is sorted in the first phase. The resulting material streams undergo further rounds of treatment at downstream facilities, yielding biogas, district heat, power and recyclable materials. A separate anaerobic digestion and composting plant is envisaged to treat biodegradable and green waste.

The layout of the different sections of the CRE allows synergies to be tapped perfectly, meaning that the CRE will make a significant contribution to expanding Hamburg's district heating network in a climate friendly manner and ensuring waste management reliability in Hamburg. The CRE is to start full operations in the middle of 2025.

1 About Stadtreinigung Hamburg

Stadtreinigung Hamburg is one of Germany's biggest municipal providers of waste and recycling services. As a public waste management authority and certified waste management company, Stadtreinigung Hamburg collects, transports and treats waste from approximately 924,000 Hamburg households and about 100,000 different businesses. SRH guarantees reliable waste management services, thereby fulfilling a key tenet of public services of general interest for people living in Hamburg. SRH's statutory tasks include:

- Management of waste in the City of Hamburg (collection, transportation and treatment)
- Cleaning paths and roads
- Winter services
- Preventative and defensive disaster prevention efforts
- Cleaning green spaces and parks
- Building and operating public toilets

SRH is also responsible for planning, building and operating facilities and infrastructure as part of its waste management duties in order to contribute towards low-cost, efficient and environmentally sound energy supply.

Since 2011, Stadtreinigung Hamburg has been focusing specifically on waste prevention, reuse, upcycling and recovery with the Hamburg Recycling Campaign. SRH offers a service-friendly four-bin pick-up system for paper, biodegradable and green waste, packaging, and residual waste. A drop-off system with locations throughout the city (12 recycling yards, depot containers and hazardous waste mobile) is also available.

The planned strategy for the CRE guarantees that waste required to be handled by the City of Hamburg and waste not under its remit can be treated safely and securely with its own plant technology in order to guarantee reliable and self-sufficient waste management services. At the same time, a significant contribution can be made towards providing the city with reliable and climate-friendly district heat.

2   Centre for Resources and Energy (CRE)

The corporate decision to close and dismantle the old Stellinger Moor waste-to-energy plant (WTE) after buying Müllverwertung Borsingstraße (MVB) placed SRH in a position to develop a sustainable waste treatment centre and to bring it to fruition within the metropolitan area.

The CRE focuses on the environmentally friendly use of material and energy contained in waste and the management of non-recyclable waste. In addition to environmentally sound waste management, the planned waste recovery plants have another key mission: climate-neutral and sustainable energy generation and waste recovery.

The CRE is made up of a set of different facilities that can be broadly grouped as follows: household waste sorting, anaerobic digestion and drying of the fine fraction from household waste, anaerobic digestion and composting of separately collected biodegradable and green waste, biogas upgrading, a biomass heating plant and a refuse-derived fuel plant. Besides recycling materials, the CRE also generates energy in the form of biogas, heat and power.

2.1   Use of existing infrastructure and building

As mentioned above, the CRE's plants and machinery are being installed on the site of the former Stellinger Moor waste-to-energy plant as well as on the property of the former Biowerk anaerobic digestion facility and parts of the current car park.

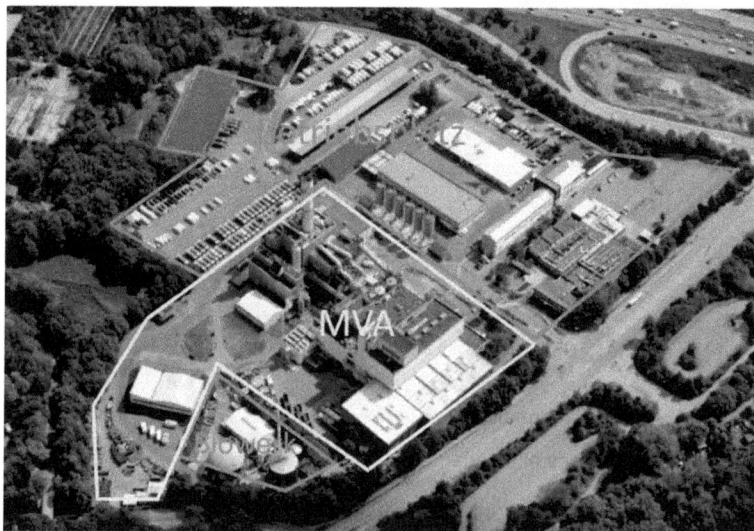

A prior study examined the structure and estimated the required refurbishment costs for the old building in order to conserve resources and reduce the costs of investing in new technology. The goal was to decide which building in the CRE plant might be used.

With regards to the waste bunker, the analysis indicated that the structure, over 40 years old, was in a good condition, meaning that it could continue to be used after refurbishment.

2.2 Creating the CRE

Once systems and machinery for the CRE were chosen and existing buildings suitable for continued use were identified, a strategic line-up, including interfaces between the different treatment phases, was developed in the first phase of planning. Figure 1 illustrates the process within the CRE in diagram form:

EBS | Hausmüll | TA 1 Sonstige Biomasse | TA 3 | Holzige Biomasse | TA 2 Bio- und Grünabfall

Sortierung | Aufbereitung | Biomethan | Aufbereitung | Aufbereitung

Wertstoffe

Ersatzbrennstoff (EBS)

Vergärung | Biogas | Biogas-Aufbereitung | Biogas | Vergärung

Gärrest | Trocknung | Kompostierung | Gärrest

TA 5

EBS-HKW | TA 4 | Biomasse-HKW | Siebüberlauf | Kompost

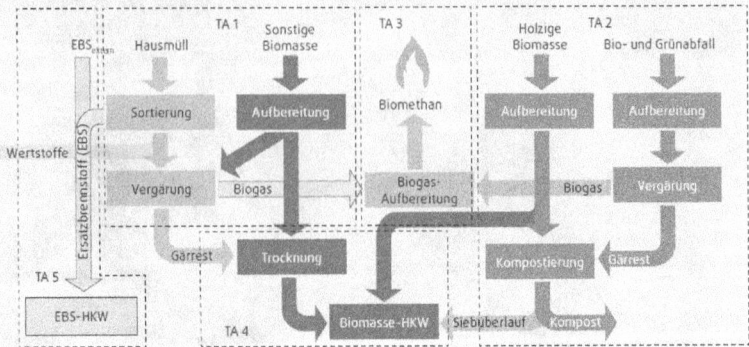

Figure 1: Diagram showing treatment processes in the CRE

The following sub-sections of the plant will be needed to bring this process to fruition:

1. Section 1.1: Sorting line for up to 140,000 Mg/a commingled municipal waste
2. Section 1.2: Anaerobic digestion of the organic-rich fine residual waste fraction
3. Section 2: Anaerobic digestion and composting of 45,000 Mg/a biodegradable and green waste
4. Section 3: Upgrading biogas from sections 1.2 and 2 into biomethane, which can be fed into the public natural gas grid
5. Section 4.1: Drying digestate from section 1.2 and other waste-based biomass to produce a biofuel for section 4.2
6. Section 4.2: Biomass co-generation plant for biomass conditioned in section 4.1, other waste-based biomass and waste wood with a thermal rating of 47 MW
7. Section 5: RDF co-generation plant for energy recovery of the RDF made in section 1.1 and of third-party RDF with a thermal rating of 47 MW

Since a relatively small amount of space is available to create the CRE (42,000 m²), the project's challenge lies in perfectly arranging the different sections of the plant on this plot. The following illustration shows the layout of the CRE's different sections on the site of the former WTE plant:

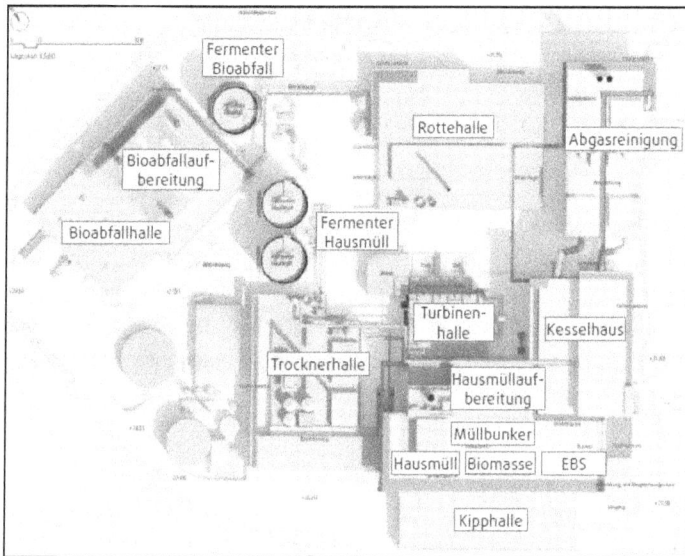

Figure 2: Layout of the CRE's different sections

2.3 Detailed description of the CRE

The former WTE plant bunker is split in a ratio of 6/14 and 8/14 for the reception of collected household waste and fuel for the biomass co-generation plant respectively. An area for RDF will also be added to the bunker.

2.3.1 Section 1 – Household waste treatment

Section 1 of the plant is broken down into 1.1 (sorting plant) and 1.2 (anaerobic digestion plant), meaning that section 1 in its entirety is made up of the following units: the bunker, mechanical pre-treatment (sorting) and anaerobic digestion of the fine fraction.

2.3.1.1 Section 1.1: Sorting plant for up to 140,000 Mg/a household waste and paper
bin waste

The mechanical pre-treatment phase pre-shreds 140,000 Mg/a household waste and paper bin waste, separates it into different fractions and extracts recyclable materials. The fine organic-rich fraction is sent for anaerobic digestion (section 1.2), with the coarse fraction from which recyclable materials have been removed going to an RDF co-generation plant (section 5).

Mechanical pre-treatment takes the form of a vertical sorting. Material is fed from the old bunker at the Stellinger Moor WTE plant.

Shredding is directly followed by screening using two screening drums connected in series that are installed in the upper levels. Screening can take place at 50 to 60 mm and at 250 mm, as needed.

The > 250 mm coarse fraction is fed back to the pre-shredder for another round of shredding.

The > 50 to 60 mm medium-sized fraction undergoes automatic sorting during which metals and valuable rigid plastics and paper and board are removed. The remainder is a high-calorific fraction and another product in the form of RDF, which is recovered in section 5 of the plant. The paper and board fraction is made up of a heavy and a light-weight fraction. The lightweight fraction is presumably much less contaminated and is recycled where possible. The heavy paper and board fraction undergoes a household waste anaerobic digestion process.

Rigid plastic and the lightweight plastic and board fraction is baled to optimise transpor-tation of recyclable materials. This feedstock will not use up all of the bale press's ca-pacity so the plan is to accept and bale recyclables collected at neighbouring recycling yards, such as plastic film and rigid plastics.

Recyclable materials (ferrous and non-ferrous metals) are extracted from the <50-60 mm fraction (household waste fine fraction). This fine fraction is then screened (ca. 20 mm), with the coarser sub-fraction undergoing combined removal of plastics and inert materials (glass and ceramics, china and stones). The product fraction extracted in the first phase undergoes a second sorting phase that removes impurities (ceramics, china, stones and plastics), creating a recyclable glass fraction. The fine fraction from which inert materials have been removed is then sent to the collecting hopper for household waste anaerobic digestion.

The reception area in section 1.1 is designed in such a way that household waste can be moved out of the bunker in an emergency. To this end, waste will be transferred from the bunker to a new automated loading station using the existing crane discharge and a new chute with a slat conveyor. This loading station can then fill walking-floor vehicles or containers. The centre can also tranship up to 850 Mg/d of household waste along-side operation of plant section 1.1

The emergency transhipment system will also regularly be used to cushion surges in demand at the SRH waste-to-energy plant, with a 'buffer volume' of up 2,000 Mg made available.

2.3.1.2 Section 1.2: Anaerobic digestion of the organic-rich fine fraction

The anaerobic digestion plant for household waste is preceded by a reception area that bridges stoppages in operations at the mechanical pre-treatment systems for household waste. This allows the anaerobic digestion plant to be constantly fed.

Another option is for the receiving bunker to be fed directly with other types of waste, such as:

− Grass cuttings, up to 250 Mg per month

− Waste from zoos and wildlife parks

− Spoiled fruit and vegetables

Conveyor belts transport the fine household waste fraction and other waste-based biomass into a fermenter. Digestate is sent for drying via a receiving bunker (section 4.1).

2.3.2 Section 2 – Anaerobic digestion and composting of biodegradable waste and green waste

On the one hand, the plant treats biodegradable waste (from the dedicated biowaste bin) from the north-western region of Hamburg that SRH is required to collect as part of its mandate and green waste from SRH recycling centres. Any wood in the green waste should be collected separately at recycling yards, where possible, allowing it to be sent directly to the biomass co-generation plant. On the other hand, the plant uses biodegradable and green waste from the district of Harburg.

Biodegradable and green waste is received in a new delivery hall that will have to be built. During construction, this hall was used to tranship biodegradable waste, green waste and road sweepings.

Biodegradable waste first undergoes a round of initial screening (screen cut of about 60 mm), with 80% of biodegradable waste separated for anaerobic digestion as undersize material (fine fraction - FF). This material is transported to an intermediate bunker. A magnet separator is located on the conveyor route to extract ferrous metals. The intermediate bunker has a volume of 2.5 days to ensure that the fermenter is continuously fed as far as possible. The intermediate bunker feeds the anaerobic digestion phase, which is to take the form of a continuous fermenter. The anaerobic digestion stage produces both biogas and digestate. The biogas is sent for upgrading. The digestate is separated into liquid and solid components using a press and decanter. Solid matter is sent for composting (see below). Liquid digestate is partly mixed with input materials prior to entering digestion (inoculation). Surplus liquid digestate is either used in agriculture, managed externally or sent to the household waste anaerobic digestion phase

(section 1.2). Two containers (with a combined storage capacity of four days) are envisaged to provide intermediate storage.

The coarse fraction from the aforementioned biodegradable waste screening and (at a
different time) green waste are first carefully pre-shredded in another processing stage
before a star screen removes impurities (especially plastics). The product serves as
structural material for composting digestate. Shredding is carried out so carefully that
plastic film is kept as large as possible and can thus be extracted from biodegrada-
ble/green waste. Solid digestate and structural material are mixed and placed in composting boxes. The final maturation phase serves to stabilise the compost at a composting degree of II – III (fresh compost).

This fresh compost then undergoes another round of post-treatment so that it meets
requirements for high-quality compost recovery.

2.3.3 Section 3 – Biogas upgrading

The biogas upgrading plant upgrades biogas from the anaerobic digestion plants for
household waste (1.2) and biodegradable waste (2) to natural gas quality (biomethane)
through amine scrubbing. The main reasons for choosing amine scrubbing lie in the low
methane slip in exhaust air and the heat available at the site. Biogas is stored in a gas
storage tank so that upgrading can take place constantly and thus efficiently. Following
upgrading, biomethane is fed into the public natural gas grid.

2.3.4 Section 4 – Biomass heating plant

Section 4 is made up of drying and the biomass co-generation plant. The core task of
this plant (section 4.2) is recover energy from digestate from household waste. A drying
stage (section 4.1) precedes the co-generation plant to ensure that the fuel, which contains both digestate and other LCV biomass, has a constant calorific value.

2.3.4.1 Section 4.1: Biomass drying and processing

The dryer and reception area for biomass to undergo drying (e.g. greenery from roadways) is located in a dryer hall, which needs to be built. The dryer hall also includes a
storage bunker for anaerobic digestion (section 1.2) and a storage bunker for the dryer.

Moreover, a reception area for other biomass requiring drying or further processing (e.g.
foliage from roadways, but also screen overflow from composting plants) will be created
and have shredding and screening capabilities.

The dryer dries digestate from section 1.2 together with foliage and other biomass. While foliage from roadways can enter the dryer without additional processing, other biomass may need to be shredded beforehand.

In turn, digestate (from section 1.2) has virtually no structure and is hard to dry. Mixing it with structural material is thus helpful, allowing digestate to be aerated.

Substances such as the fine fraction from waste wood treatment (waste wood FF) and a fine fraction from processing green cuttings can be used as structural material. Moreover, other types of biomass such as green cuttings (see above) can be added to the agitator.

2.3.4.2 Section 4.2: Firing system, exhaust gas treatment and energy offtake

The biomass co-generation plant is designed as a single line with a thermal rating of 47 MW. It uses waste-based biomass to generate power, process heat and district heat. LCV biomass, which until now has not been eligible for higher-quality use, is also used in the biomass co-generation plant. One significant input stream is dried digestate from the biogenic fraction of household waste from section 4.1.

The plant is designed in such a way that the old tipping hall and 8/14 of the waste bunker can continue to be used by installing a partition wall for the biomass co-generation plant. Its storage capacity of 10 to 15 days is enough to bridge longer pauses in incoming deliveries (e.g. over Christmas or Easter). Dried digestate and biomass are transported to the bunker using a conveyor belt. Waste wood and other waste-based biomass that does not have to be dried is delivered to the tipping sites in the biomass bunker. The bunker has a shredding machine in case any arriving waste wood has to be shredded.

Air extracted from the bunker is discharged into the surrounding air using the central exhaust air treatment system via dust filters or, just like highly contaminated exhaust air streams from sections 1 to 3, is sent to the incineration plant as combustion air.

Combustion air from the biomass co-generation plant is pre-heated to a minimum temperature of 170°C using a steam air pre-heater. The combustion air's dust content thus has to be reduced to a permissible level for the air heater using a filter.

Grate firing is the envisaged firing technology. The parameters for live steam before the turbine should stand at 40 bar(a) / 400°C. Steam is converted into electricity in a back-pressure turbine and used to heat district heating water to a maximum of 136°C. A tapping point is envisaged in the turbine to be able to deliver steam with a saturated steam temperature of about 170°C (ca. 8 bar(a)) to supply section 3, even when working in partial load. Moreover, extraction at 4 bar(a) will supply the heat condensers in order to

supply district heating water at a maximum temperature of 136°C, and supply the digestate dryer. The district heating water is heated in two stages. Exhaust steam from the
turbine undergoes condensation in another heat condenser at around 1.2 bar(a).

The heat condenser is to be used jointly by the biomass co-generation plant and the
RDF co-generation plant (section 5). The heat condensers for both plants will be sized
in such a way that they guarantee significant flexibility. The flow temperature should be
kept at a constant 136°C. The temperature will be able to be changed to the flow temperature required by the district heating network through admixture upstream of the
heating condensers. During the transition phase (where low flow temperatures are required by the district heating grid), the corresponding heat condenser can also be by-
passed to the 4-bar(a) track.

The plan is to use remaining heat from exhaust gases and to cool it to 80°C before the
flue to improve energy efficiency.

When heat is not consumed by the district heating grid, exhaust stream from turbines is
condensed in an air condenser at approximately 800 mbar(a). Heat condensers for the
district heating grid are not in operation then.

All condensate is returned to the water-steam cycle.

Exhaust gases are treated so that they definitely meet the requirements of German
emission control legislation. A few levels are even below the limits, meaning that the
biomass co-generation plant achieves better-quality exhaust gas than traditional biomass plants. The use of a catalytic exhaust air denitrification (SCR) system is envisaged to further improve exhaust air quality.

2.3.5 Section 5 – RDF co-generation plant

A single-line RDF co-generation plant with a thermal rating of 47 MW is to also be built
at the site. This plant will use both the RDF extracted in section 1.1 (ca. 52,000 Mg/a)
and plastic-rich impurities from section 2 (3,000 Mg/a) together with third-party RDF to
generate power, process steam and district heat. The plant will recover around
100,000 Mg/a of RDF (SRH's own RDF plus third-party RDF).

A bunker with a tipping hall will be built directly next to the existing bunker to store the
RDF. The crane in the new section of the bunker is to be connected to the old bunker.
RDF enters the bulker via a conveyor belt from the waste sorting area (section 1.1) and
from the tipping areas. The boiler is then fed using an automatic crane system.

Otherwise, the RDF co-generation plant is essentially subject to the same considerations as the biomass co-generation plant. The only difference lies in the fact that the air

heater only pre-heats combustion air to 100 °C because the calorific value of fuel used by the RDF co-generation plant is much higher than that of the biomass co-generation plant.

Its exhaust gas treatment system is also identical to that of the biomass co-generation plant. Exhaust gas from both boilers is treated in multi-stage drying process based on natrium hydrogen carbonate and lime combined with catalytic sulphur removal. An exhaust gas heat exchanger is installed in the boiler of each exhaust air treatment line. This exchanger cools exhaust gases and provides heat for heating purposes.

2.4   Project timeline

The CRE will be planned and built in five planning stages.

Table 1: Planning stages for construction of the CRE

Planning stage 1	Design phase (completed)
Planning stage 2	General contractor award and selection of a preferred option (completed)
Planning stage 3	Development of the construction and procurement tendering documents with awards (in progress)
Planning stage 4	Official permit process under German emission law (the scoping date has already taken place), the permit application with the related reports and the EIA is currently being drafted
Planning stage 5	Implementation

The CRE strategy (planning stage 1) was developed by SRH and has been completed. SRH has tasked an expert planning consortium with serving as the general contractor for the following stages. Planning stage 2 (selection of the preferred option) has also been finalised (since September 2017). This phase entailed refining the plans for the plant strategy once again, identifying potential for optimisation and carrying out definitive layout and design of the different sections.

The general contractor has been working on planning stage 3 (development of the construction and procurement tendering documents with award to technology providers) since October 2017. A manufacturer-neutral application is being filed for the first partial permit affecting the following sections of the plant: household waste sorting, co-generation plants, rebuilding work on existing buildings, infrastructure and overarching electrical technology for the CRE. The second partial permit will add manufacturer-specific information about the suppliers of the anaerobic digestion technology.

According to current plans, planning stage 5 (implementation) is to begin in mid-2020.

The five sections cannot be created in an existing technology park at the same time due to the space needed for assembly and construction site equipment. A decision was thus made to build the sections anti-clockwise at the construction site and at different times. The final section to be constructed will be the biodegradable waste anaerobic digestion plant, which is slated to commence regular operations at the start of 2025.

The commissioning of the CRE will give Hamburg one of Germany's most modern waste treatment centres. The CRE will be capable of providing more than 20% of the heat needed by the western portion of the city in winter. During the summer, no additional heat will need to be generated for urban heat supply. Moreover, the separation and recovery of recyclable elements will be maximised. The combination of sorting and biological and thermal treatment methods at a single site will reduce the required number of journeys, significantly relieving the burden on city-centre refuse collection traffic and reducing air-borne vehicle and noise emissions. Emissions reductions will also be supported by the fact that both co-generation plants have optimised exhaust gas treatment systems, which will result in emissions well below legal limits. Furthermore, the plant will also be highly energy-efficient, minimising in-house consumption and allowing as much residual heat as possible to be used (e.g. heating exhaust gas before the flue).

New Material-From-Waste facility in PARIS 17

Christophe CORD'HOMME

CNIM Group, Paris, France

Abstract

In 2015, the Syctom awarded to CNIM consortium the contract to design, build and operate its brand new waste sorting facility in Paris (France). Located in the new Clichy-Batignolles quarter (17th arrondissement North of PARIS), this new large-capacity plant will process the household packaging waste from more than 900,000 people. It will recover packaging materials coming from the selective collection in preparation for recycling. This latest-generation sorting centre in Paris 17 will be able to process up to 15 tonnes an hour of municipal waste. Thanks to its 13 optical sorting machines, it will be super-efficient and fully automated. Capable of sorting new plastics types and so implementing the extension of waste sorting regulations, it will represent a step forward for the recycling of household waste and help achieve the target of recycling 75% of packaging as defined by legislation.

Keywords

Material-from-Waste, sorting, recycling, packaging, resources, household, circular economy...

1   Introduction

The Syctom, the PARIS metropolitan agency for household waste treatment, is the public body responsible for the processing and recycling of the municipal solid waste produced by the 5.8 million inhabitants in its 84 member towns (Paris and nearby suburbs), representing almost 10% of the population of France.

In 2017, 2,313,363 tons of municipal solid waste were processed by SYCTOM. In the context of the increasing scarcity of raw materials and the energy transition, all this waste should be considered as resources. In 2017, the 6 existing sorting centres of Syctom (see figure 1) have treated 184,600 tons of household paper and packaging waste recovered by selective collection (yellow bin).

NANTERRE
capacité:
40 000 t/an

PARIS XVII
capacité:
45 000 t/an

SEVRAN
capacité:
17 000 t/an

ROMAINVILLE
capacité:
45 000 t/an

PARIS XV
capacité:
20 000 t/an

ISSÉANE
capacité:
22 500 t/an

IVRY/PARIS XIII
capacité:
30 000 t/an

Figure 1 : Syctom sorting facilities

This is a daily challenge for SYCTOM, which is constantly seeking innovations to optimize its facilities' performances (increasing energy efficiency, enhancing sorting and recycling processes) and to find optimal solutions for the processing of the various waste flows.

2   A new Material-from-Waste facility in Paris

2.1   New selective collection instruction

The fight against wastage not only urges us to consume better, but also to recycle the waste we produce more effectively. It is this objective that Syctom strives to achieve by modernising its sorting centres and by organising a network of waste reception centres, which are receiving the different flows coming from the selective collection of Parisian municipal waste (see figure 2). In 2022, plastic sorting will no longer be limited to bottles. It will cover all plastic packaging: jars, trays, films, etc. To comply with the Law for Energy Transition for Green Growth, which requires this expansion of the sorting instruction (see figure 3), Syctom is transforming its   industrial tool.

Figure 2 : selective collection in Paris

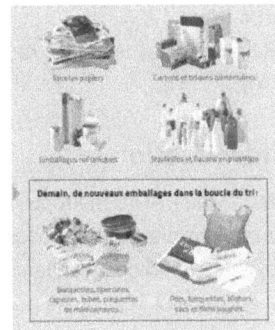

Figure 3 : Evolution of "Yellow bin" content

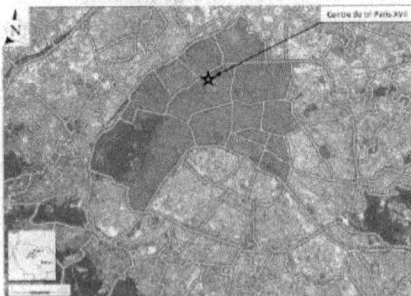

Figure 4 : waste supply zone

Sort more and better: this is the course set by Syctom in response to ever greater environmental challenges. The new Paris XVII sorting centre responds fully, thanks to its large capacity, efficient equipment and an optimized organization. It will prepare for recycling household packaging waste from selective collection by 900,000 inhabitants of Paris and 4 neighboring municipalities (see figure 4). it is the second one in Paris intra-muros., which gives Syctom a new generation tool capable of handling 45,000 tonnes of waste per year and implements the sorting instructions for all plastic and metal packaging.

2.2   A waste sorting centre in Paris intra muros

Situated on the northern edge of central Paris, the railway wasteland of the 17th arrondissement gave way to the new eco-district Clichy-Batignolles, transforming the image of the North-West of the French capital. At the heart of this 54-hectare "ZAC" (mixed development zone), this area is including the largest law courts complex in Europe with the new architectural attraction of the skyscraper "Palais de Justice" (Paris courthouse), designed by star architect Renzo Piano.

Credit :Ateliers Monique LABBE

Figure 5 : Palais de Justice and Syctom sorting

In 2015, Syctom awarded to a consortium led by CNIM the contract to design, build turnkey and operate its brand new waste sorting facility in this zone. In addition to CNIM in charge of the design-and-build responsibility and of the operation and maintenance, this group includes Ateliers Monique Labbé, for the architecture, "Urbaine de travaux" for civil works, "Ar Val" for equipment, Ingerop and Segic for engineering. Established rue de Douaumont on a site of 1.1 hectares, the new Syctom sorting centre is well placed near the locations where waste is produced.

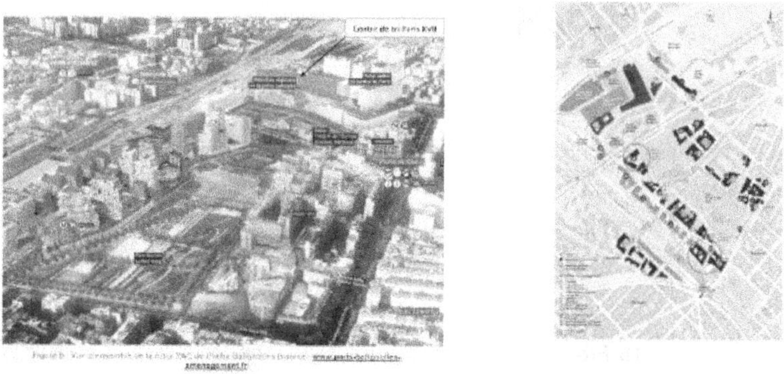

Figure 6 : Clichy Batignoles development zone

It is a clear response to the challenges of the energy transition and the circular economy to improve recycling of different types of packaging waste.

3   A "new generation" centre adapted to new challenges

3.1   Principle

This latest-generation sorting centre in Paris 17 will be able to process up to 15 tonnes an hour of municipal packaging waste. Capable of sorting new plastics types and so implementing the extension of waste sorting regulations, it will represent a step forward for the recycling of household waste and help achieve the target of recycling 75% of packaging as defined by European legislation. It will receive recyclable waste from selective collection trucks and also from the pneumatic collection terminal located nearby.

The received waste will be the clean and dry conventional, packaging, paper, cardboard, plastics, metals, but also small waste electrical appliances and electronic equipment (WEEE). With this new sorting centre, the sorting instructions can be extended to food trays, polystyrene and plastic films, as it will be asked by 2022 for the sorting to the residents.

Credit :Ateliers Monique LABBE

Thanks to its 13 optical sorting machines, it will be super-efficient and highly automated to limit manual gestures and guide the activity of agents towards quality control; this is designed to improve working conditions and staff safety.

3.2   Input

As required by the 2015 French law on energy transition, the extension of sorting instructions for recycling is requiring by 2022, that residents will sort all plastic packaging and metal packaging at home. The Paris XVII sorting centre, with its high capacity and state-of-the-art equipment, will be able to welcome and process these new flows in the best conditions. The sorting guidelines apply to the following waste:

- Today :
  - o All papers
  - o Food cartons and bricks
  - o Metallic packaging
  - o Plastic bottles and flasks
- Tomorrow, new packaging in the sorting loop:
  - o Trays, caps, capsules, tubes, medicine blisters
  - o Pots, trays, blisters, bags and soft films

3.3 The different steps of process for material sorting

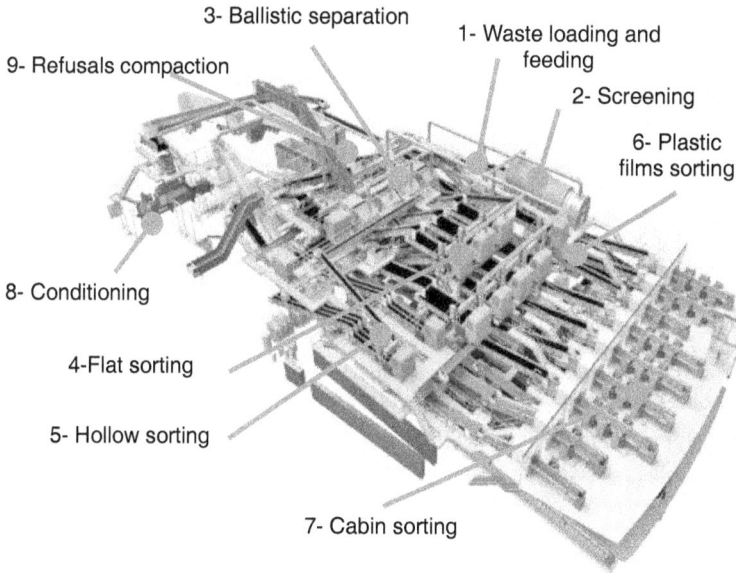

Figure 7 : sorting process

With a waste reception 18 / 24h and an operation 6/7 days, the sorting horary capacity of 15 t/h will allow to treat 45,000 t / year with a recovery rate of 95%.

3.3.1 Waste loading and feeding

The wheel loader unloads the collected waste from the storage area into a hopper with an Alternating Moving Floor ("FMA") followed by the conveyor, which is feeding the sorting process.

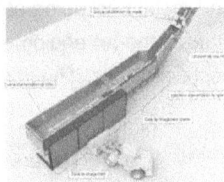

3.3.2 Waste Screening

The trommel screen, a rotating cylinder with larger and larger holes, separates waste according to its size: below 90mm, over 350mm and in between (intermediate).

3.3.3 Ballistic separation

3D or "hollow" "heavy" bodies bounce on the ballistic separators and fall by gravity at the bottom. The 2D "flat" pieces, so-called "light" are blown by fans and evacuate on the top.

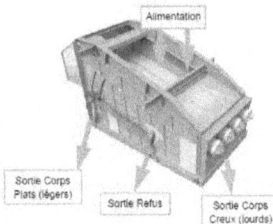

3.3.4 Flat bodies sorting

8 optical sorting machines (7 binary and 1 ternary) separate with positive and negative sorting Newspapers, Journals, Magazines (JRM), Recyclable Household Packaging (EMR), plastics and films.

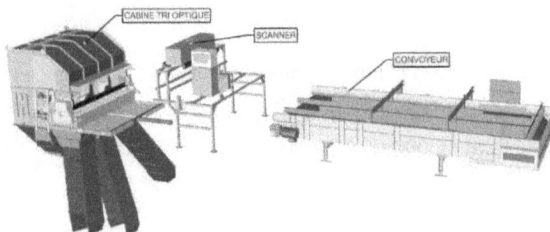

3.3.5 Hollow bodies sorting

4 optical sorting machines (1 binary and 3 ternary) separate PE-PP-PS (Polyethylene - Polypropylene - Polystyrene) + Packaging for Food Liquid (ELA), light and dark PET (Polyethylene terephthalate).

3.3.6 Plastic films sorting

1 ballistic separator and 1 optical separator allow to isolate the films that feed the PEBD Films sorting table.

3.3.7 Cabin sorting

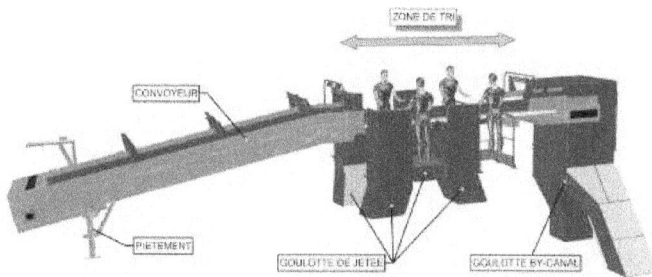

On 8 sorting conveyors, the sorting staff controls the quality of each product before baling:

- Dark PET
- Clear PET
- PP-PE-PS Polyethylene - Polypropylene - Polystyrene
- Food Fluid Packaging (ELA),
- Recyclable Household Packaging (EMR)
- Cartons
- Plastic films
- Newspapers, Journals, Magazines (JRM)
- Store cardboard (GDM)
- Aluminium
- Steel
- Small household appliances (PAM)
- Glass.

3.3.8 Conditioning

The sorted materials are packaged using two balers and a packer.

3.3.9 Compaction of process refusals

Alternating Moving Floor ("FMA") compactors are used for process refusals.

4   Lay out & Circulation

4.1   Lay out principle

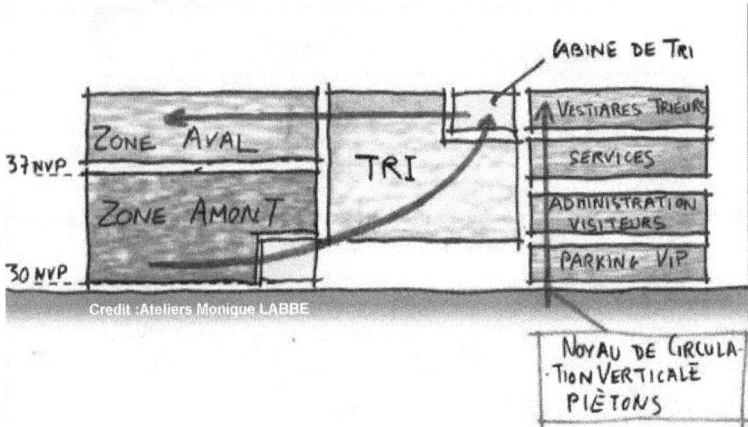

Figure 8: lay-out

A simple principle of functional layout has been adopted to achieve strict separation of functional areas and allow for the scalability of the sorting area

Figure 9 : sorting zone during erection of equipment

4.2   Waste inlet

A traffic of 50 to 80 trucks a day is planned for the loading of unsorted waste coming

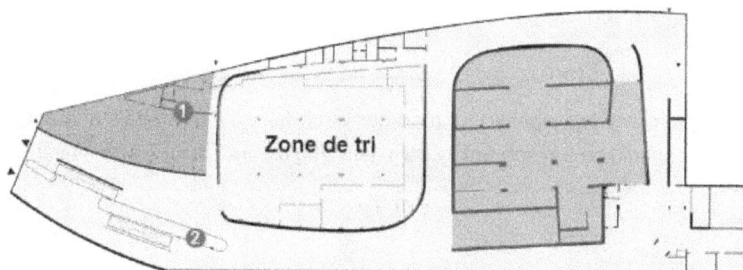

directly from selective collection. This waste flow will arrive on the low level of the plant to be stored in the dedicated area for the waste reception on the right side after radioactive detection. It will be completed by pneumatic collection flow.

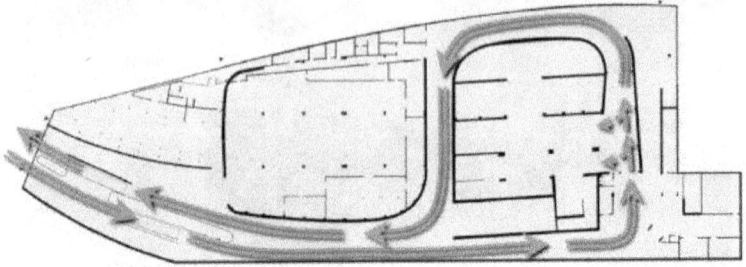

4.3   Material outlet

After sorting, recyclable materials will leave the plant from the dedicated area for storage of baled materials (on the right side) by a dedicated access on top level.

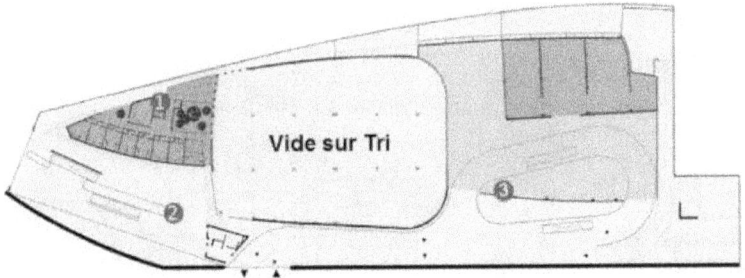

Vide sur Tri

25 to 30 shipments are planned by road per week for plastics, metals and cardboard. Paper bales should be transported by train, avoiding the traffic of 12 trucks every week.

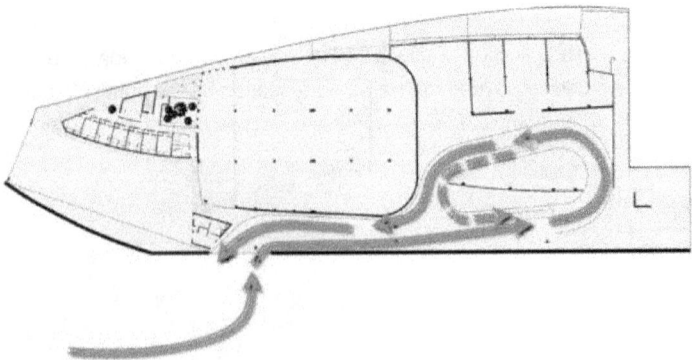

5   Exemplarity desire

Connected to the ring road and the railway, the sorting centre in Paris XVII is fully integrated into its environment. It took also up the challenge of blending the installation's architecture into the urban environment of this development zone and in the vicinity of the skyscraper "Palais de Justice", which is emblematic of the renewal of this district. The project is characterized by large vegetated areas, with suspended gardens and terraces, and a wide use of renewable materials. Equipped with photovoltaic panels, the roofs offer a landscape of streams using a wide variety of species, in favour of a bio-diversified environment.

Renewable material are used for the building facade and structure (wood).The North facade of the centre benefits from a specific treatment to limit the reverberation of traffic noise as shown on the following figures of noise mapping in day time.

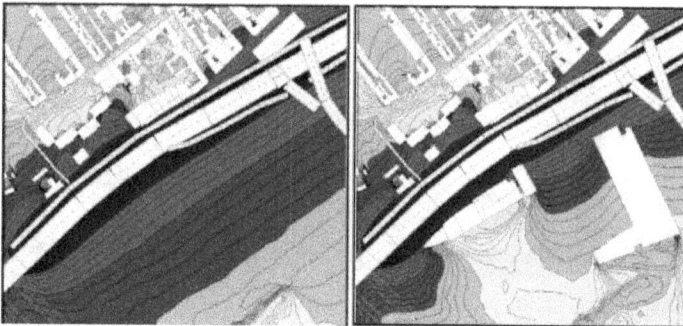

Figure 10 : Noise mapping :
Current status (left) and with absorbent facility facade (right)

With the help of careful insulation of the building, privileged natural lighting, luminaires with systematic presence detectors, the energy needs of the building will be moderate. The building heating is provided by 85% renewable energy with the connection to the CPCU district heating. In the operating phase, the dust generated will be directly sucked by an internal dust collection system.

The generated traffic by the plant operation will be lower than 1% of neighbourhood traffic The sorting centre will receive, in particular, recyclable waste from the nearby automated vacuum collection terminal. The nearby railway line will enable half the material sorted at the centre to be transported in goods wagons to recycling plants. This will help improve air quality by reducing heavy trucks traffic in the city.

6   Construction

Started in 2007, the construction is almost complete and the plant start up is planned in 2019.

7   Literature

Credit photo and architecture Ateliers Monique LABBE
Credit SYCTOM and CNIM photos and documents
Website : https://www.syctom-paris.fr/installations-et-projets/projets/paris-xvii/centre-de-tri.html
Video : https://www.youtube.com/watch?time_continue=2&v=fvxbvFGgx9Q

Author's address

Christophe CORD'HOMME
CNIM Group (Constructions Industrielles de la Méditerranée)
 35, rue de BASSANO – 75 008 PARIS – France
Tel : +33(0)1.44.31.11.00
e-mail : ccordhomme@cnim.com

1.5 years experience with 1,000 t/d MBT in Mumbai (India)

Aurel Lübke and Bernhard Gamerith

Compost Systems, Wels, Österreich

Abstract

A project with a final life span of nearly 10 years has to overcome many obstacles in order to become operational in the end of day. The plant, that has now been operating for 1.5 years, has the capacity to process 1,000 t of household waste per day on a secure 24/7 base via a sheltered, emission reduced windrow composting process, followed by a sheltered and aerated maturation phase. On the one hand this project can be seen as a model especially for India or other regions in South-East Asia, where this technology can not only eliminate many tons of CO_2, but can also create numerous jobs and pave the way to a modern, sustainable future. On the other hand, the produced high quality compost can be sold at a prime rate of approx. 55 US dollars per ton, required to meet the highest standards in India, under a regulated governmental fertilizer program.

Keywords

MBT, composting, India, Mumbai, effective waste management, GHG reduction

1 Background

The project started back in 2008 and would create the largest mechanical biological treatment (MBT) plant in South-East Asia to Western standards, for the city of Mumbai. Almost ten years later, the finished plant was presented to the public as part of the IFAT India 2017 trade fair. It is the largest waste treatment centre with Western technology in all South-East Asia. You could also say that "Things could only get worse". About 1,000 t of waste can be treated per day at the new plant. Unlike in Europe, waste is transported on relatively small trucks in India. This means that this volume of waste corresponds to some 300 to 500 truckloads a day.

The waste is first sorted in a mechanical treatment plant. The throughput with a high organic content is fed into the biological treatment system, the part of the plant which Compost Systems designed. In India, waste is relatively wet when delivered due to the rain and consumer behaviour. As a result, the organic part may already contain over 60 % moisture. The system itself consists of a roofed aeration platform, which is powered by a compost turner. Compost is produced using intensive aeration for four weeks followed by another four to eight weeks of curing. This compost has high market value in India. Provided the limit values for contaminants are also met, the compost can be sold at a profit thanks to the promotion of fertiliser and a high demand for natural fertiliser. Refuse Derived Fuel (RDF) is sold, recycling products are also sold on to the recycling industry. The presentation in 2017 showed how the plant could be operated all year round, even in India. More than 2,000 mm of precipitation falls in Mumbai over about two months with volumes increasingly fluctuating because of climate change.

This project can be seen as a model especially for India or other regions in South-East Asia, where this technology can not only eliminate many tons of CO_2, but can also create numerous jobs and pave the way to a modern, sustainable future.

2   The situation in India

The Global Ambient Air Quality Database of the World Health Organisation (WHO) shows that of the 10 most polluted cities in the world, nine are in India. While air pollution influences nearly every place in the world, people in low-income cities are mostly affected of it. Furthermore, data show that 97 % of cities in low- and middle- income countries with more than 100,000 inhabitants exceed WHO air quality limits. In high-income countries, this percentage is only 49 % (WHO, 2018).

Increases in the Gross Domestic product (GDP) usually leads to a significant gain in consumption, waste generation and the ecological footprint. The same applies for India, which has been explored by different studies. For instant Delhi had a municipal solid waste production of 400 t/day in 2000, where only 15 years later the daily waste production has exploded to the enormous summary of 8,700 t/day. This reflects an increase of over 2,000 % in only 15 years. BARUA & HUBACEK (2009) found significant relationships between water pollution and per capita income for 12 regions of India. It is evident that many Indian states face similar transitions of initial high per capita pollution followed by improvements of per capita pollution levels and finally additional increases of pollution levels with further economic growth.

Studies have shown that air pollution is cutting life expectancy in India significantly. Following VALENTINE (2015), India's high air pollution is diminishing live expectancy of certain Indian regions inhabitants significantly. In the last 2 decades, Indian cities have seen a rising tide of waste that's disposed of in open dumps. Such landfills, often on fire, are only aggravating the pollution problem, contaminating both air and groundwater. Uncontrolled landfills in a warm climate like India also release significant amounts of methane into the atmosphere. Methane, 25 times more toxic to the climate than CO_2, signing responsible for up to 30 % of the greenhouse-gas emissions in a country. Uncontrolled landfills are emitting between 1 and 1.7 tons of CO_2 equivalents of CO_2 per ton of household waste into the Atmosphere and are considered the second highest emitter of methane after the cow, on the planet.

Bangalore, as the Silicon Valley of India, is considered to become inhabitable by 2025 by the WHO, due to garbage being dumped uncontrolled into lakes, known now as the burning lakes of Bangalore.

These environmental issues are challenging India in many ways, yet for the rest of the world leading to the conclusion, that the world's economy and ecosystems are intertwined and a clean and resource efficient India is essential for the sustainability of the planet.

3   Development Phase

The initial step of the BMC (Bombay Municipal Corporation) releasing a tender for the treatment of 1,000 t/day of solid household waste per day, triggered a consortium of the local partner Antony Lara Enviro Solutions Pvt. Ltd. and Compost Systems to enter into a bid in 2008. Finally, the project was released to the consortium, together with a landfill project, taking further 3,000 t of waste per day in the same location.

During the planning phase, some legal obstacles had to be overcome, including a law case in the connection with land use, taking the case all the way to the high court. After clearing all legal obstacles, the financial structure had expired and knock on effects of the world economic crisis required a complete new financial structure for the financing of the project. This together with the challenge, having to bring the building plot, that used to be a salt mine over 100 years ago, from a level of minus 2 m to a safe building level of plus 4 m, and building a surrounding wall of over 5 km around the plot, caused a delay of the project for over 7 years. Finally, in 2016 the project was able to proceed and was opened in September 2017 to take its first loads of solid municipal waste.

4   Technology

The plant, that has now been operating for 1.5 years, has the capacity to process 1,000 t of household waste per day via sheltered, emission reduced windrow composting process, followed by a sheltered and aerated maturation phase.

Figure 1: The Kanjur Mumbai MBT plant, 50 % already built

On the facility, the following equipment is being used:

- Four open newEARTH technology buildings, each 40 m x 110 m, with 8 traingle-shaped negative-aerated windrows, each windrow containing approx. 500 to 600 t of organic fraction waste.

- Wireless real-time process monitoring with multi temperature probes.

- Automatic computer controlled irrigation system.

- Four roofed, positive-aerated maturation buildings, also with wireless tempera-ture probes, for a maturation of up to 8 weeks.

- The newEARTH process was founded by Compost Systems, based on extensive research and development. By creating and maintaining the perfect environment for aerobic micro-organisms, the process will maximize the yield of compost, while reducing the ability of contaminants to enter the final product.

- It further is proven to stabilize a fully aerobic process within hours and maintains that process throughout the processing time, securing a process that is close to 100 % Methane free and therefor contributing a significant reduction of GHG emissions.

- The real-time monitoring of the process together with the work regime secures a removal of pathogens according and approved by European Regulators to com-ply with the Animal Byproduct Regulation.

- As a result, composting time is reduced which in turn enhances economic value. The process is "emission reduced" by using a negative aeration process, which has the ability to remove unpleasant emissions from the process by an efficiency rate of over 80 % from the process, while keeping operation on a simple level.

- The plant is under an enormous economic pressure. While the tipping fee is only approx. 10 € per ton of waste, the yield of compost has to contribute a significant value to the project economy.

- Knowing the differences of management and other skills in India compared to Europe, it was important to design the plant in a step by step process, making it easy to detect and erase operation defaults in a quick and effective way. Only then it was possible to secure a 24/7 quality process that is securing a high quality product on continuous base.

- The mechanical splitting line is based on a 0/80 mm split, sending the small fraction to the composting plant, while the oversize fraction is further separated into valuables, RDF fraction and Inert fraction going to landfill. Currently the valves for the energetic fraction are currently being explored in detail and temporary contracts for the offtake are in place. Missing governmental regulations on the use of RDF are leaving yet a gap of uncertainty to the Industry at current time.

Figure 2: Mechanical treatment plant

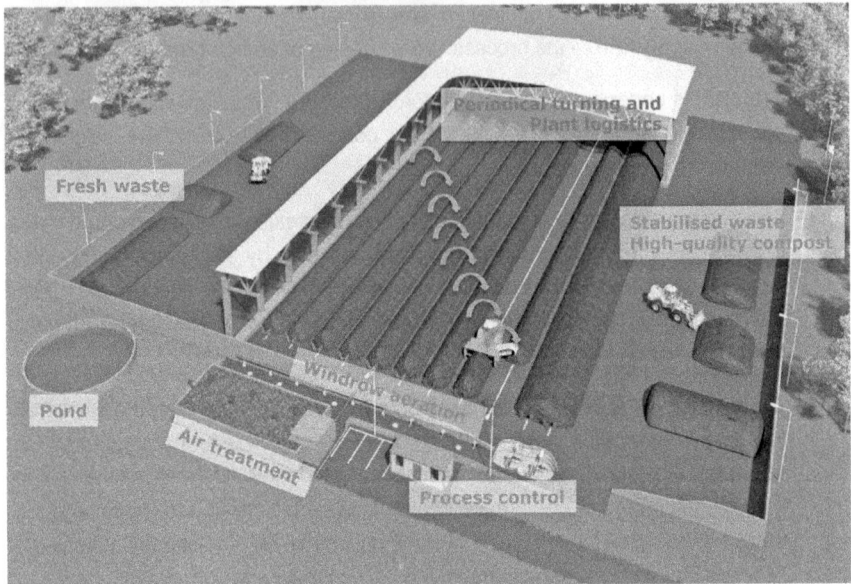

Figure 3: The concept of the newEARTH system in detail

Figure 4: Sheltered windrow composting with turner TracTurn

- The TracTurn is a tractor connected compost turner, operated with a reverse drive system. The side stacking ensures a constant material flow which saves logistic costs, but also secures that fresh and finished material have no physical connection and recontamination is prevented. With a maximum capacity of up to 2,000 m³ per hour, the windrows are mixed in a very gentle way with a slow turning rotor, to prevent plastic or glass from becoming milled and impossible to be removed in the final refinery.

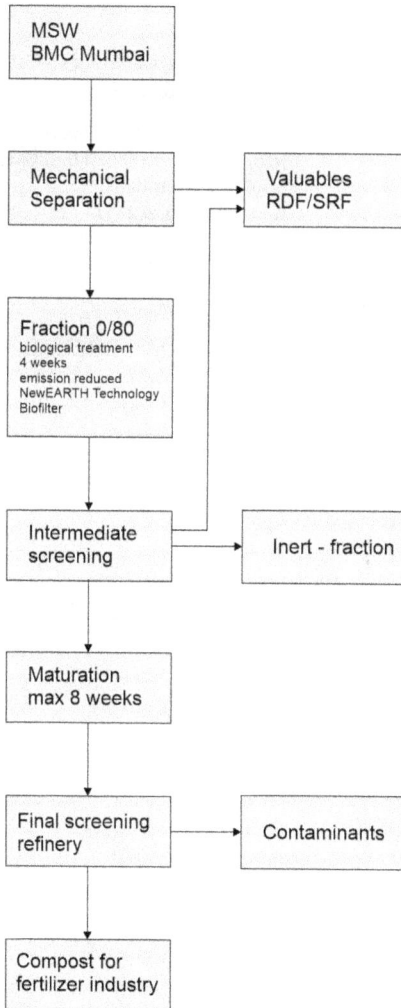

Figure 5: Material flow on the Mumbai plant

5   The finished product

The finished compost is currently sold for approx. 55 US dollars per ton. This is prime rate for high quality compost, required to meet the highest standards in India, under a regulated governmental fertilizer program, applying subsidy to compost and chemical fertilizer. Here a current analyses of the final product, showing that the requirements are being achieved.

Table 1: Results of the recent compost analysis of the finished product (ANA LABORATO-RIES, 2019)

Test	Standard as per Fertilizer Control Order 1985 of the Indian government, amended up to 2017	Observation
Moisture percent by weight	Max. 25	13.50 %
Colour	Dark brown to black	Black
Odour	Absence of foul odour	Absence of foule odour
Particle size	Min. 90 % material should pass through 4 mm IS sieve	95.45 %
Bulk density	< 1.0	0.9776 gm/cm³
Total organic carbon percent by weight	Min. 12	32.47 %
Total notrogen (as N) percent by weight	Min. 0.8	1.707 %
Total Phosphorus (as P_2O_5) percent by weight	Min. 0.4	0.6035 %
Total potash (as K_2O) percent by weight	Min. 0.4	0.5975 %
C : N ratio	< 20	19.02 : 1
pH (of 50 % suspension)	6.5 to 7.5	7.10
Conductivity (as d sm^{-1})	Not more than 4	3.3 µs/cm^{-1}

Table 2 Values for pathogens and heavy metals of the recent compost analysis (ANA LABOR-ATORIES, 2019)

Test	Standard as per Fertilizer Control Order 1985 of the Indian government, amended up to 2017	Observation
Pathogens		
E-coli	Nil	Absent
Salmonella	Nil	Absent
S-aureus	Nil	Absent
Pseudomonas	Nil	Absent
Listeria monocytogene	Nil	Absent
Faecal streptococci	Nil	Absent
Heavy metal content (mg/kg) Maximum		
Arsenic (as AS_2O_3)	10	0.22 mg/kg
Cadmium (as Cd)	5	3.94 mg/kg
Chromium (as Cr)	50	10.75 mg/kg
Copper (as Cu)	300	39.05 mg/kg
Mercury (as Hg)	0.15	0.14 mg/kg
Nickel (as Ni)	50	4.17 mg/kg
Lead (as Pb)	100	16.08 mg/kg
Zinc (as Zn)	1000	40.57 mg/kg

6   Lessons learned

India is a complicated country. Starting with a strongly varying lifestyle, different man-agement styles all the way to a society and habits that are hard to understand for Euro-peans, it is even more important to build an INDIAN plant with Western Technology in India, verses importing a European plant into India, which will likely never work!

India is a very competitive market, and it must be said that time has a different meaning in India than in Western countries. This again underlines the requirement to keep the supply of western technology only on the level of technological design and leave the execution of the detailed design and building to the local force. So the supply of knowledge and products was reduced to the absolute minimum, while the use of local force was maximized, yet securing a safe, environmentally sound technology that is able to work in India in a secure 24/7 base.

Flexibility in a building schedule is nothing uncommon in India. In this case a pause of 6 years in the building execution had to be overcome and can challenge a supplier - customer relationship in many ways.

India has yet a need for many environmental solutions, including the requirement for a solution to treat still more than 500 Million tons of Household waste. It must be considered as a fact, that only in terms of cost, any sort of incineration is only viable for contaminated or hazardous waste. Labour being relatively cheap still in India, the solution will have to work towards a recycling scheme, with the requirements to build many more waste recycling and treatment plants in India. With the ability to provide a significant contribution to the reduction of GHG emissions, conserving natural resources and reducing significantly local emissions to water and atmosphere, while preventing pests or disease risk from uncontrolled dumpsites.

Finally, it must be said, that India has a strong will to move up as a global player in the world economy, targeting China as a model. It is the responsibility of the western world to keep supporting India in its ability to secure a safe environment for the local people, but also in terms of global interests especially in terms of GHG emissions. Concluding that the world will not be able to hold the global worming to 2°C, without integrating India, a society that will pass China by inhabitants by 2023.

Figure 6: Processing 1,000 t of MBT per day

7   Literature

World Health Organisation (WHO)	2018	WHO Global Ambient Air Quality Database (update 2018), At: https://www.who.int/airpollution/data/cities/en/ (08.03.2019).
Barua, A. & Hubacek, K.	2009	Water Pollution And Economic Growth: An Environmental Kuznets Curve Analysis At The Watershed And State Level. Journal of Ecological Economics and Statistics, 2008, Volume 10, Number W08.
Valentine, K.	2015	India's Air Pollution Cuts Life Expectancy by 3 Years. At: https://ourworld.unu.edu/en/indias-air-pollution-cuts-life-expectancy-by-3-years (08.03.2019).
ANA LABORATORIES (Mumbai)	2019	Certificate of Analysis. Report Number 19/A/354. City Compost Sample – Location: ISWM Project, Kanjur.

Deriving lessons for solid waste management sector in India from experiences of Germany

Shivali Sugandh, Dirk Weichgrebe

Institute for Urban Water Management and Waste Technology, Gottfried Wilhelm Leibniz University Hannover, 30167 Hannover, Germany

Abstract

With the growing pressure to reduce greenhouse gas emissions and climate change impacts, India is striving to improve Municipal solid waste management (MSWM) practices. Developed nations like Germany joined this stride way earlier due to early onset of industrialisation and subsequent environmental and public health degradation in 80's and 90's. During this time, Germany adopted a spectrum of measures and solutions to deal to address the challenges linked to improper waste management (WM) and implemented a modern, future-oriented waste management system, driven by European Waste Framework Directive. As many developing countries have relatively new understanding, laws & infrastructure in WM sector, this study presents lessons that can be learnt from experiences of Germany. This study bases these lessons on an analysis of key contrasting charactersitics between India & Germany: Demographic (population density), economic condition (GDP per capita) and social status (literacy rate & skilled population) and key convergences: Historical MSWM situation, generation pattern and development of legislations, commitment to achieve global sustainability goals.

Based on the analysis, the study showcases potential lessons which can be adapted in India specifically linked to economic instruments (market-based), legislative, technological and social instruments adopted by Germany. These include legislative reforms linked to ambitious waste laws emphasising resource efficiency & circular economy principles, extended producer responsibility, quality standards for waste derived products and data recording & monitoring regulations. Economic Instruments linked to monetary obligations on waste generators like 'Pay as you throw' system, landfill ban & tax and monetary subsidies which encourage industry to innovate solutions instead of creating a cut throat competition that determines quality. Technological guidance on Mechanical Biological Technologies as best suitable technology for transitioning waste conditions, supplemented by a SWOT analysis supporting a business case scenario. Social instruments linked to Germany's capacity building & training efforts such as the Dual Vocational Training and research & development funding also hold an important take as way as employment generation is a key developmental goal in developing countries. Lessons from negative interaction between policies have also been highlighted to prevent nations from pitfalls in future.

As MSWM sector is in nascent stages in India, the study showcases a positive opportunity for India to shift from the conventional reductionist approach of SWM to an Integrated and synergistic approach by applying combination of approaches like circular economy, systems thinking and Indian smart city approach, which is even new for Germany. However, this would require using an integrated policy approach -regulatory, economic, and social policy instruments coupled with Inter- departmental communication (waste, water, food & energy) & cooperation.

Keywords: Municipal solid waste management, lessons, climate change, instruments, Germany, India

1.1 Copyright

The paper is under review in a scientific journal and cannot be published.

Author's address(es) :

Shivali Sugandh
Businesss Development Associate
Umwelttechnik und Ingenieure GmbH
Wöhlerstraße 42
Telefon +49 15175313005
E-Mail: s.sugandh@qualitaet.de

PD Dr.-Ing. habil. Dirk Weichgrebe
Gottfried Wilhelm Leibniz Universität Hannover
Institut für Siedlungswasserwirtschaft und Abfalltechnik
Welfengarten 1, 30167 Hannover
Hauptgebäude 1101, Raum e231
Tel +49 511 762-2899
E-mail: weichgrebe@isah.uni-hannover.de
http://www.isah.uni-hannover.de

Hydrothermal Carbonization (HTC) and Vaporthermal Carbonization (VTC). Key technologies in global waste management

Alfons Kuhles

GRENOL GmbH, Ratingen, Germany

Abstract

Hydrothermal carbonization (HTC), and the related process of vapothermal carbonization (VTC), refer to the natural process of coalification and the related industrial process of converting biomass waste into coal and water. The article explains carbonization in contrast to other procedures, it explains the closed material cycle of waste disposal and decentral energy production, the basic functioning of the GRENOL base module and, finally, the economic view and profitability with an empirical example.

Inhaltsangabe

Hydrothermale Karbonisierung (HTC) und das verwandte Verfahren der Vapothermmalen Karbonisierung (VTC) sind auf den Prozess der Inkohlung bezogen, mit dem seit Jahrmillion hat die natürlichen Vorkommen an Torf, Braunkohle und Steinkohle entstanden sind, sowie auf das Verfahren, mit dem organische Abfälle in wenigen Stunden zu Kohle und Wasser umgewandelt werden. Der Beitrag erläutert Karbonisierung im Vergleich zu herkömmlichen Verfahren, den geschlossenen Stoffkreislauf aus Entsorgung und dezentraler Energiegewinnung, Grundlagen der Funktionsweise des HTC-Basismoduls, sowie die wirtschaftliche Dimension und Profitabilität an einem empirischen Beispiel.

Keywords

Karbonisierung, Bioabfallentsorgung, Biokohle, dezentrale Energiegewinnung, Carbonization, organic waste disposal, biocoal, decentral energy production,

1 Introduction

The procedure presented here, and its newly developed technical implementation were already invented by Friedrich Bergius in the 1920s (BERGIUS 1932). Together with Carl Bosch, he was awarded the Nobel Prize for Chemistry in 1931 for the discovery and development of chemical high-pressure processes. The technology and process currently used by GRENOL were then forgotten. Fossil raw materials such as coal, oil, and natural gas could then be mined and used more cheaply. It was not until 2006, when the public debate on climate change and global warming began, that Prof. Markus Antonietti of the Max Planck Institute for Colloid Research rediscovered the process and its application as a regenerative form of energy (ANTONIETTI 2006). In a selected consortium, to which the managing director of GRENOL belonged, the technical possibilities of hydrothermal carbonization (HTC) were taught and passed on to the participants by Prof. Antonietti. This was the foundation stone for the foundation of GRENOL GmbH.

2   Know-how and technological idea source - Who is GRENOL?

After the meeting with Prof. Markus Antonietti in 2006, GRENOL was founded. The company is the world's oldest company on the market dealing with hydrothermal processes and the associated patented technologies including all peripheral components required for a complete operation. One of the two company founders, besides Lothar Hofer, is the managing director Alfons Kuhles (Dipl. Agrar engineer). Over the years, various reactor types for different, primarily industrial applications have been developed and assembled (see figure 1). In addition to technical development, economic factors and market-relevant influences were also considered and adapted concepts further developed.

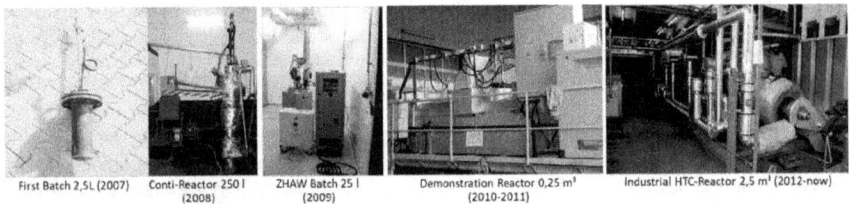

First Batch 2,5L (2007)   Conti-Reactor 250 l   ZHAW Batch 25 l   Demonstration Reactor 0,25 m³   Industrial HTC-Reactor 2,5 m³ (2012-now)
                          (2008)   (2009)   (2010-2011)

Figure 1: Timeline of different developments from the GRENOL company

In addition to further technical and economic developments, the company has been awarded several environmental and energy prizes over the years. The competent team of GRENOL impresses with its professional knowledge from different fields, a holistic approach to customer requirements, a very open-minded and friendly approach to the tasks set by the customer. The company also has a large and extensive international network of support partners and a portfolio of diverse peripheral components for the entire system.

3   The Hydrothermal and Vapor-thermal Carbonization

Hydrothermal carbonization is the conversion of biomass into biocoal and water within a few hours in a closed system under pressure and temperature, as in the natural coalification in the ground for millions of years. This means that it is now possible to produce a storable energy source from currently unused biomass. The hydrothermal process takes place at temperatures of approx. 200-230 °C and approx. 20-25 bar pressure in a pressure vessel. In contrast to other processes, hydrothermal carbonization is not a biological process; the HTC process is a physical-chemical process. This means that the process can be interrupted and restarted at any time without having to observe long dwell times. By splitting the carbon compounds in the biomass, heat energy is released.

This is called the exothermic process. The released heat varies depending on the composition of the biomass: the more carbohydrates are contained in the biomass, the higher the released heat energy is in the process. This exothermic energy is fed back into the HTC process via a sophisticated heat management system. Depending on the input material, the carbonation process only takes a short time (2-6 hours). This means that large quantities of unused biogenic residues can be converted into coal products in a continuous process.

The HTC process is mainly used for wet and liquid waste materials with a dry matter content (DM-content) of max. 30 % dry matter. Materials with a higher dry matter content can be treated with the "sister process", vapor thermal carbonization (VTC). This process uses steam to carbonize lumpy, solid materials as opposed to hydrothermal carbonization, which works in a liquid medium. The process parameters of pressure and temperature do not differ from hydrothermal carbonization. At present, VTC technology is only operated in batch mode, this means in a discontinuously process.

Organic waste   HTC / VTC-process   Bio-Coal   Process water

Figure 2: Hydrothermal / Vapor thermal carbonization (HTC/VTC) and its products

4 Differences and advantages compared to conventional methods: Why should HTC/VTC technology be used?

The use of hydrothermal carbonization involves the closure of waste management worldwide, ideally in closed regional material cycles. This means that with the HTC/VTC process, all organic residues can be usefully recycled (see Figure 3). At present, organic waste is not yet being used to its full extent. In many processes that utilize biomass, the carbon efficiency is too low. The production of regenerative energy from biomass often releases climate-damaging carbon dioxide (CO_2). The HTC process, on the other hand, converts almost 100% of the carbon contained in the biomass into biocoal (see Figure 4). The energy stored by photosynthesis in the form of carbohydrates (e.g. sugar ($C_6H_{12}O_6$)) with the help of the sun, water and CO_2 from the atmosphere is converted into 6 C + 6 H_2O in the HTC process. In this process, $2/3$ of the energy from the biomass is retained and $1/3$ is released as exothermic thermal energy.

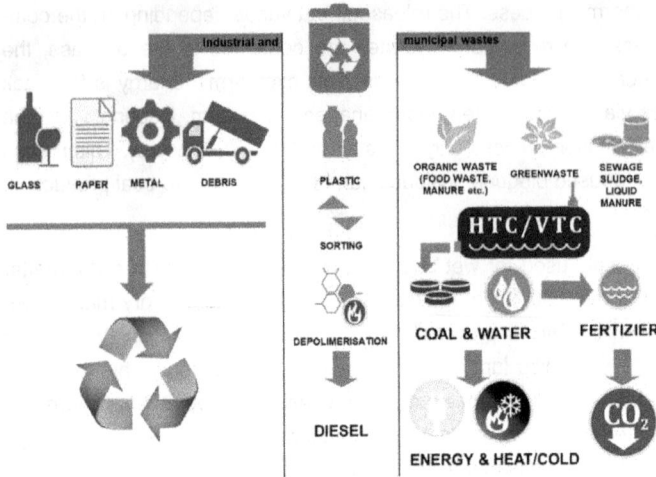

Figure 3: The closed waste management cycle

If the HTC bio-coal is stored, used in materials (activated carbon, insulating material, etc.) or used as a soil optimizer, a CO_2 sink occurs because the carbon dioxide removed from the atmosphere by the plants has been fixed in the carbon. When bio-coal is used as an energy source, the CO_2 balance is neutral because the carbon dioxide, which was recently removed from the atmosphere by the plants, is released again by combustion.

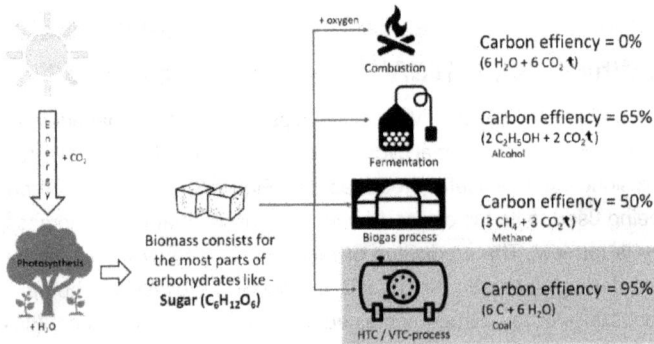

Figure 4: Carbon Efficiency of different processes for biomass utilization in the production of different energy sources

The biocoal is characterized by a higher pressability as compared to the input material. Using the example of sewage sludge, which can be mechanically pressed with up to 30 % dry matter content, after the HTC-process the sewage sludge bio-coal can be pressed up to 60-70 % dry matter content. Moreover, the biocoal can be easily stored and transported. With the use of a wood/coal gasifier and a CHP unit, the bio-coal can be used to produce electricity for feeding into the grid and heat for on-site use - an economical and environmentally friendly use of the energy still contained in the used input material.

Additional advantages of hydrothermal carbonization include the possibility of processing any type, wet or dry, and any mixing ratio. The process does not have to be changed for this. The quantity of coal is determined by the organic dry matter content of the input material, the quality of the coal by the quantity of carbohydrates of the input material. These parameters are again reflected in the calorific value and ash content of the coal produced (see Figure 5).

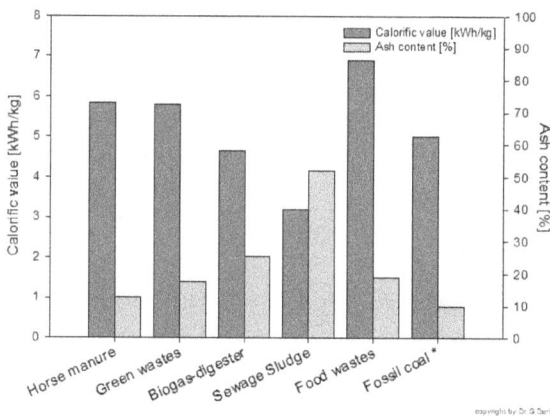

Figure 5: Calorific value comparison and ash content of different input materials in comparison with fossil brown coal*

The biocoal produced in this way can be stored, it can be used for decentral energy production and it has a high energy density (up to 7 kWh/kg Bio-coal). Furthermore, in contrast to fossil fuels (such as crude oil, natural gas and hard coal/brown coal), coal is CO_2-neutral when burned, because the biomass has only recently grown and has not been produced over a longer period of millions of years, so there is no additional emission of carbon dioxide into the atmosphere.

The pressure and temperatures in the HTC-reactor hygienize all pathogens, fungi and bacteria contained in the biomass. Moreover, most antibiotics, hormones and even pesticides are decomposed (WEINER ET AL. 2016). Depending on the plastic composition and melting point behavior, certain plastic fractions are also converted into coal or an oil precursor in the process. This also applies to microplastics, which is currently the subject of in-depth scientific studies.

The resulting process water can be used in different ways depending on the input material. Firstly, it is possible to return the process water to a fermenter (e.g. fixed-bed fermenter or biogas plant) in order to increase methane gas production (WIRTH 2015a, 2015b, 2016). Secondly, the minerals containing the biomass can be separated from the process water by means of vacuum distillation. In this process, the nitrogen content is extracted into a commercially available ASL fertilizer and the remaining nutrients are processed into a basic fertilizer. A third possible application is the evaporation of the process water in a greenhouse with special aquatic plants. This enables an additional increase in biomass for coal extraction.

In addition to generating energy from unused residues, the HTC plant is easy to handle. The concept is space-saving and can be used decentral all around the world, wherever biomass (to be disposed of) accumulates.

Summary of the advantages over conventional techniques:

➢ Hygienization: Antibiotics, pesticides, bacteria and fungi are destroyed and parts of microplastics are converted into coal.

➢ Efficient and environmentally friendly process

➢ Low-emission and economical use

➢ Storable coal, CO_2-neutral for energy generation, CO_2 sink for storage, as a carbon building block or for use in soil optimization

➢ Higher compressibility of the carbonized solid; efficient storage and transport

➢ Production of synthesis gas, increase of methane in biogas production by returning the process water to the biogas plant or the fixed-bed fermenter.

➢ Added value for agriculture through the separation of nitrogen and phosphorus

➢ Easy handling and high applicability due to mixing of input materials

➢ Carbon efficiency: approx. 95%.

5   The GRENOL HTC base module

The patented GRENOL HTC base module operates continuously, with a permanent feed of wet biomass up 30 % dry matter content and a permanent discharge of mixed HTC coal and water. Figure 6 shows the simplified technical process diagram of the HTC basic module (15 t/d input or 5,000 t/a). From a storage tank, the biomass is pumped into the pressure tank via inlet pump. The still cold biomass is heated to approx. 170 °C by a heat exchanger system. In the pressure vessel, the biomass is further heated to operating temperature via thermal oil heating sleeves and forced through the reactor via an internally mounted screw. After a certain dwell time, the biomass is converted into a coal/water suspension and leaves the reactor via a discharge system. The coal/water suspension cooled by the heat exchanger can then be further processed. The residence time of the biomass in the HTC reactor is controlled by the feed system and the screw drive. The coal/water mixture can be separated, the coal pressed into briquettes and dried for storage. The separated process water can be further processed depending on the intended use. The system is controlled over a SPS and operates 24h/7d per year. The GRENOL HTC base module is constructed on a 40-foot container for better transportability, by truck, train or ship. To prevent an uncontrolled increase of pressure, the GRENOL HTC base module is equipped with two electrical and one mechanical security systems. Alarms or messages from the HTC-reactor controlling system can be sent to the operators in charge automatically via E-Mail or SMS, as soon as the alarms occur.

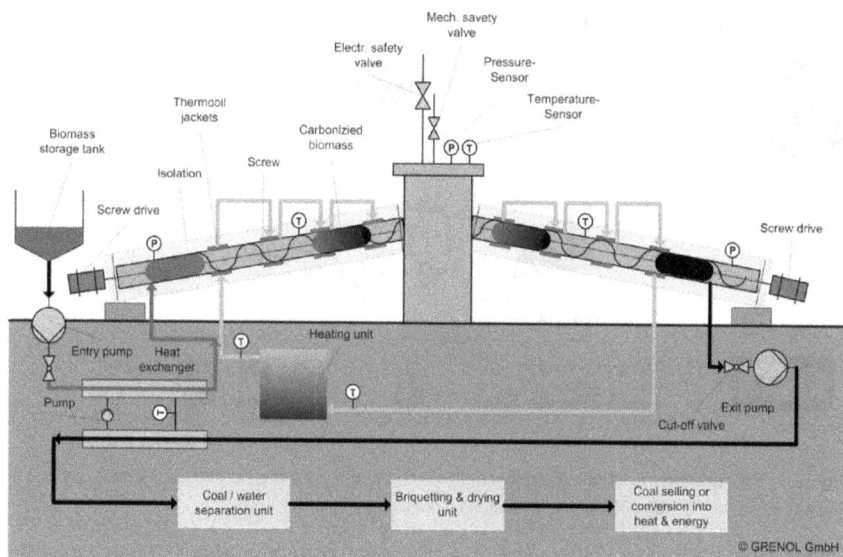

Figure 6: Functional diagram of the HTC reactor at the Rheinmühle site in Chur/Switzerland

6   The Economic view

Using the example of the use of biogas digestate, it will be shown how hydrothermal carbonization can be a sensible addition and value creation to an existing biogas plant. Figure 7 shows an exemplary utilization of 5,000 t/a of pressed digestate (22-27 % dry matter content) from an existing biogas plant.

Depending on the region in Germany, the disposal costs for digestate amount to between EUR 10-25 per ton. These costs are saved when the digestate is recycled by means of hydrothermal carbonization. In order to refine the input materials, approx. 100 MWh/a electricity and approx. 600 MWh/a heat must be fed into the HTC reactor including peripheral equipment (coal/water separation, wood/coal gasifier, water treatment etc.).

Due to the given process parameters of hydrothermal carbonization, the digestate is hygienized and 60% CO_2 savings are achieved in relation to the status quo. From the assumed 5,000 t/a with a dry matter content of 22-27 %, approx. 750 t/a of bio-coal with a calorific value of approx. 5 kWh/kg can be produced. From the CO_2-neutral biocoal, approx. 1,125 MWh of electricity and 2,250 MWh of heat can be produced annually via wood/coal gasification. In addition to generating energy, the process water also produced can be used to optimize methane production. In this process, a compact fixed-bed fermenter is used to increase the methane yield by 60%. The calculation does not include the additional methane yield.

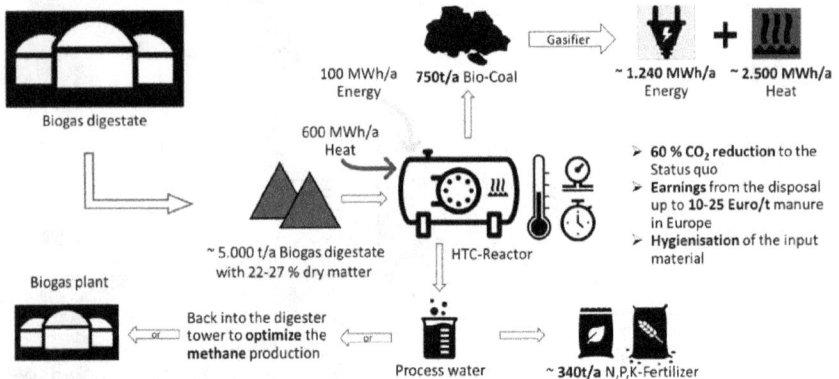

Figure 7: Exemplary representation of biogas digestate utilization by means of hydrothermal carbonization and its value creation potential at an existing biogas plant

On the other hand, up to 340 t/a N, P, K fertilizer can be produced from the process water of the refined fermentation residue by means of further processing steps. All this leads to a value creation of the digestate from the existing biogas plant. The costs for the refining of an assumed 5,000 t/a are approx. 1.48 million euros for a turnkey complete system (HTC reactor, peripheral equipment, wood/coal gasifier, water treatment, etc.). The costs can vary slightly depending on the location and use of the products. Before building a plant, GRENOL always carries out a planning/feasibility study to ensure that all relevant local conditions and the customer's wishes have been considered. For the example presented, the annual operating costs amount to approx. 150,000 Euros. This includes electricity and heat consumption of the system, personnel costs, property rent, maintenance and repair costs. On the other hand, revenues from disposal savings, electricity and heat sales, as well as fertilizer sales and a possible CO_2 certificate trade total approx. 424,500 Euro/a. If expenditure and income are offset against the investment costs, a return of investment (ROI) of approx. 5.4 years results for a complete system for the recycling of 5,000 t/a biogas fermentation residues including fertilizer and electricity/heat production.

7   Conclusion

The GRENOL HTC/VTC technology is a new alternative process to create an ideally regionally closed material cycle consisting of organic waste disposal, decentralized energy generation (electricity and heat) and the use of recyclable nutrients. Hydrothermal carbonization is a convenient and inexpensive option for the "processing" of all unused organic waste generated worldwide and decentral. The HTC/VTC reactor is the heart of an overall concept for the decentralized generation of electrical and thermal energy and enables the return of plant nutrients to agriculture.

8   Literature

ANTONETTI, M.	2006	Zauberkohle aus dem Dampfkochtopf, MaxPlanck-Forschung Multimedial Heft 2.
BERGIUS, F.	1932	Chemical reactions under high pressure. *Nobel Lecture, S. 33.*
WEINER, B., RIEDEL, G., KÖHLER, R., PÖRSCHMANN, J., KOPINKE, F.-D.	2016	Hydrothermaler Schadstoffabbau: hydrothermaler Schadstoffabbau unter besonderer Berücksichtigung von Medikamentenrückständen und PVC. *In: Thrän, D., Pfeiffer, D., Klemm, M. (Hrsg.).*
WIRTH, B., KREBS, M. & ANDERT, J.	2015a	Anaerobic degradation of increased phenol concentrations in batch assays. *Environmental Science and Pollution Research, Vol. 22, Issue 23, 19048-19059.*
WIRTH, B., REZA, M., T., MUMME, J.	2015b	Influence of digestion temperature and organic loading rate on the continuous anaerobic treatment of process liquor from hydrothermal carbonization of sewage sludge. *Bioresource Technology 198, 215–222.*
WIRTH, B. & REZA, M. T.	2016	Continuous Anaerobic Degradation of Liquid Condensate from Steam-Derived Hydrothermal Carbonization of Sewage Sludge. *ACS Sustainable Chemistry & Engineering. 4, 1673-1678.*

Author's address

Dipl. Agar. Ing. Alfons Kuhles
GRENOL GmbH
Artzbergweg 6,
D-40882 Ratingen
Telefon + 2104 2145153
Mail: info@grenol.de

How to REACH Circular Economy –

Liquid Resources from SRF to Feed BASF Production

Nicole Karpensky and Christian Haupts

RECENSO GmbH, Remscheid, Germany

Abstract

Catalytic tribochemical conversion (CTC) is a direct liquefaction process for high-molecular substances of organic origin. The technology converts waste materials into a liquid, hydrocarbon-based resource known as CARBOLIQ CLR, thereby closing the resource cycle. However, its intermediate status creates technical and legislative challenges, both on the product side and with the process definition. A clear distinction must be made between a waste and a product. For a chemical product to gain product status, it must be registered in accordance with the rules of the REACh Regulation. RECENSO took this step for its use during product and process-oriented research and development (PPORD) activities. CLR is therefore qualified as product and no longer considered waste.

RECENSO is working together with BASF in the ChemCycling project. CLR, a resource made out of solid recovered fuel (SRF), was fed into the existing production network in Ludwigshafen in October 2018. The 'plastic cycle' was thus demonstrably closed for the first time. The project's expansion is planned for 2019.

Keywords

Liquefaction, circular economy, plastic, legislation, technology

1 The CTC method

1.1 The process

Catalytic tribochemical conversion (CTC) is a direct liquefaction process for high-molecular substances of organic origin. This process is characterised by the simultaneous use of thermal, catalytic and mechanochemical (tribochemical) mechanisms. As a single-step process, it differs from other methods that extract liquid products by synthesising gases that are generated (e.g. Fischer-Tropsch synthesis).

CTC works in moderate conditions, meaning that operations take place at approximately atmospheric pressure and at low temperatures (< 400 °C). The process uses a mineral catalyst. A rotating energy input unit (the tribo-chemical reactor or TCR), which is connected to the reactor vessel, is the centrepiece of the CTC plant (see Figure 1). It creates the mechanochemical effect and simultaneously heats the system and the medium itself, avoiding wall carbonisation processes. This combination of mechanisms hydrocarbon-based materials to be converted efficiently.

The reactor vessel contains a liquid reactor medium that serves as a base collector vessel for the TCR. A pre-heated blend of feedstock (e.g. shredded biomass, plastic waste and mattresses), catalysts and, if necessary, neutralising agents is constantly fed into this reactor medium. At the same time, constantly rotated hydrocarbons are released from the reactor system.

Zeolites are used as the catalyst, with 1-6% by weight used depending on the feedstock in question. When halogenated polymers are used as feedstock, both calcium oxide and the catalyst are fed into the system. This calcium oxide aims to neutralise any acids that form and is able to bind halogens.

1.2   Design of the CTC plant

RECENSO GmbH carried out the first industrial-scale use of the CTC technology in Europe at its waste management centre in Ennigerloh. (ECOWEST). Known as DIESEL-WEST, this plant essentially consists of the following (see Figure 1):

- a feedstock pre-treatment system (A) to remove materials that are not suitable for processing,
- the CTC core reactor (B), including the TCR and reactor vessel, and
- a product collection unit (C), which condenses and collects liquid CARBOLIQ.
- It also includes a system to remove solid residues from the reactor system (D), which allows the liquid reactor medium to regenerate and facilitates continuous operations.

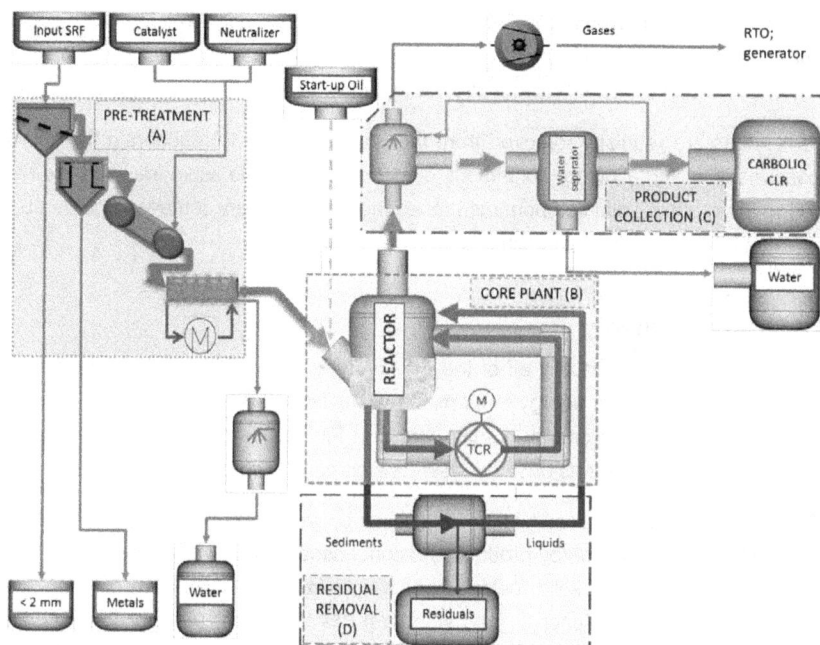

Figure 1: Diagram illustrating the CTC system in Ennigerloh. It shows the pre-treatment system (A), the core CTC reactor (B) and the product collection unit (C). The plant also has a way of removing solids from the reactor system (D).

The CTC pilot plant in Ennigerloh uses SRF as a feedstock to extract crude CTC oil (CARBOLIQ CLR). This feedstock is made out of household and commercial waste in accordance with the RAL-GZ 724 seal of approval. Before entering the CTC plant, SRF undergoes additional pre-treatment to remove materials that cannot be processed by the CTC plant. Pre-treatment (A) aims to guarantee the quality of material entering the reactor (B). Feedstock has to be as free as possible from inert materials and water. Moreover, the plant has requirements relating to the fragmentation of material to undergo liquefaction. The catalyst is added after this pre-treatment. To reduce moisture, material is then dried with the help of a heated auger system.

Pre-heated and pre-dried feedstock is then fed into the reactor, which is the centrepiece of the CTC plant (B). This reactor contains a liquid reaction medium, which is kept at a temperature of < 400 °C during the process. The simultaneous use of temperature, catalysts and friction forces cracks hydrocarbons, generating steam. This steam escapes from the reactor and is captured (C). The high-performance TCR – a unique feature of the CTC technology – is at the heart of the reactor system.

The product (C) is generated by condensing steam flowing from the reactor. Once water arising as a by-product is separated off, crude CARBOLIQ CLR oil is captured. This oil can then be used in the chemical industry, for instance.

Solid residues are regularly removed from the reactor system to regenerate the liquid process medium and allow for continuous operations (D). While liquid fractions are returned to the reactor, solid components that cannot be turned are removed and can be used as an energy source, for instance.

1.3   Mass and energy

The CTC process uses virtually all of the energy input to the TCR for the conversion process due to its efficient energy input mechanism. Therefore, its energy efficiency is beneficial. The mass and energy results shown in Figure 2 were estimated and confirmed by independent experts based on test runs carried out with SRF.

As Figure 2 shows, the process yields < 5% gases by weight, 55% CARBOLIQ CLR by weight, 18% water by weight (by-product, and condensed water from pre-treatment (see (A) in Figure 1)) and 19% solid residues per weight from the conversion of 100% by weight of SRF. Moreover, energy yield stands at 1% for gases, 95% for CARBOLIQ CLR, 0% for water and 4% for removed residues based on the SRF used. The process consumes about 11% of the SRF's energy content, meaning that a total of 84% of the SRF's energy content can be stored in CARBOLIQ CLR.

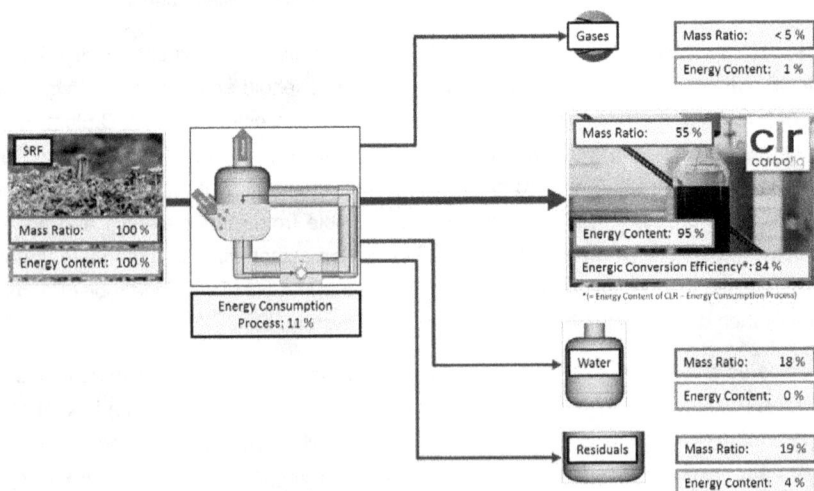

Figure 2: Estimated mass and energy results, based on test runs with SRF as feedstock (Diagram shown without reference to pre-sorting).

1.4   Recycling pathways

Almost 100% of the 6 million tonnes of plastic waste generated in Germany each year is recovered, although the lion's share ends up in energy recovery processes. Figure 3 shows the mass flows. The amounts recycled relate to the quantity of plastic waste generated in Germany in 2017, regardless of whether it was recycled in Germany itself or in other countries. When it comes to energy recovery, the use of plastic waste as SRF still accounts for a significant proportion (~18% based on total plastic waste volumes) [3].

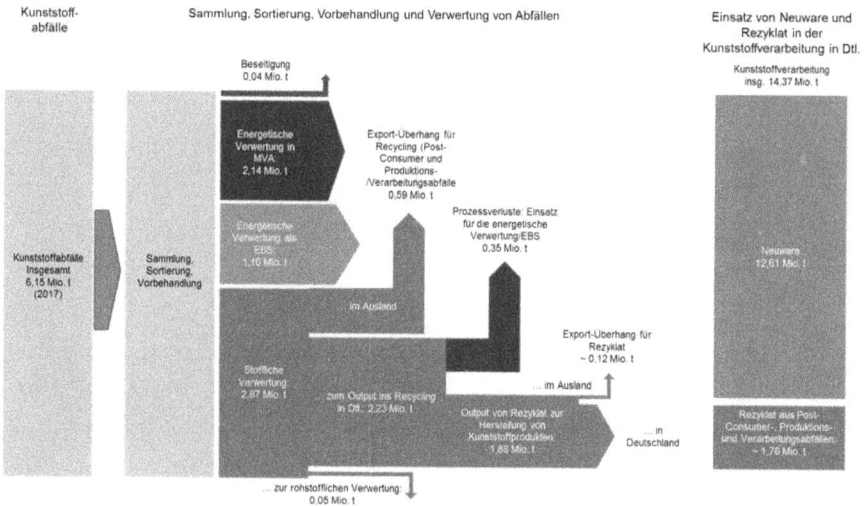

Figure 3: Material flows for processing plastic
waste and reuse in plastic conversion in 2017; based on [3].

It is noteworthy that a proportion of the waste in the material recovery stream is actually exported or is ultimately incinerated to generate energy. The need for alternative recycling technologies is evident. When leveraging the potential held by SRF, the CTC method allows for use of its intrinsic energy and materials.

2   The CARBOLIQ CLR conversion product

2.1   Characteristics of CARBOLIQ CLR

CARBOLIQ CLR is the main product from the CTC process. It is a liquid resource that can be stored and used in a variety of applications. The pilot plant in Ennigerloh primarily liquefies ECO20 SRF made by ECOWEST. CLR is a blend of different hydrocarbons.

Its characteristics mainly depend on the feedstock. Analyses of CLR generated from ECO20 indicate a distillation range of 70 – 352 °C and a density of 803 – 863 g/cm² (at 15 °C). Its kinematic viscosity (at 40 °C) stands at 0.011 – 0.035 cm²/s. CARBOLIQ CLR is mainly made up of aliphatic and aromatic hydrocarbons. The following illustration shows the basic composition of CARBOLIQ CLR, as illustrated with the help of GC-MS analyses. It also contains a typical breakdown of the elements contained in crude oil from the Kuwait region [2].

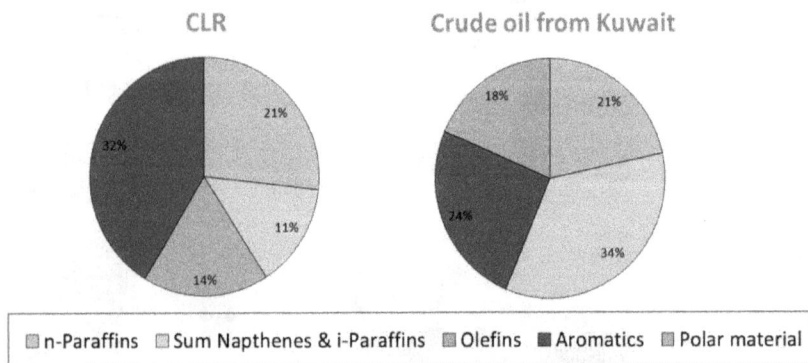

Figure 4: Chemical characteristics of CARBOLIQ CLR (left) and crude oil from the Kuwait region (right) [2]

It should be noted that this image should give an initial impression because any potential variations in analysis methods have an impact on the detection of substances, which may lead to distortions. Nonetheless, initial profiles are created, together with an idea of common features and differences. For example, there are similarities in the n-paraffins and aromatics content, but variations in naphthene/i-paraffins, olefins and polar materials.

2.2   Use of the CTC product

CARBOLIQ CLR is a blend of different hydrocarbons. In many of its key characteristics, CLR is similar to products obtained from crude oil and can hence substitute fossil resources. It can be used as both a liquid source of energy (and energy store) and as a base material for making base chemicals. There is empirical

Figure 5: A photo of CARBOLIQ CLR.

evidence backing both options. When using it as fuel in an engine, upgrading (desulphurisation) is recommended to adapt the product to meet trade standards.

In addition to scaling CTC technology, a series of legal and bureaucratic hurdles must be surmounted in order to use CLR commercially. It is key that CLR be classified as a marketable product since the manufacturing sector can only use this new resource if its end-of-waste status is clear. In summer 2018, RECENSO registered liquid CLR made out of standard SRF in accordance with Regulation (EC) No. 1907/2006 of the European Parliament and Council of 18 December 2006 concerning the Registration, Evaluation, Authorisation and Restriction of Chemicals (REACH) as a substance within the parameters of PPORD activities.

2.2.1 CLR as an energy source and energy store

Several major petrochemical companies have explored using CLR in existing production facilities. From a technical perspective, the results were classified as unproblematic and as "to be preferred" when looking at the outcome of the analysis of pyrolysis oils.

The petrochemical industry has strategic interest in sourcing CLR to meet global demand for advanced fuels. However, the legal position regarding the treatment of waste-based energy sources is unclear in this regard. EU Directive 2018/2001 of 11 December 2018 on the Promotion of the Use of Energy from Renewable Sources ('RED II'), which prescribes rising targets for 'climate-neutral' energy sources, does not govern the allocation of waste-based energy sources. The document now to be transposed into national law defines *recycled carbon fuels* as "liquid and gaseous fuels that are produced from liquid or solid waste streams of non-renewable origin which are not suitable for material recovery in accordance with Article 4 of Directive 2008/98/EC". The European Commission is now required to add a *methodology for assessing greenhouse gas emission savings from recycled carbon fuel* to the Directive by the end of 2021 for transport fuel. A medium-term reliance on "fuels of non-biological origin" is only made explicit with reference to fuels for aviation.

2.2.2 CLR as a circular resource for the polymer industry

High-tech polymers are an essential basis for modern industry and account for a large proportion of the mass of consumer goods whose service lives are constantly becoming shorter and shorter. Polymer recycling opportunities are limited. The growing quantities of plastic-based waste around the globe pose a major environmental policy challenge and the management of 'end-of-use plastics' is a crucial aspect on the path to an efficient circular economy. Innovative technologies are needed to reclaim the maximum amount of raw materials contained in waste streams and return them to the material cycle. Two legal hurdles need to be surmounted: relating to the product definition and process definition.

The first legal difficulty relates to the product definition. Secure 'end-of-waste' status is imperative for conversion products to be used in chemical engineering. Applicable law is not a reliable basis for determining this status. Burocratic responsibility for this classification is not clear, either. Consequently, RECENSO opted to register CLR as a substance in accordance with rules set forth in the REACh Regulation. CARBOLIQ CLR has been listed with the European Chemicals Agency (ECHA) under EC 948-684-8 since 7 September 2018 and is approved for PPORD.

Following successful REACh/PPORD registration, the first significant amounts of CLR were delivered to BASF. In October 2018, this CLR was fed into a steam cracker as part of BASF's ChemCycling project in Ludwigshafen. The steam cracker is the point of origin for integrated production and cracks raw material at temperatures of around 850 °C, essentially producing ethylene and propylene. A large number of chemical products are subsequently made from these base chemicals. Working together with ten customers from different industries, BASF is already developing pilot products of this kind – the first time that the polymer cycle has been closed on an industrial scale.

Figure 6: Images from BASF's ChemCycling image video [1]

Products with quality and hygiene challenges, such as food-contact packaging, can be made out of waste-based feedstock since ChemCycling products manufactured by BASF have exactly the same characteristics as products made out of fossil raw materials. Converting different solid hydrocarbon-based feedstocks into a liquid eliminates contamination and traces of use. Any identifiable 'fingerprint' of the original material entering the CTC plant is eliminated when it is used in a steam cracker.

BASF is working towards the goal of being able to connect recycled raw material's share of total material use in the production network with each end product made and certified at the site by employing a mass balance method. By contrast, other major polymer manufacturers are interested in drawing a 1:1 tie between specific products and the conversion product. Analyses are currently progress in this area.

Another legal and regulatory hurdle lies in the process's classification since neither applicable law nor bureaucratic responsibility are entirely clear. In actual fact, there seems to be a gap in the definition when it comes to classifying recycling operations, which prevents the polymer cycle from being closed reliably at present.

Following successful REACH registration of the liquefaction product, an application was filed with the competent permit authority for the DIESELWEST plant seeking confirmation that:

- it should be viewed as a recycling plant in accordance with Section 3(25) of the German Closed Substance Cycle and Waste Management Act (KrWG) for all types of waste approved for processing in compliance with the applicable permit under the German Federal Immission Control Act (BImSchG),
- processing plastic-rich waste from the preparation of waste electrical and electronic equipment (WEEE) is to be viewed as materials recycling in keeping with the KrWG and the German Electrical Equipment Act (ElektroG),
- processing plastic-rich waste from the preparation of packaging is to be viewed as equivalent to mechanical recycling under the German Packaging Ordinance (VerpG).

The company also requested clarification that liquid products
- made using the DIESELWEST plant,
- for which a market for use exists for certain purposes,
- that fulfil all technical and legal requirements for these purposes (product, environmental and health protection as well as harmlessness for humans and the environment)

should no longer be classified as waste under the KrWG and such rules no longer apply.

The permit authority had yet to respond at the time that this paper was published.

3   Summary and outlook

Catalytic tribochemical conversion (CTC) is a direct liquefaction process for high-molecular substances of organic origin. Since the method makes a liquid, hydrocarbon-based resource out of waste (CARBOLIQ CLR), it essentially lies at the interface between both areas so it can close the resource loop. However, its intermediate status creates technical and legislative challenges, both on the product side and with the process definition. For instance, a substance or commodity must be broadly classified as either a waste or a product. For a chemical product to gain product status, though, it must be registered in accordance with the rules of the REACh Regulation. In this vein, RECENSO has had the liquid product from CTC direct liquefaction of SRF registered under the CARBOLIQ CLR name in keeping with REACh rules for PPORD activities. CLR is thus deemed to be a product and no longer holds waste status. Certain classification of the CTC method as a recycling operation is a prerequisite for investments in this technology.

Mounting interest in circular technologies over easily the past two years is slowly being followed by the first real investment decisions. RECENSO is working together with BASF in the ChemCycling project. Within the parameters of the project, CLR was fed into the existing integrated production site in Ludwigshafen in October 2018, demonstrably closing the plastic cycle for the first time. An expansion of production and continuous CLR deliveries are planned for 2019.

4   Literature

[1] BASF   2018   Video on the ChemCycling project, https://www.youtube.com/watch?v=NlyVO-q2Jg0

[2] National Research Council   1985   *Oil in the Sea: Inputs, Fates, and Effects.* Washington, DC: The National Academies Press, 1985. https://doi.org/10.17226/314

[3] Conversio Market & Strategy GmbH   2018   A short version of a substance flow chart for plastics in Germany in 2017. This study was carried out on behalf of BKV in partnership with and the support of the project's supporting organisations – PlasticsEurope Deutschland, bvse, IK, VDMA, BDE, AGPU (for PVC), KRV, GKV – with its associations pro-K, TecPart, AVK and FSK, as well as IG BCE

Contact details of the author(s)

Dr Nicole Karpensky   Christian Haupts
RECENSO GmbH   RECENSO GmbH
Karlstr. 8 b   Karlstr. 8 b
D-42897 Remscheid   D-42897 Remscheid
Germany   Germany

Tel. +49 2191 4 22 75 14 | E-mail: info@recenso.eu | Website: www.recenso.eu

Real-time analytics to determine the quality of input in waste pre-treatment plants.

Thomas Weißenbach, Roland Pomberger, Renato Sarc

Chair of Waste Processing Technology and Waste Management, Montanuniversitaet Leoben, Austria

Abstract

In the framework of a larger project (ReWaste4.0), research work is carried out to characterise input into waste pre-treatment plants by means of real-time analysis. Commercial waste was selected as input material for the experiments. A number of samples were pre-processed (shredding, sieving) and sorted into a number of fractions. The main experiments will be carried out with individual waste objects, which are taken from nine sorting fractions. Regarding the real-time analysis, two approaches will be used, i.e. sensor-bases analysis (NIR-sensor/RGB-camera) and a deep learning approach (image classification system). The produced data are related to data, which are generated by manual measuring (object size, weight) and laboratory analysis (heating value, water and chlorine content). By using regression analysis, the data of the real-time analysis are related to the data of standard laboratory analysis.

Keywords

Waste characterisation; real-time analysis; sensor-based analysis; image classification system, commercial waste.

1   Introduction

1.1   Motivation

The European Union aims at transforming its economy into a circular economy. For this purpose, the European Commission has adopted a Circular Economy Package in order to increase the efficiency of resource use and to reduce the negative environmental impacts of waste generation and its management. The legally binding measures of the Circular Economy Package are implemented by amending six waste directives, which entered into force in June 2018. The measures which are foreseen by the Circular Economy Package comprise – among others – the introduction and/or increase of recycling rates for different waste streams, such as municipal waste in the Waste Framework Directive (European Union, 2018a) and packaging waste in the Packaging Waste Directive (European Union, 2018b). The higher recycling rates will have to result in remarkably increased quantities of recycled waste in the upcoming decades.

1.2 State-of-the-Art of waste characterisation

A proven method to increase the recycling quantity is the separate collection of recyclable materials. Still, mixed waste (such as municipal or commercial waste) contains a large amount of recyclables which represent an additional recycling potential. The classic method for identifying the share of recyclable materials in mixed waste is the sorting analysis. Usually, sorting fractions are defined by material classes, such as metals, paper, glass, etc. Depending on the objective of the sorting analysis, the fractions are further subdivided according to a number of criteria, e.g. by sub-classes of materials (FE vs. NFE-metals), by function (packaging vs. non-packaging waste) or by shape (2D- vs. 3D-materials). The disadvantage of these manual sorting analyses is that they are carried out only sporadically due to the high efforts and costs. Therefore, this method can only provide a limited picture of the reality, because only a small share of the input material can be investigated, which includes the challenge of taking representative samples, and results are usually available only after a number of days. Desirable would be a real-time and online (on the conveyor belt) analysis system for the identification of the waste composition and for obtaining of relevant parameters.

1.3 State-of-the-Art of Sensor-based Analysis of mixed waste treatment

State-of-the-art regarding sensor-based methods is currently the detection and separation of certain waste fractions. Well known examples of this method are the removal of recyclable material fractions (e.g. PET bottles) or contaminants (e.g. Chlorine-carriers like PVC) in waste pre-treatment plants using NIR and VIS sensors and air ejection units. (Flamme et al., 2018)

A recent approach has been the use of robotic technology to sort selected waste types. Examples of the use of robotics in recovering valuable materials include the Eberhard Group facilities in Switzerland and Fortum Waste Solutions in Finland, which treat mixed construction waste. To identify the fractions to be sorted, a combination of different sensors ("Sensor Fusion") is used: RGB camera, NIR spectrometer, metal detector and 3D laser scanner. (Baldt, 2018)

In addition, there are a number of research approaches that determine various parameters for the characterization of waste and solid recovered fuels (SRF) produced by means of sensor technology (Vrancken, 2017). Particularly well developed is a methodology for quality assurance of SRF, in which the surface area and the material type of individual particles of the SRF are measured with a NIR sensor. These data are used to calculate the calorific value as well as the water, ash and chlorine content by means of regression analysis and a database. (Krämer, 2017)

Approaches to analysing waste input with the help of sensors are so far rather limited. Wünsch et al. (2015) used NIR-sensors to identify waste fractions and to determine material specific basis/surface weights. In addition, a project for the real-time analysis of specific metals is carried out at the moment and is planned to be finalised in 2019. The so-called ARGOS-system has the task to identify target metals in residual materials by means of a multi-sensor-concept. The core unit is an X-ray fluorescence sensor which is developed in the course of the project (BMBF 2016).

1.4 ReWaste4.0

The present research work is part of the K-project Recycling and Recovery of Waste 4.0 (ReWaste4.0), which is managed under the Austrian COMET programme. The Re-Waste4.0-project consists of two scientific-technical areas with seven specific sub-projects. The main objective of the project is the development of a smart waste factory of the future, based on Industry 4.0 approaches. It includes the real-time collection of data about waste flows at several positions in the waste treatment plant, the sensor-based monitoring of the machinery and the communication of all obtained data in a networking system.

1.5 Objectives of the research

The main objective of the present research work is the real-time collection of data about the waste input into a waste pre-treatment plant. For this purpose, the following tasks have to be reached:

- Selection of a suitable input waste stream;

- Characterisation of the selected waste stream regarding its composition and regarding a number of quality parameters by means of manual sorting and laboratory analysis;

- Sensor-based analysis (incl. deep learning approach) of the selected waste streams;

- Identification of correlation between the two approaches in order to generate correction factors by means of statistical regression analysis.

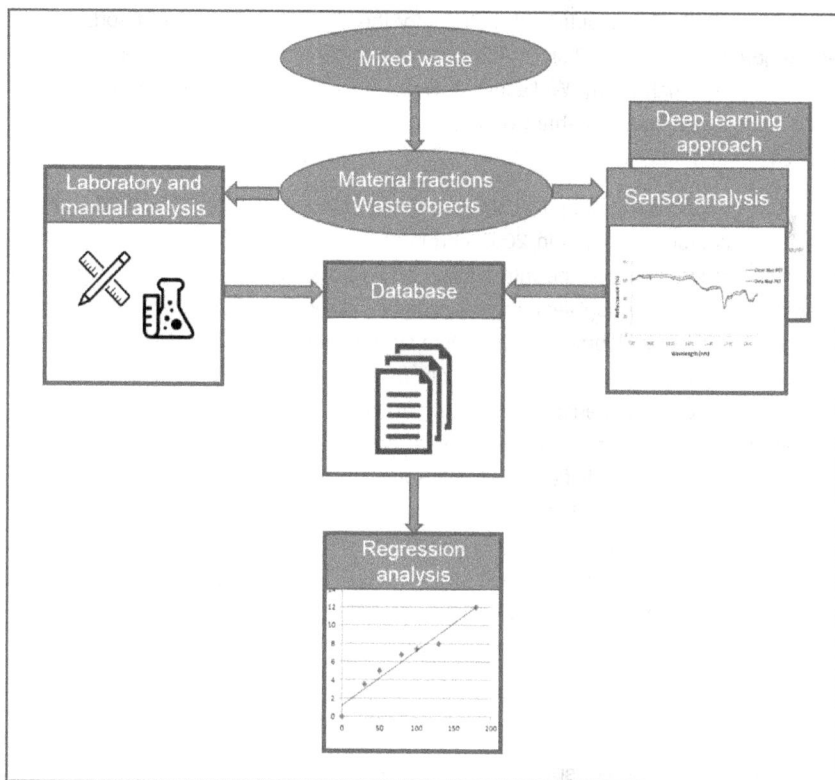

Figure 1   Overview of research work

2   Investigated waste input

2.1   Selection of input waste stream

In general, the scope of the ReWaste4.0-project comprises non-hazardous, solid, and mixed waste. Additional criteria for the selection of a suitable input waste streams are:

- The waste stream must be relevant in terms of generated waste quantities.

- The composition of the waste stream must ensure a sufficient share of recycling materials.

Based on the mentioned criteria, the waste stream of mixed commercial waste was selected for further investigation. In Austria, the generated quantity of commercial waste is estimated to be around 1 Million tonnes (Hauer 2008). According to a personal communication (Hauer, December 5th 2018), this quantity has increased to about 1.1 Million tonnes until 2017. This is sufficient material for the operation a number of sorting facilities. Regarding the composition of Austrian commercial waste, not many data are available. A sorting analysis by Wellacher & Pomberger (2017) of a specific commercial waste sample resulted in a share of selected recycling materials (paper, metals, wood and PET) of about 22%.

The situation concerning data availability is better in Germany. The "Gewerbeabfallverordnung" was first introduced in 2002 and in 2017 a major revision took place, which included - among others – the introduction of a recycling rate of 30% for waste pre-treatment facilities. The legislation was accompanied by data collection activities in order to monitor the development of the waste stream. In addition to waste quantities and treatment plants, the composition of mixed commercial waste was investigated in a number of studies. Experiences with sorting analyses, however, showed that the composition of commercial waste streams varies remarkably between different commercial and industrial branches. Helftewes (2012) sorted waste samples of the five branches Catering (e.g. restaurants, hotels), Healthcare (e.g. nursing homes, hospitals), Food (e.g. bakeries, butchers), Craft (e.g. construction, various trades) and institutions (e.g. universities, banks). The aggregation of the main recycling fractions (paper, plastic, metals and wood) varied between 41% and 58% with an average of about 50% (Helftewes 2012).

2.2   Preparation of samples

In order to obtain a broad spectrum of diverse waste materials, samples of different commercial and industrial branches have been taken at an Austrian waste pre-treatment facility.

Before the sorting started, the waste samples were pre-processed in two steps. In the first step, the waste material was shredded. The aim of the shredding is the comminution of the biggest waste objects and the opening of bags in order to make the waste accessible. On the other hand, the grain size should not be too low in order to make the separation of recyclables more effective and efficient. Preferred grain size would be around $d_{95} = 500$ mm. The second step served for screening out the fine fraction below 80 mm in order to facilitate the sorting process. For this purpose, a drum sieve was used. The conveyor belts carrying away the grain fractions from the drum sieve were used to take representative increments, which were later combined to composite samples.

The composite samples were manually sorted into 18 fractions. The two fractions 2D-plastics and 3D-plastics were subsequently split up in 15 material sub-fractions each (e.g. PE, PET, PVC). In addition, the grain size distribution of the composite samples will be determined.

For the further investigations, the following fractions will be used as a source for taking sample materials for experiments: HDPE, LDPE, PET, PVC, PP, PS/EPS, paper/cardboard, wood, and Tetra Pak. For the laboratory analysis, mixed samples will be taken; the manual and sensor-based analysis will work with individual waste objects.

3 Analysis of samples

3.1 Sensor-based analysis

For the sensor-based analysis, a test rig has been set up by the project partner BT Wolfgang Binder. In the first step, a NIR-sensor and an RGB-camera are used. The sensors have the task to identify the type of the waste material and the size (surface area) of the investigated waste objects in terms of. Regarding the material type, a test sample has been used to develop a teach-in for the sensor system. The generated data, which can be exported by the sensor system, comprise the raw data (NIR-spectra), the classified NIR-data, an RGB-picture and a file for the metadata.

The waste objects, which are generated by the pre-processing and sorting procedure described in chapter 2.2, are scanned by the test rig. The recorded data are fed into a database, in which the regression analysis will take place.

3.2 Manual and laboratory analyses

The manual and laboratory analyses represent the usual approach of waste investigation and are regarded as standard for the comparison with the sensor-based analysis.

- **Manual analysis**

The same waste objects which are subject to the sensor-based analysis of chapter 3.1 are measured manually. The measuring comprises the size (surface area) as well as the weight of the waste objects. For documentary purposes, a digital picture is taken of each waste object.

- **Laboratory analysis**

It is not possible to carry out laboratory analysis of each waste object for financial considerations. Therefore, a representative sample of each investigated waste fraction will be taken. In order to level out variations, up to five parallel samples will be investigated. The laboratory analysis comprises the following parameters, which are mainly relevant for the use of the waste as raw material for SRF-production:

o   Water content

o   Heating value

o   Chlorine content.

The results of the manual and laboratory analysis are imported into the above-mentioned database, as well.

3.3   Deep learning approach

Parallel to the sensor-based analysis of chapter 3.1, a deep learning approach will be used to characterise the waste input material. The aim is to develop an image classification system (ICS) which provides four services:

o   Differentiation between different waste types (residual waste, commercial waste),

o   Particle size distribution,

o   Detection of some selected brands/product groups,

o   Waste composition (materials).

First work has been carried out regarding the identification of certain waste types. Several thousand pictures of residual, commercial and bulky wastes have already been taken and are now in the status of labelling.

4   Regression analysis

All data, which are produced during the diverse analyses of chapter 3, will be imported into a common database. As the real-time methods (sensors, image classification system) are only able to measure material type and area size, all other relevant parameters have to be calculated by regression analysis. The first step is the estimation of the weight of the waste objects by applying fraction-specific area to weight ratios. Based on the calculated weight, further correlations can be identified with regards to heating value and chlorine content. The regression analysis will result in correction factors, which adapt the measured values to the results of the standard laboratory analyses.

5   Literature

Baldt, T.   2018   Robotersortierlösungen von ZenRobotics. In: Pom-
                                           berger, R. et al. [Ed.], Proceedings 14. Re-
                                           cy&DepoTech-Conference, Leoben.

BMBF (Bundesministe-rium für Bildung und Forschung)	2016	ARGOS: Mit Multisensorik analysieren Forscher Recycling-Metalle in Echtzeit. Download (5.4.2019): https://www.r4-innovation.de/files/ARGOS.pdf
European Union	2018a	Directive 2008/98/EC of the European Parliament and of the Council of 19 November 2008 on waste and repealing certain Directives, amended by Directive (EU) 2018/851 of the European Parliament and of the Council of 30 May 2018.
European Union	2018b	European Parliament and Council Directive 94/62/EC of 20 December 1994 on packaging and packaging waste, amended by Directive (EU) 2018/852 of the European Parliament and of the Council of 30 May 2018.
Flamme, S., Hams, S., Zorn, M.	2018	Sensortechnologien in der Kreislaufwirtschaft. In Pomberger et al. [Ed.], Proceedings 14. Recy&DepoTech-Conference, Leoben.
Hauer, W.	2008	Sammlung und Behandlung von Sperr- und Gewerbemüll – Analyse und Ausgangslage in Österreich, Vortrag bei der Veranstaltung „Sperrmüll und Gewerbeabfälle: Sammlung, Zwischenlagerung und Behandlung" am 2. Oktober 2008, Wien.
Helftewes, M	2012	Modellierung und Simulation der Gewerbeabfallaufbereitung vor dem Hintergrund der Outputqualität, der Kosteneffizienz und der Klimabilanz. Dissertation, Rostock.
Krämer, P.	2017	Entwicklung von Berechnungsmodellen zur Ermittlung relevanter Einflussgrößen auf die Genauigkeit von Systemen zur nahinfrarot-gestützten Echtzeitanalytik von Ersatzbrennstoffen. Dissertation, Shaker Verlag, 1. Auflage.
Vrancken, C., Longhurst, P.J., Wagland, S.T.	2017	Critical review of real-time methods for solid waste characterisation: Informing material recovery and fuel production. Waste management (New York, N.Y.) 61, 40–57.
Wellacher, M., Pomberger, R.	2017	Recyclingpotenzial von gemischtem Gewerbeabfall. Österreichische Wasser- und Abfallwirtschaft (ÖWAV) 69, S. 437-445.

Wünsch, C., Ilinykh, G., Borisov, D. 2015 Determination of waste compositions by the identification of fraction specific material surfaces by near infrared-technology and the allocation of appropriate basis/surface weights. Proceedings SARDINIA 2015 Symposium: "15th International Waste Management and Landfill Symposium", Sardinia, Italy.

Acknowledgements

Partial funding for this work was provided by: The Center of Competence for Recycling and Recovery of Waste 4.0 (acronym ReWaste4.0) (contract number 860 884) under the scope of the COMET – Competence Centers for Excellent Technologies – financially supported by BMVIT, BMWFW, and the federal states of Styria, managed by the FFG.

Author's address(es)

Dipl.-Ing. Thomas Weißenbach*
Dipl.-Ing. Dr. Renato Sarc
Univ.Prof. Dipl.-Ing. Dr. Roland Pomberger
Chair of Waste Processing Technology and Waste Management
Montanuniversitaet Leoben
Franz-Josef-Strasse 18
A-8700 Leoben, Austria
*Telefon +43 3842 402-5137
*E-Mail: thomas.weissenbach@unileoben.ac.at

Production of alternative fuels from municipal solid waste in Latin America – Possibilities and Challenges

Rafaela Craizer

BlackForest Solutions GmbH, Berlin, Germany

Abstract

Several countries in Latin America are beginning to explore the implementation of waste to energy (WTE) as part of their integrated solid waste management plans. Even though WTE technologies are well developed worldwide and have proved its efficiency along the last 20 years, project development is still challenging due to the multidisciplinary character of the venture and different initial conditions from each site. The aim of this paper is to briefly discuss common waste management practices in Latin America, specifying waste to energy technologies and current approaches, focussing on the production of refuse derived fuel (RDF) from municipal solid waste (MSW) and its challenges.

Keywords

waste to energy; refuse derived fuel; Latin America; municipal solid waste; waste management

1   Introduction

Waste crisis, geopolitical and global economic dynamics have induced governments to re-examine how basic infrastructure services might be more effectively and economically acquired, without degradation of the quality of service and the environment. Some nations are already developing legislation which enables different types of business development from the historically publicly owned and operated infrastructure services - such as electricity generation, drinking water provision, as well as waste to energy plant ownership and operation.

Waste to energy is a general term to describe the process of generating energy in the form of electricity and/or heat from the primary treatment of waste, or the processing of waste into a fuel source. The WTE industry is gaining growing acceptance worldwide as an important part of the waste treatment hierarchy: "reduce, reuse, recycle, recover and dispose", as being considered part of the step "recover". According to ISWA GUIDELINES (2013), it is only applicable when certain criteria are fulfilled:

- Existence of a mature and well-functioning waste collection and management system for several years;

- A minimum and stable supply of combustible waste;
- A minimum average lower calorific value (at least 6-7 MJ/kg);
- A community that is willing to absorb the increased treatment cost via paying a stable gate fee, and/or a government which is willing to pay respective subsidies in form of gate fees and feed-in tariffs;
- Skilled staff that can be recruited and maintained;
- Solid waste disposal at controlled and well-operated landfills, meaning that collection and disposal are already regulated;
- A constant planning environment for the community – project development and conditions must be from at least 15 years duration.

Selection of the appropriate WTE technology is not a simple task, as the generation of solid waste is influenced by seasonality and socioeconomic level of producers. Policy instruments for sustainable waste management also have a significant impact on the selection of technology. They can be classified into biochemical and thermochemical processes while municipal solid waste (MSW) can be classified as biodegradable and non-biodegradable, suitable for biochemical and thermochemical processes, respectively. Biochemical processes are inter alia related to anaerobic digestion technologies to produce biogas and thermochemical processes are, among others, related to pyrolysis, gasification and incineration technologies. There is also the possibility of combining biological and thermal processes, to produce alternative fuel which will be more suitable and efficient for energy generation (GIUGLIANO, M. AND RANZI, 2016).

Within the Latin American context, it is difficult to fulfil all listed requirements. In fact, the management of solid waste is complex and has evolved in parallel with the urbanization process, economic growth, and industrialization.

To address this subject, it is not sufficient to recognise the technical aspects of the collection, street sweeping and final disposal. It is also necessary to apply new concepts in finance, decentralization, private sector participation, health issues, environmental standards, education, and community participation. Therefore, the introduction and expansion of WTE programs become a real challenge in the region.

Capacity building, market creation and law enforcement, for example, play a major role in the success of such a project. Therefore, choosing a technology based on these initial parameters is not trivial and major importance should be addressed to the feasibility study for such an investment.

2　Waste Management in Latin America

Municipal solid waste management in Latin American and Caribbean (LAC) countries can be compared with eastern Europe, Central Asia, Middle East and North Africa, with an MSW generation of 1.09 kg/capita/day (Figure 1).

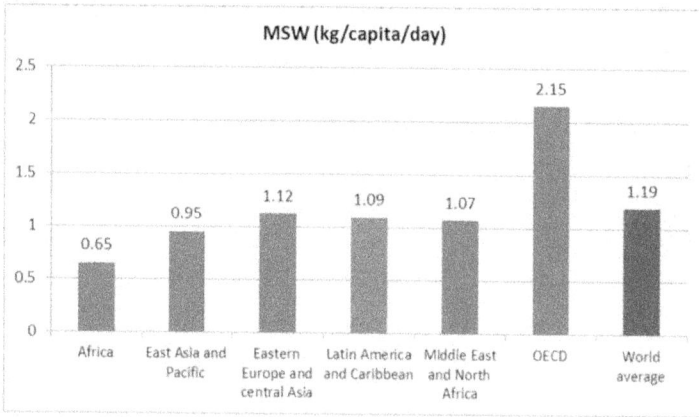

Figure 1 - MSW generation by various regions in the world (Worldbank, 2012).

As shown in Table 1, organic content is often higher in waste from developing/emerging countries compared to developed ones.

Table 1 – Percentage of Organic Waste in various regions in the world (Worldbank, 2012).

Region	LAC	Africa	East Asia Pacific	Eastern Europe & Central Asia	Middle East & North Africa	OECD
Organic Waste (%)	54	57	62	47	61	27

Among all developing regions, LAC has the highest food loss per capita in the world (CHAINEY, 2015). Besides the high organic content in urban waste, a large volume of agricultural waste also contributes to the wet fraction.

A third of all waste generated ends up in open dumps or in the environment, according to UN ENVIRONMENT (2018). Even though collection rates are considerably high, in comparison with other regions - Africa (46%), South Asia (65%), and the Middle East and Northern Africa (approximately 85%) – following the report from INTER-AMERICAN DEVELOPMENT BANK (IDB, 2015), landfilling practices (controlled or sanitary) are considered "adequate disposal" and inadequate disposal rates are still very high in some countries (Figure 2) .

Figure 2 - Collected Data on the Status of Solid Waste Management in LAC (IDB, 2015).

Country	Per capita generation (kg/hab/day) HSW	USW	Municipalities with solid waste management plans (%)	Collection coverage (%)	Collection service frequency in LAC (%) Daily	2 to 5 times a week	Once a week	Types of final disposal (number of inhabitants) (%) Total adequate disposal	Total inadequate disposal	Unit costs (US$/ton) Collection	Final disposal	Types of collection in LAC as a percentage of population (%) Property tax	Electricity	Drinking water & sanitation	Regular customer billing	Recycling rate
Argentina	0.77	1.15	74	99.8	71.9	27.9	0.2	64.7	35.3	54.02	17.63	68.2	3.9	0	27.9	-
Barbados	0	0.9[10]	-	90[10]	-	-	-	82[10]	18[10]	-	-	-	0	0	-	9[10]
Belize	0.68[a]	1.07[11,a]	21.9	85.2	0	88	12	34[a]	66[a]	15.27	-	100	0	0	0	2[11]
Bolivia	0.46	0.49	9.8	83.3	5.4	94.6	0	44.8	55.2	-	7.89	0	95.6	0	4.4	-
Brasil	0.67	1.04[21]	1.6	90.4[11]	44.7	54.5	0.8	58.3[21]	41.7[21]	42.46	31.48	79.1	0	9.2	11.8	1[20]
Chile	0.79	1.25	53.4	97.8	22.3	77.6	0.1	82.2[10]	17.8[10]	23.34	11.43	58.6	0	0	41.4	10[17]
Colombia	0.54	0.69[a]	-	98.9	0	98.6	1.4	93.18[b]	6.82[b]	34.12	23.31	0	34.5	65.5	0	17.2[b]
Costa Rica	0.63[a]	0.88	57.1	90.4	0	68.8	31.2	67.4	32.6	22.65	18.81	31.8	0	16.3	68.2	0.3[a]
Ecuador	0.62	0.73[10]	-	84.2	57.3	42.7	0	30.3	69.7	30.05	5.61	7.1	75.9	0	0.8	-
El Salvador	0.5	0.89	41.3	78.8	20.9	79.1	0	78.9[10]	21.1[10]	30.42	21.02	0	40.9	0	59.1	-
Guatemala	0.48	0.61	28.5	77.7	1	86.5	12.5	15.5	84.5	10.84	-	0	0	0	100	-
Guyana	0	1.5[a]	-	89[a]	-	-	-	-	-	-	-	-	-	-	-	19.3[10]
Haiti	0	0.7[10]	-	11[10]	-	-	-	0[10]	100[10]	-	-	-	-	-	-	-
Honduras	0.51	-	26.7	64.6	5.4	75.7	19	11.3	88.7	20.81	8.16	62.6	0	10.5	26.9	-
Jamaica	0.71	-	0	73.9	0	35.3	64.7	0	100	-	-	-	-	-	-	-
Mexico	0.58	0.94	35	93.2	71.6	28.4	0.1	65.6	34.4	26.39	10.56	0	0	0	0	9.6[a]
Nicaragua	0.73	-	1.2	92.3	0	94.2	5.8	0	100	-	-	3	0	0	100	-
Panama	0.55	1.22	43.1	84.9	13.1	79.5	7.4	55.9[10]	44.1[10]	6.59	5.88	15.1	0	69.4	27.7	-
Paraguay	0.69	0.94	18.8	57	16.1	79.8	4.1	36.4	63.6	15.02	5.98	85.1	0	4.1	80.8	-
Peru	0.47	0.75	57.2	84	55.7	43.5	0.8	43.5	56.5	-	-	0	0	0.2	14.7	14.7[a]
Dom. Rep.	0.85	1.0[a]	5.1	97	55.2	37.1	7.7	33.9[a]	66.1[a]	-	-	0	0	8.8	91.2	-
Suriname	0	1.4[a]	-	80[a]	-	-	-	0[a]	100[a]	-	-	-	-	-	-	-
Trinidad & Tobago	0	1.8[a]	-	100[10]	-	-	-	0[10]	100[10]	-	-	100	-	-	-	8.2[10]
Uruguay	0.75	1.03	73.9	98	18.6	81.4	0	13.7[20]	86.3[20]	47.85	9.19	0	90.9	0	0	-
Venezuela	0.55	0.86	33.4	100	58.2	41	0.8	13	87	-	-	0	90.9	0	9.1	-
LAC Average	0.6	0.9	19.8	89.9	45.4	52.7	1.8	55.4	44.6	34.22	20.43	52.0	15.3	12.4	20.2	-

From the same picture, it is noticeable that collection charges account for at least 60% of waste management costs and that waste tariffs are collected through variable levies – property tax, electricity, drinking water and sanitation and regular customer billing - in these countries.

3 Waste to energy in Latin America

There are several approaches and technologies to enhance solid waste treatment practices, which range from reducing the generation of waste by better design of products and packaging, to recycling usable materials, composting green wastes, and combustion with energy recovery. Globally, over 80% of the total urban post-recycling MSW is landfilled (about 1 billion tons per year), and only 20% of this waste is disposed of in sanitary landfills (THEMELIS ET AL., 2013).

Waste to energy is a broad concept and includes from biogas production out of landfills and anaerobic digestion until direct incineration, co-processing and gasification, among other technology assimilations. Some WTE technologies have been broadly used for waste management practices in Latin America while others are still in progress.

3.1 Proven Technologies

The most common WTE technologies which have already been technically and economically proved to be feasible in LAC are biogas production from landfill methane gases, anaerobic digestion, mainly for agro-industrial wastes, refuse derived fuel from industrial waste and some successful cases of pyrolysis from waste mono fractions.

The production of biogas is a reality in many countries in LAC. The gas is produced from landfills, wastewater treatment plants, agriculture residues, organic waste among others. Bioenergy Nuevo Leon, SA de CV (BENLESA) is the first project in Mexico and in Latin America using biogas as renewable energy produced from a landfill. The project began in the 1990s and proved to be so successful that other cities have since replicated the approach (LFGCONSULT, 2007). The city of Maldonado, in Uruguay, is capturing biogas from landfills, capable of producing more than 4,818 MWh/year (UNSD, 2011). In São Paulo, Brazil, there are two landfills where landfill gas has been used to produce electricity (DE SOUZA ET AL., 2014). Other studies found out that the biogas obtained from landfills can also significantly contribute to the energy matrix of Brazil (LINO AND ISMAIL, 2011). It is predicted for the next years a major role from biogas and biomethane in the energy matrix of the country (FERREIRA-COELHO,2018).

The first small-scale biodigesters were inaugurated in the region in the '70s. This process was accelerated on the '90s and, and in the following years most of the Latin

American countries had experiences with these technologies (MARTÍ-HERRERO ET AL., 2016). Biodigesters have spread successfully mainly in rural zones, to address not only waste management issues but also to enhance energy production due to lack of connectivity to the grid (LOPEZ AND BORZACCONI, 2017).

Besides the handling of the organic matter, there are some initiatives to thermally treat industrial waste by converting the streams into an alternative fuel, which substitutes fossil fuels in cement plants. The biggest cement producers in Latin America are Brazil, Mexico and Argentina, figuring on the list of the 20 biggest producers of cement in the world (CEMBUREAU, 2012). The major Latin American cement company is the Mexican multinational Cemex, which has several factories worldwide and accounted for 8% of use of alternative fuel (CEMEX, 2009). The company attributes this restricted substitution amount to local regulations of waste management. According to Cemex "in many countries, our alternative fuels substitution rate is low, far below its real potential. The reason is that our technical know-how must be matched by appropriate waste management regulations" (CEMEX, 2013). In Brazil, the largest company is Votorantim, which has been practicing co-processing since the early '90s and more than 90% of the facilities are authorized to receive wastes to co-processing (VOTORANTIM, 2014). In 2007, 800.000 tonnes of industrial wastes were co-processed in the country, corresponding to 30% of all the total industrial wastes per year. The main kind of wastes used are contaminated soil, tires, oily sludge, used catalysts, adhesives, resins, latex, rubberized and contaminated materials as paper plastics and woods (CONSTRUÇÃO TOTAL, 2008).

3.2   Ongoing Initiatives

Besides the above-mentioned technologies, there are several waste to energy projects which are currently being studied and/or conducted, with different incineration technologies and framework conditions. Veolia has won a public tender, published by the government of Mexico City, and signed a contract to Design, Build, and Operate the first waste to energy facility in Latin America. The operation concession was granted for 30 years and the plant shall treat 1.6 million tonnes of household waste per year, producing 965 GWh of electricity, which will power the Mexico City Metro (VEOLIA, 2017). In Barueri, Brazil, another waste to energy facility is being developed, with 825 tons per day input capacity of municipal solid waste. It is a public-private-partnership headed by Foxx Haztec and will generate power capacity of 17.5MW (CITVARAS, 2016). In Chile, the Ministry of Energy has conducted a feasibility study to determine the technical, economic, social and environmental viability for the installation of a waste to energy plant in the Metropolitan Region (PÖYRY, 2018).

3.3   Refuse Derived Fuel from Municipal Solid Waste

While Latin American countries are developing laws and strategies regarding industrial waste co-processing, many European countries are co-processing not only industrial, but also municipal wastes (ARANDA USÓN ET AL., 2013).

In most countries, urban waste is limited to handling at the source, collection, and disposal at landfills without any pre-treatment. Energy recovery from MSW in Latin America is almost non-existent (UN, 2004).

Regarding the current management perspectives and volume of municipal waste, co-processing of MSW could be an alternative to properly destinate such stream.

Some matters must be addressed when developing such a project. Mixed municipal waste must be pre-sorted in order to obtain regular waste streams of known quality and remove potential contaminants. Due to this, co-processing municipal waste should be regarded as an integrated part of municipal solid waste management (GTZHOLCIM, 2006).

In this constellation, the three main topics are ecological, technical and financial components. Due to a lack of source sorting in most of the countries, municipal solid waste is highly diverse and might contain hazardous substances like batteries, thermosets, lamps. Alternative fuels that have a high amount of chloride like PVC should be used in limited amounts and fuel mix optimization is very critical in terms of enough heat value in kiln and cement quality (MURRAY, 2008).

From the technical point of view, introducing RDF from MSW into a cement plant requires some adaptations on the plant. They can include feed-in and kiln amendments, as well as gas treatment upgrades with strict monitoring on the amount of dusts, sulfur dioxide (SO_2), nitrogen dioxide (NO_2), carbon monoxide (CO), volatile organic compound (VOC), hydrochloric acid (HCl), hydrogen fluoride (HF), ammonia (NH_3), dioxins/furans ($PCDD_s$/PCDF), benzene, mercury (Hg), thallium(Tl), cadmium (Cd) and other heavy metals from waste gas (SCHÄFER, 2017).

As mentioned in chapter 2, the financing of municipal waste collection systems is highly variable in the continent. Argentina, Brazil and Chile use a mix of property taxes and regular customer billing. Bolivia and Venezuela have most of the population paying through electricity billing. 100% of Guatemala is covered through regular customer billing and 100% of Uruguay is enclosed through a property tax (IDB, 2011).

Gate fees also have deep variations depending on the region, country or municipality. A review of indicative waste tariffs is shown in Table 2.

Table 2 - *Indicative Gate Fees in Latin America (a.Themelis et al.,2013; b. Centro de Economia y Administracion de Residuos Solidos,2011; c. Barriga Rodríguez,2011; d. BFS, 2019)*

Country	Location	Status	Gate Fee (USD/Ton)
Argentina	Buenos Aires	Sanitary	15[a]
Brazil	Rio de Janeiro	Sanitary	40[d]
Brazil	Piratininga	Sanitary	22[d]
Brazil	Sao Paulo	Sanitary	31[d]
Brazil	Florianopolis	Sanitary	28[d]
Chile	Valparaiso	Sanitary	14[b]
Chile	Villa Alemana	Non-sanitary	6[a]
Chile	San Pedro	Non-sanitary	8[a]
Mexico	Toluca	Sanitary	13[a]
Mexico	Hidalgo	Non-sanitary	5[d]
Uruguay	Montevideo	Sanitary	16[c]

With financial decentralization, common in many countries, municipalities have had to overtake funding of services previously funded by central government. This reduces the resources available for other municipal services. How municipal waste management systems are financed has direct, concrete implications for the ability of municipalities to develop sustainable, quality services and to promote decent employment conditions in the sector (PSIRU, 2017).

Considering that gate fees from industrial waste are usually higher than from urban waste, business models for RDF production from MSW must be settled accordingly. A simplified financial scheme is shown in Figure 3, highlighting key cost and income drivers for such an investment.

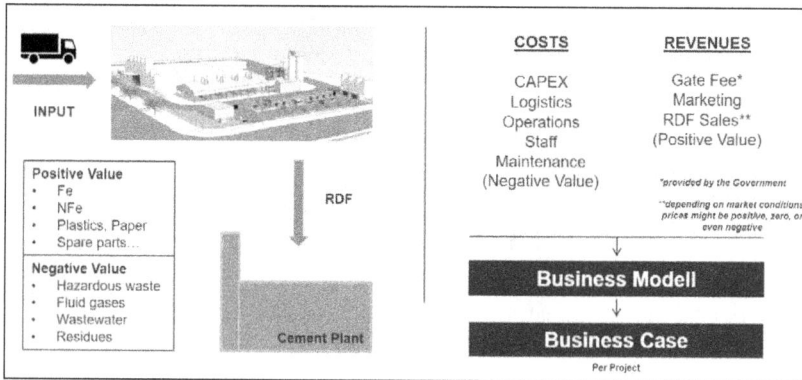

Figure 3 - Simplified financial analysis (BFS, 2019).

Besides government support on guaranteeing long-term agreements for waste receipt and gate fees oriented on the market situation, electricity and fossil fuel costs play a major role in financing such a system.

4   Challenges and Conclusions

The challenges, opportunities and required support for private sector, especially from waste management and cement industry, must be considered through extensive stakeholder engagements. From the recycler perspective, the uncertainty of take-off and price are major concerns, while from the receiving party (for example a cement factory), uncertainties regarding supply and quality, heterogeneity and inadequate infrastructure are main barriers (CPHEEO,2018).

Specially in the Latin American context, logistical issues due to megacities and urban complexes play a major role in designing such a structure. Besides, countries are large and almost a hundred percent of cargo is transported per truck. This highly increases operational costs for any waste management investment. As a result of the large landfill history in the region, landfill companies have strong political representation and compete with cement factories to receive waste input (LEME, 2019). Along with this, divergences between political spheres are common and it is unusual to obtain long term contract with municipalities.

Apart from technical, financial and political topics, MSW management is a critic environmental problematic common to almost all cities in Latin America. As co-processing is already occurring in cement kilns, RDF from MSW is a feasible option for dealing with the urban waste challenge. However, it should be correctly implemented, otherwise it could result in a quality loss in cement production and potential environmental liabilities. If properly handled, it is a win-win situation for parties involved.

5 Literature

Barriga Rodríguez, M. E. D.   2011   Cost-benefit analysis of a waste to energy plant for Montevideo; and waste to energy in small islands

BlackForest Solutions Gmbh (BFS)   2019   Research from various sources and BFS networks

CEMBUREAU   2012   Activity Report - The cement sector: a strategic contribute to Europe's future

CEMEX   2009   Local Reports - Sustainable Development Report

CEMEX   2013   Alternative Fuels, http://www.cemex.com/MediaCenter/files/ CEMEX_POSITION_on_Alternative_Fuels.pdf

Central public health and environmental engineering organisation (CPHEEO)   2018   Guidelines on Usage of Refuse Derived Fuel in Various Industries

Centro de Economia y Administracion de Residuos Solidos, Universidad Federico Santa Maria   2011   Plan de Manejo Integral de Residuos Solidos Region de Valparaiso

Chainey, R.   2015   Which Countries Waste the Most Food? Available online: https://www.weforum.org/agenda/ 2015/08/which-countries-waste-the-most-food/

Citvaras, A.   2016   Foxx Haztec. 5th Annual World Waste to Energy City Summit. *London*

Construção Total   2008   Co-processamento - uma iniciativa para o futuro

De Souza, S. N. M., Horttanainen, M., Antonelli, J., Klaus, O., Lindino, C. A. and Nogueira, C. E. C.   2014   Technical potential of electricity production from municipal solid waste disposed in the biggest cities in Brazil: Landfill gas, biogas and thermal treatment. Waste Management Res., vol. 32, no. 10, pp. 1015–1023

Ferreira-Coelho, J. M.   2018   Artigo perspectivas para o setor de biocombustiveis do Brasil em 2018

Giugliano, M. and Ranzi, E.   2016   Thermal Treatments of Waste. Waste to Energy (WtE)

GTZ-Holcim	2006	Guidelines on co-processing Waste Materials in Cement Production
InterAmerican Development Bank (IDB)	2011	Regional Evaluation of urban solid waste management in LAC 2010
ISWA	2013	Waste to Energy in Low- and Medium-income Countries. ISWA Guidelines
Leme, F.	2019	Presentation "Co-processing activity in South America" - Global CemFuels Conference
LFGConsult	2007	Case of CDM Landfill Gas Projects Monterrey, Mexico (BENLESA)
Lino, F. A. M. and Ismail, K. A. R.	2011	Energy and environmental potential of solid waste in Brazil. Energy Policy, vol. 39, no. 6, pp. 3496–3502
López, I. and Borzacconi, L.	2017	Anaerobic Digestion for Agro-industrial Wastes: a Latin American perspective. International Journal of Engineering and Applied Sciences (IJEAS)
Martí-Herrero, J., Pino, M., Gallo-Mendoza, L., Pedraza, G. X., Rodríguez, L. and Víquez, J.	2016	Oportunidades para el desarrollo de un sector sostenible de biodigestores de pequeña y mediana escala en LAC
Murray A, L. P.	2008	Use of Alternative Fuels in Cement Manufacture: Analysis of Fuel Characteristics and Feasibility for Use in the Chinese Cement Sector
Pöyry	2018	Pöyry presents results of feasibility study for first Waste to energy plant in Chile. https://www.poyry.com/news/poyry-presents-results-of-feasibility-study-for-first-waste to energy-plant-in-chile
Public Services International Research Unit (PSIRU)	2017	Municipal Solid Waste Management Services in Latin America
Schäfer, S.	2017	Untersuchung des Einflusses alternativer Brennstoffe auf die Ansatzbildung in Dreh-rohröfen der Zementindustrie. Vereins Deutscher Zementwerke e.V.
Themelis, N.J, Diaz Barriga, M. E., Estevez, P. and Velasco, M.G.	2013	Guidebook for the Application of Waste to Energy Technologies in Latin America and the Caribbean
UN Environment	2018	Waste Management Outlook for Latin America and the Caribbean

UN Habitat   2004   The state of the world's cities 2004/2005. Globaliza-
                                      tion and Urban Culture Settlements Program

UNSD   2011   Turning waste into resources: Latin America´s waste
                                      to energy landfills

Usón, A. A., López-Sabirón,   2013   Uses of alternative fuels and raw materials in the
A. M., Ferreira, G., Sastre-   cement industry as sustainable waste management
sa E. L.   options. Renewable and Sustainable Energy Re-
                                      views, 23:242–60

Veolia   2017   Waste to energy Renewable Energy Mexico.
                                      https://www.veolia.com/en/news/waste to energy-
                                      renewable-energy-mexico

Votorantim   2014   Votorantim Cimento

World Bank   2012   WHAT A WASTE: A Global Review of Solid Waste
                                      Management. Urban Development Series
                                      Knowledge Papers; Paris, France

Sorting process and energy recovery possibilities of hospital waste in Brazil by the hospital association in Porto Alegre

Silvia Kohlmann

envitecpro Gmbh, Rostock, Germany

Abstract Treatment of hospital waste in Rio Grande do Sul
Waste landfills are often not properly secured in Brazil. Secondary raw materials are not or only partially recycled. Contaminated leachate flows into the groundwater and pollutes the drinking water, which leads to serious consequences for the nature and health of the population and the flora and fauna. Industrial, commercial and household waste is hardly treated or recycled. 90 % is intended for final disposal to landfills. As an example the hospital waste of the association in Porto Alegre was analysed to derive treatment measures for local energy recovery.

Keywords Treatment of hospital waste, waste treatment, Brazil, hospital waste

1 Situation in Brazil

Landfills are often not properly secured in Brazil. Secondary raw materials are not or only partially recycled. Contaminated leachate gets into the groundwater and pollutes the drinking water, which leads to serious consequences for the nature and health of the population, the flora and fauna.

In the capital Porto Alegre of the southern federal state of Rio Grande do Sul with approximately 1.4 million inhabitants, the association with 62 members such as hospitals and medical facilities represents about 6,600 beds. The currently existing waste management system is able to separate the waste into different categories. The infected, medical waste is just only sterilized by a monopoly supplier from neighbouring Uruguay with an autoclave (gas-tight sealable pressure vessel for the thermal treatment of substances in the overpressure area) and then transported to a landfill near Porto Alegre. The hospitals, such as other facilities as well, are bound to this monopoly provider for the sterilization of medical waste. This has repeatedly led to incalculable price increases in the past. In addition, after the sterilization the waste is just transported to the landfill without any material recovery or thermal utilization.

The German company IBS Technik Gmbh from Neubrandenburg has many years of experience in the planning, realization, operation, maintenance and dismantling of thermal utilization power plants. Together with the German company envitecpro which is responsible for the project management measures will be developed to improve the separation and treatment of medical waste and the material recovery and for the local production of electricity and heat.

2   Project plan

IBS and envitecpro collaborate with the regional hospital association to develop measures for the environmentally friendly use of hospital waste in order to reduce the volume of waste to be landfilled, to use the secondary raw materials and reduce the costs of waste management. The decentralized produced electricity and heat will be available to the hospitals for the own operations. The professionals of the hospitals and other medical institutions are trained in regional environmental protection and sustainable resource management. The hospital association represents about 6,600 beds, of which as many as possible are included in the data collection.

To achieve the project goals, the following steps are required:

In the **first phase** of the project (work package I) the project organization will be established and the project goals and target groups will be identified jointly. The kick-off event with the Brazilian project partners, including workshops and developing the cooperation agreements, will form the basis for a successful project implementation.

In the **second phase** of the project (work package II) detailed data on the actual situation of hospital waste management will be collected together with the Brazilian project partners. These include the relevant legal framework conditions (city / municipality, state and state), the local requirements of the hospitals (location analysis) as well as the collection, treatment and utilization of the current work flow. The German project partners will provide a questionnaire. The quality of the data collected and the evaluation will be crucial for the project result.

At the same time, the state of the art in the utilization of hospital waste in Germany will be analyzed and the transfer to the Brazilian conditions will be examined.

The results will be discussed and evaluated in a workshop in Brazil.

In the **third phase** (work package III) based on the values and data of phase 2 the framework and the basic data for a concept development are defined. Measures for an optimized collection and presorting of waste will be proposed. The concept for a thermal utilization of waste will be developed as well. In addition to the technical, ecological and economic aspects, the licensing of such a plant will be constantly coordinated with the responsible environmental authorities. It is also necessary to examine which technologies are available and to what extent they are suitable for the purpose of the application and the local conditions, or how they have to be adapted. The initial eligible processes are incineration or gasification. For combustion various methods (e.g. grid, fluidized bed, etc.) and for plant concepts (e.g. heat generation, cogeneration, etc.) are available. The aim is to provide energy for the central laundry of the hospitals. In addition, reasonable alternatives for the use of energy can be considered.

The proposed technologies are also dependent on the legal framework in the state of Brazil and the federal state of Rio Grande do Sul.

For educational purposes, Brazilian decision-makers will travel to Germany, focusing not only on a workshop for presenting and discussing the concept, but also on visiting reference facilities. These are mainly members of the management staff and employees of the hospitals and other medical institutions on the subject of environmental protection, work safety and the sustainable use of resources.

3   Analysis

The respected association has 16 hospitals. 13 provided information to values regarding quantity and quality of the waste according to the standardized questionnaire. To "group 1" belongs 4 hospitals. One hospital was eliminated as the values were not consistent. The respected 12 hospitals with 5.159 beds out of 16 hospitals with 6.121 demonstrate 84,3 %. See the distribution of waste per hospital per year in the diagram. The total amount is 10.243,6 t per year.

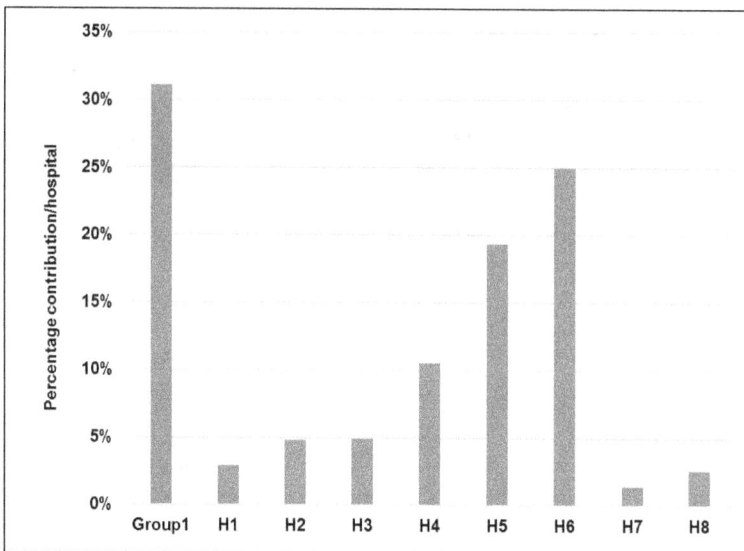

Figure 1   Distribution of waste per hospital per year (total of 10.243,6 t per year)

For the participating hospitals an average of 169,7 kg per bed and month of all classes of waste was calculated according to the data. In the State Plan of Solid Waste of Rio Grande do Sul (PERS-RS) an average of 127 kg per bed and month of the hospitals of Porto Alegre in 2014 was identified.

The waste is collected by five categories:

Class A – biological waste
Class B – chemical waste
Class C – radioactive waste
Class D – common and recyclable waste
Class E – sharp objects as needles

Figure 2 shows the average distribution of waste of the participating hospitals according to the classes.

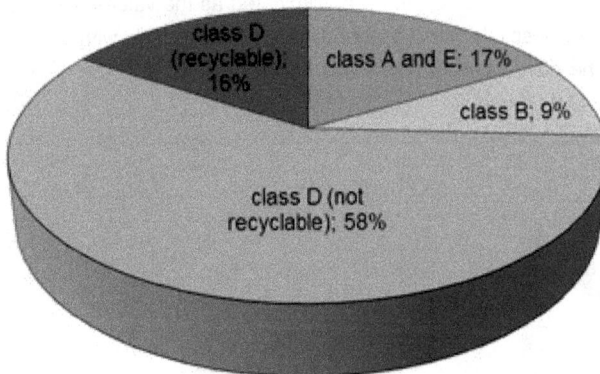

Figure 2   Distribution of waste of the participating hospitals according to the classes

A verification of the analyzed values by the comparison with the data of the State Plan of Solid Waste of Rio Grande do Sul shows that class D is matching where class A and E and B varies from the data in 2014. One reason is, that the information about the method of data analysis is lacking. Figure 3 shows the value of the state plan of 2014.

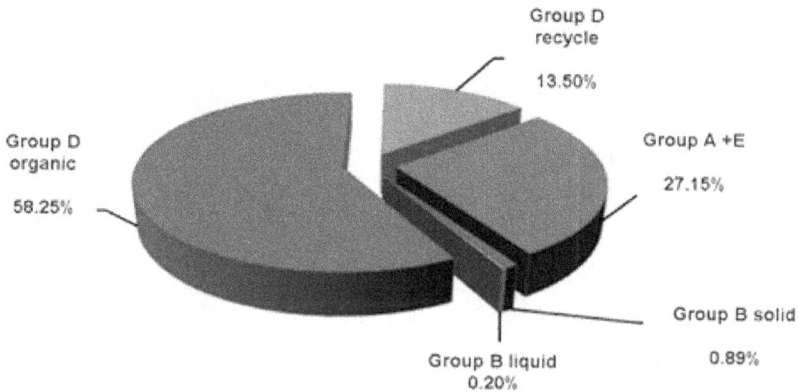

RSS Composition in hospitals of Porto Alegre

Group D
recycle

13.50%

Group D
organic

58.25%

Group A +E

27.15%

Group B solid

0.89%

Group B liquid
0.20%

**Figure 3   Distribution of waste of hospitals in Porto Alegre according to the state plan
of Rio Grande do Sul**

Figure 2 shows as well that in the respected hospitals class C (radioactive waste) is not produced. The amount of all classes per bed and month has a big range from 69 kg to 248 kg. Some hospitals do not count or evaluate class D (recyclable waste) as this is collected without charge by the community. The relevant classes for a thermal treatment are A, E and D (non recyclable). Figure 4 shows the diagram of the different classes by hospital.

The relevant class A and E per bed as well shows a big range. This allows the participating hospitals from the association to analyze the amounts for a better separation and a reduction in the future. The benefit is to compare the results among each other. This offers a great potential for a reduction of the cost of the waste management. H8 with 809 kg per bed and year for class A and E is considered to be inconsistent. The amounts are displayed in figure 5.

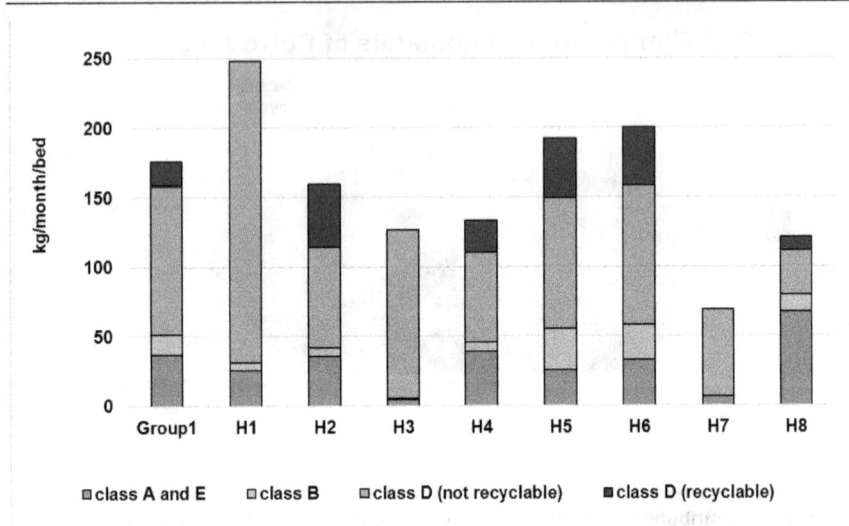

Figure 4   waste per class by participating hospital per month

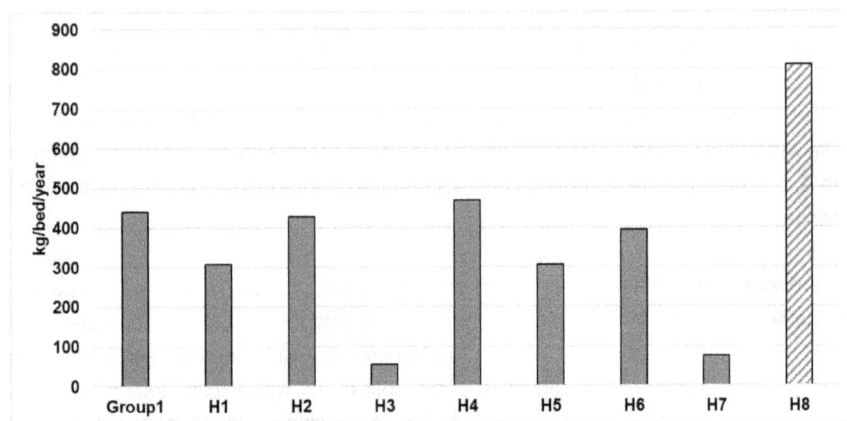

Figure 5   class A and E per hospital in kg per bed and year

The amounts for class D per hospital is shown in figure 6. As well H1 was considered to be inconsistent.

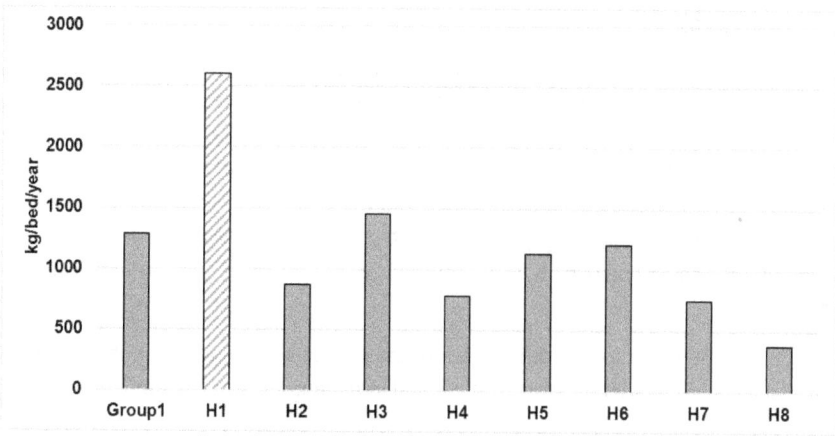

Figure 6   class D non recyclable per hospital in kg per bed and year

One of the analysis during the project additionally to the quantitative analysis is the documentation of the waste management of the different classes. This allows the optimization of the waste handling as well as training of staff on waste handling. Figure 7 shows the flow chart including the process within the hospital as well as the final disposal.

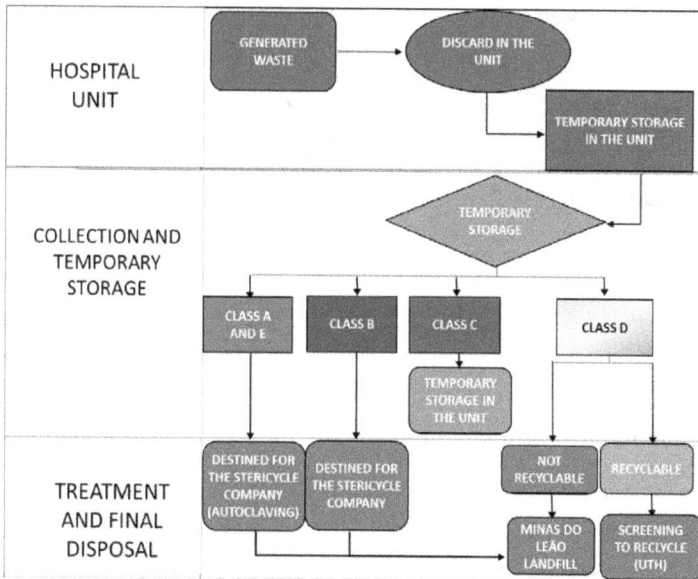

Figure 7   flow chart of hospital waste according to the classes

4   Outlook

The first and second phase of the project is finished. The results were analyzed and summarized. They show a great potential for the optimization of the waste management in the hospitals. That means the gathered information will be used for training purpose of the staff as well as for the improvement of the waste handling process. This will lead to a more secure and cost efficient process. The information as well will allow the hospitals to compare the single results between each other. Their position for negotiation with the waste treatment companies for a reduction for the handling costs will be strengthened.

Now the requirements for the third phase of the project are fulfilled. With the analysed data the technical possibility for a local thermal treatment will be calculated. Therefore a precise and reliable data base was necessary.

The project is funded by the DEG (German development bank) with the program developpp.

5   Literature

PERS-RS   2014   State Plan of Solid Waste of Rio Grande do Sul

Prof. Dr. Jeane Estela   2019   Qualitative and quantitative analysis of hospital
Ayres de Lima, Prof. Me.   waste of the association of Porto Alegre
Adriana Moura, Me.
Gabriela Messias,
Leonardo Siviero, Maria
Vitória Lisot, Dr. Maria-
Elisabete Haase-Möllmann,
Dr. Dirk Bludau, Dipl.-Kffr.
Silvia Kohlmann

Author's address:

Dipl.-Kffr. Silvia Kohlmann
Envitecpro GmbH
Luise-Reuter-Ring 8
18069 Rostock
Germany
M: +49 (0) 172 – 77 171 90
E: silvia.kohlmann@envitecpro.de

Biological Drying of MSW for RDF production in MBT Plants with membrane covered boxes plus biofilter

Axel Huber

STRABAG Umwelttechnik Gmbh, Dresden, Germany

Abstract

Mechanical biological treatment (MBT) plants are widely found, also in countries and regions with low level of solid waste treatment, in which a first step of organised treatment shall improve the overall situation of the collected waste. This shall be the reduction of uncontrolled emissions from landfills, raising the recycling quota and finally depositing off of mostly inactive waste. Sometimes these plants are realised without being embedded in an overall strategy of the regional waste management, and therefore most of the positive impacts are lost. Heron under an MBT plant is introduced in which the main target (loss of weight before depositing) was expanded to the production of RDF to be used in cement factories. This is not new, but the innovation of this plant are composting boxes with semipermeable membrane roof in combination with a regular open biofilter. This very cost-effective technology allows safe and organised operation of a plant and shorter retention times compared to regular membrane covered windrows or boxes.

Keywords

MBT, membrane cover, biofilter, RDF, cement factory, composting box, cost effective.

1   Introduction

Beside the mechanical treatment ("splitting") of MSW or residual waste as initial step in an MBT plant, there are numerous technologies on the market for the biological treatment. They differ in factors like energy balance, level of automation and as a result of that, investment costs. Biological treatment with anaerobic digestion as well as automatically filled and/or discharged composting systems are not further considered herein. That brings us to the reputed technology of composting boxes with forced positive aeration and waste air treatment in biofilters.

The requested target of this plant was the quick reduction of weight of the screened <80mm fraction to keep landfill costs low and certainly cheap invest cost. This versatile MBT plant could be realised at extraordinary low investment costs of ca. EUR 75 per ton of annually treated waste (turn key MBT).

Beside the guarantee parameters of the plant which were proven in the commissioning phase, the production of RDF was a trial which was realised together with the public client and a leading cement company.

2   The biological process

For the biological degradation process, each box is aerated by pressure aeration via floor ducts. In order to control the system and to maintain optimal process conditions, temperature is measured inside the material with manual probes as well as the oxygen concentration in the waste air to the biofilter in order to control the aeration system.

For aeration each box is equipped with a separately controlled ventilator and aeration pipework. Depending on the season and process, the air can partly be recirculated before it is fed to the biofilters.

In order to control humidity, a sewerage network to drain leachate resulting from the composting process is installed. The leachate is captured in a collection basin. The resulting concentrate can be re-irrigated to the boxes, if degradation is the only target, not biological drying or a combination of both.

The boxes have reinforced concrete walls and a roof made of a semipermeable three-layer membrane. The purpose of this membrane is to allow vapor to pass to get the water out of the box but to keep the odor-, pollutant- and dust emissions through the membrane roof as low as possible. From outside the membrane protects the box from rain and snow. Structurally it must withstand the loads of wind and snow.

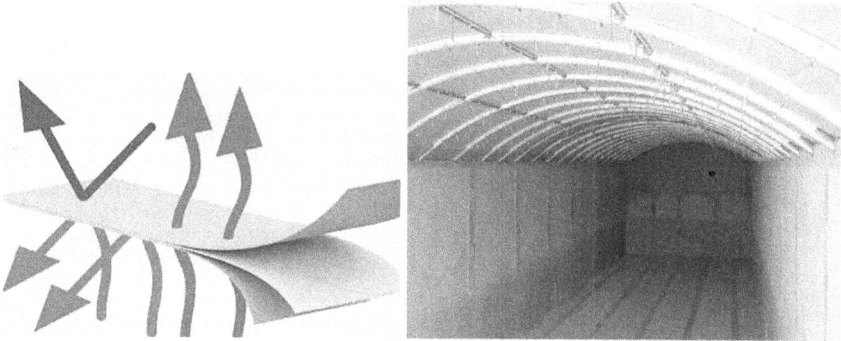

Figure 1   left: Function of 3-ply membrane. right: Stainless steel roof structure with fixed membrane

Figure 2   Large concrete box with plot of 10m x 50m and membrane roof

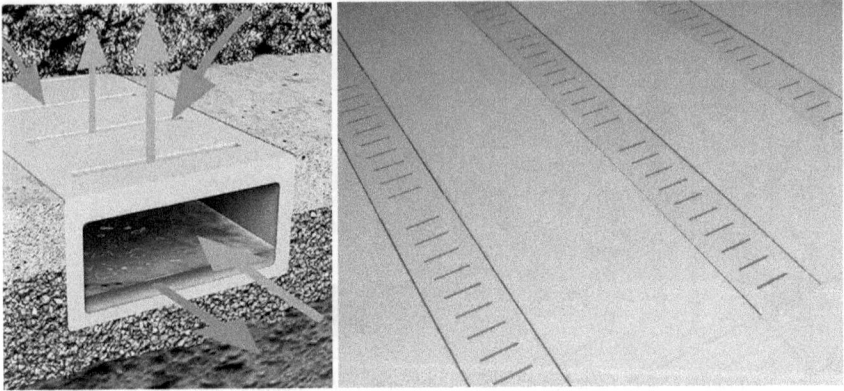

Figure 3   STRABAG uses own, simple and cost effective ducts for aeration and collection of leachate. Robust cover plates can be easily removed for cleaning.

2.1 Two way system

The three-ply membrane is limited concerning the amount of vapor (water) which can pass through in a defined time. The membranes on the market can be adjusted to the targets regarding their permeability, but if the pores of the membrane would be designed too large, also odor and other compounds could pass and therefore its functionality would not be given anymore. Therefore the idea was that additionally to the membrane, exhaust air from each box can be routed to biofilters which are installed on the back side of the boxes.

Figure 4   13 boxes and adjacent biofilter segments on backside

Each box has a gross area of 500m² (10m x 50m) and therefore the time of feeding and discharging with wheel loader takes certain time. To reduce uncontrolled emissions during this manipulation phase, the box is not aerated and the waste air is sucked off and fed to the biofilter.

After 28 days, stabilized material will be removed from the cells using a wheel loader and transferred to the maturation and finally transported to the landfill.

2.2 Shifting between Boxes

At plants which are fed and discharged manually by wheel loaders, the manipulation of feedstock has to be optimized. Experience shows that it is necessary to shift the feedstock between boxes at least once in the process cycle, to achieve the optimum performance of the chosen technology. Otherwise the material is not homogenous anymore and the necessary retention time in the box would increase unproportional for the same result. Inhomogeneous material shows zones which are very dry with higher permeability and better aeration as well as zones which are still more humid and aerated less due to lower permeability. Shifting the boxes "mixes" the waste and the homogenous waste body can be aerated uniformly throughout the large box.

Figure 5   Picture shows inhomogeneous drying results after 4 weeks if the boxes are not shifted - dark (wet) and light (dry) zones inside waste body during discharge

3 Dimensioning of the biological treatment process

3.1 General

Mass reductions achievable during the intensive composting and bio drying certainly depends on the composition of the MSW and its biodegradable fraction <80mm.

For this plant 60% of the incoming waste is of native organic origin. The dry matter content in the fraction <80 mm prior to the first stage of composting is approx. 43%.

3.2 Intensive composting

Design basis:

Plant input:	150.000 t/a MSW
Residual waste - Composition:	
Native organic fraction within residual waste	Min. 60%
Dry matter (DM):	Appr. 43%
Organic dry matter (oDM)	Appr. 65% of DM
Retention time intensive composting	28 d
Shifting of material within the retention time	Once (optionally)
Resulting height of pile based on max. area load of 1,5 t/m² and density of 0,6 t/m³	2,5 m
Resulting total mass degradation (<80mm)	Ca. 39,5 %

Figure 6   Key data of the Ploiesti plant

Figure 7 Box during feeding with <80mm fraction

4   Modified process

Romania has a huge demand for RDF, especially in the cement industry; but despite this demand, almost no actual MBT plant in Romania is operating in a way, that it fulfills these demands. In order to allow the utilization of the heating value of all residues of the Ploiesti plant, the following trials were done with the existing plant. Also the fraction >80mm was fed into boxes for biological drying.

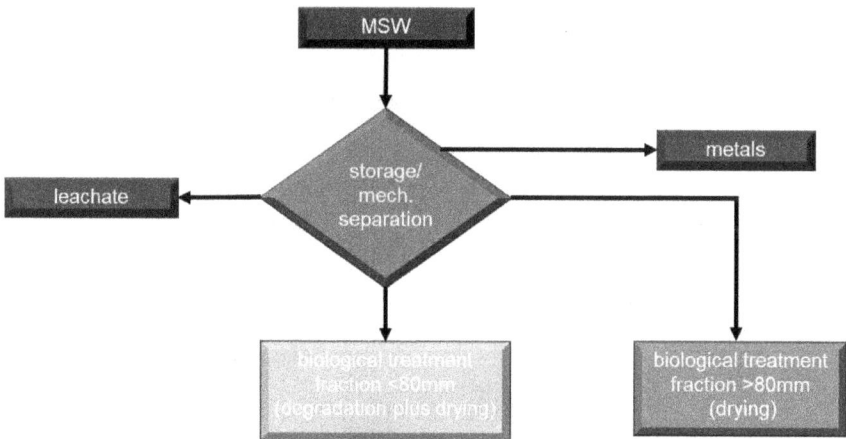

Figure 8   mechanical biological process of MSW and fractions <80mm and >80mm

Figure 9   Box filled with >80mm fraction for biological drying

To improve the quality regarding impurities, a mobile ballistic separator was used for classification of both fractions after biological treatemnt (<80mm and >80mm).

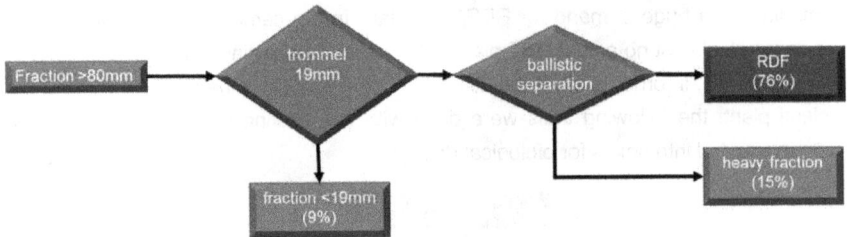

Figure 10 mechanical treatment of stabilized >80mm fraction

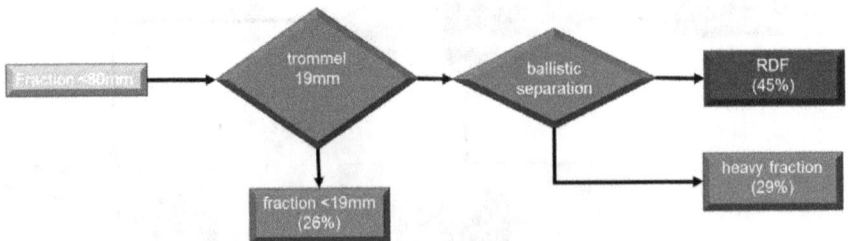

Figure 11 mechanical treatment of stabilized <80mm fraction

Finally, the achieved RDF was sent to the cement factory and the analysis showed the following results:

Samples at the reception			Wi (%)	Ash (%)	GCV (Gj/T)	NCV (Gj/T)	S (%)	Cl(%)
07.08.2018 09:58	CP18028245	STMB Ploiesti	9,55	9,13	20,69	19,33	0,17	0,41
07.08.2018 10:02	CP18028246	STMB Ploiesti	9,3	8,97	25,33	23,97	0,11	0,48
10.08.2018 08:40	CP18028867	STMB Ploiesti	12,47	15,33	21,49	19,98	0,11	0,86
10.08.2018 08:42	CP18028868	STMB Ploiesti	12,29	17,47	20,2	18,69	0,17	1
10.08.2018 10:04	CP18028440	STMB Ploiesti	27,84	9,52	17,62	15,95	0,13	0,57
11.08.2018 08:43	CP18028869	STMB Ploiesti	8,55	13	21,16	19,69	0,26	0,61
11.08.2018 08:45	CP18028870	STMB Ploiesti	19,62	12,56	18,42	16,83	0,23	0,53
13.08.2018 09:59	CP18029060	STMB Ploiesti	24,56	10,86	17,65	16,02	0,19	0,78
21.08.2018 10:06	CP18030346	STMB Ploiesti	21,63	12,19	17,27	15,67	0,18	0,6
28.08.2018 10:33	CP18031469	STMB Ploiesti	39,56	8,77	13,05	11,25	0,12	0,73
30.08.2018 08:37	CP18031795	STMB Ploiesti	16,63	15,19	17,18	15,63	0,12	1
07.09.2018 08:56	CP18033123	STMB Ploiesti	36,53	9,92	14,14	12,38	0,18	0,55
MIN			8,55	8,77	13,05	11,25	0,11	0,41
MAX			39,56	17,47	25,33	23,97	0,26	1
AVG			19,88	11,91	18,68	17,12	0,16	0,68
COUNT			12	12	12	12	12	12

Figure 12   Results from cement factory laboratory (mixture of both fractions after process as indicated)

For the entire MBT, the overall balance for the modified process, looks as following:

MBT Ploiesti	
Type of waste - MSW	
Mechanical Treatment	**Percent (%)**
Total Input	100.00%
Total Output	
Fraction <80mm	58.80%
Fraction >80mm	37.50%
Amount from magnetic separation	0.79%
Water loss	2.91%
Biological Treatment (biodegradation and biodrying)	
Fraction <80mm	
Input	100.00%
Output	59.24%
Water loss	40.76%
Fraction >80mm	
Input	100.00%
Output	51.50%
Water loss	48.50%
TOTAL outlet after Biological Treatment	51.58%
Quantity (RDF) to GEOCYCLE	30.44%
Quantity to Landfill	21.14%
TOTAL mass reduction	47.63%

Figure 13 Summary: balance of Ploiesti plant

5   Outcast

The concept of the combination of composting boxes with semipermeable membrane roof and biofilter to allow flexible, safe operation of biological degradation and biological drying of various fractions of MSW, passed all performance tests and showed promising results for the production of RDF for different applications. Depending on the legal requirements and technical standards, this plant concept will be suitable in many regions of the world to establish a very cost effective MBT concept.

Author's address

Dipl.-Ing. Axel Huber

STRABAG Umwelttechnik GmbH

Lingnerallee 3

D-01069 Dresden

Telephone +49 351 26359-0

E-Mail: axel.huber@strabag.com

Use of SRF (Solid Recovered Fuel) and RDF (Refuse Derived Fuel) in Europe – RECORD project

Literature review and administrative situations encountered in the field

LE BIHAN Mathilde, MICHEL Frédéric, DE CAEVEL Bernard

RDC Environment, Brussels, Belgium

Abstract

The 2014-2025 French waste plan reckons that the preparation of SRF will contribute to the reduction of landfilled non-inert NHW by 2.5 Mt/year (out of a foreseen 10 Mt/year reduction). France must therefore acquire SRF production and consumption capacities beyond the current capacity (250-300 kt/year).

Within this context, RECORD association launched a study that aims to improve the understanding of the European RDF market and lead to recommendations for the development of the French sector.

RECORD

The study was divided into three parts: A global analysis of the RDF/SRF market in 10 European countries; an analysis of the situations in the field by means of 13 site visits; and the development of recommendations for France.[1]

KEYWORDS

Refuse Derived Fuel, Market, Regulations, Subsidies

---

[1] Deliverables are accessible here: https://www.record-net.org/catalogue/205

1 Context and objectives

A Refuse Derived Fuel (RDF) is a solid non-hazardous waste (NHW) prepared for energy recovery. The 2014-2025 French waste plan reckons that the preparation of RDF will
contribute to the reduction of landfilled non-inert NHW by 2.5 Mt/year (out of about 10
Mt/year reduction. France must therefore acquire SRF production and consumption capacities.

Within this context, RECORD association[2] launched a study that aims to improve the
understanding of the European RDF market and lead to recommendations for the development of the French sector.

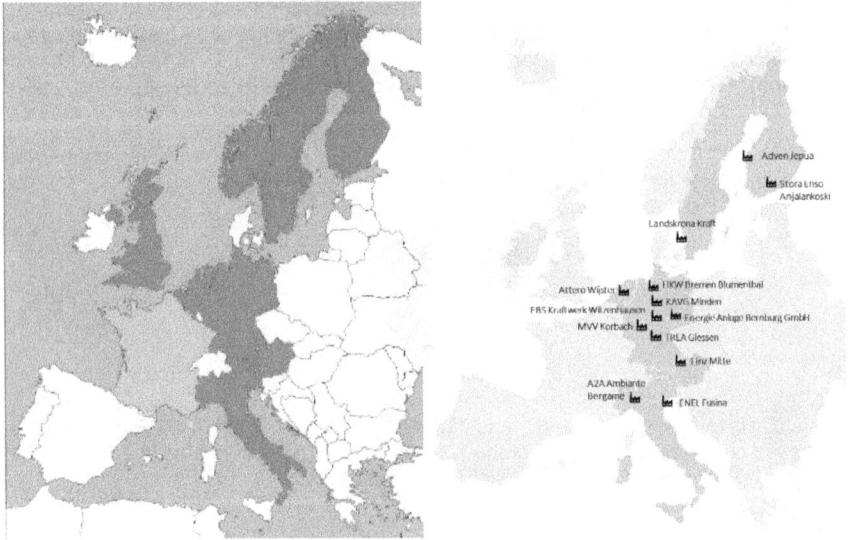

Figure 1: European countries and facilities analysed (RECORD, 2018)

---

[2] https://www.record-net.org/

2   RDF in Europe

2.1   RDF production

2.1.1   Market

Non-standardised RDF is mainly intended for dedicated facilities (heat production) and standardised RDF for cement kilns and coal-fired plants.

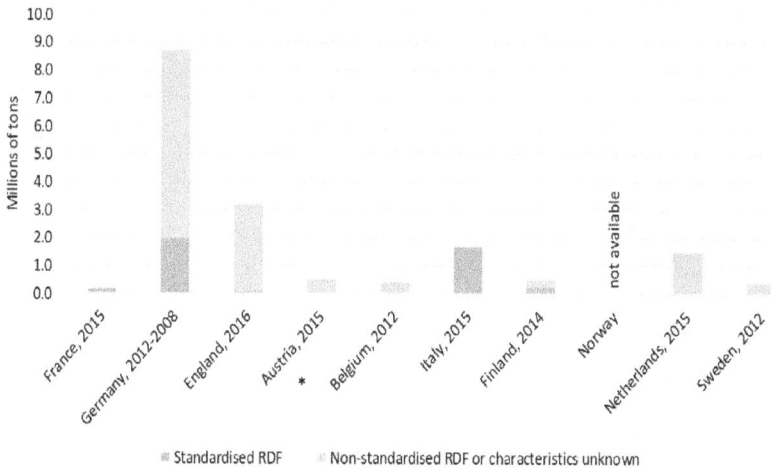

Sources : ADEME, ISPRA, UBA, BLFUW, Environmental Agency, RECOMBIO...
* Excluding I&CW

Figure 2: Production of non-standardised RDF and standardised RDF (DGE, RECORD, 2018)

2.1.2   Technical conditions for the preparation of RDF

The LHV of RDF prepared from I&CW varied between 11.5 and 16.5 at the sites visited; and between 8.4 and 20.4 for RDF prepared from MSW.

In Germany, RDF was mainly prepared using mechanical treatment (biological treatment only applied to the organic fraction). In Italy, mechano-biological treatment was the principal production route, applying biodrying to the whole input stream.

2.1.3   SRF regulations and quality standards

More often than not, **the quality of the SRF** is governed by **agreements between producers and users according to technical criteria** and in order to respect emission limit values. In all countries studied, but in France, the quality of RDF is not covered by the regulations. In any case, the local authorities are free to apply extra requirements in the environmental permits. In Germany, Austria and Sweden, RDF energy recovery plants are not required to only recover RDF, they can recover other non-hazardous waste if the technical conditions allow it.

2.2   Use of RDF

2.2.1 Market

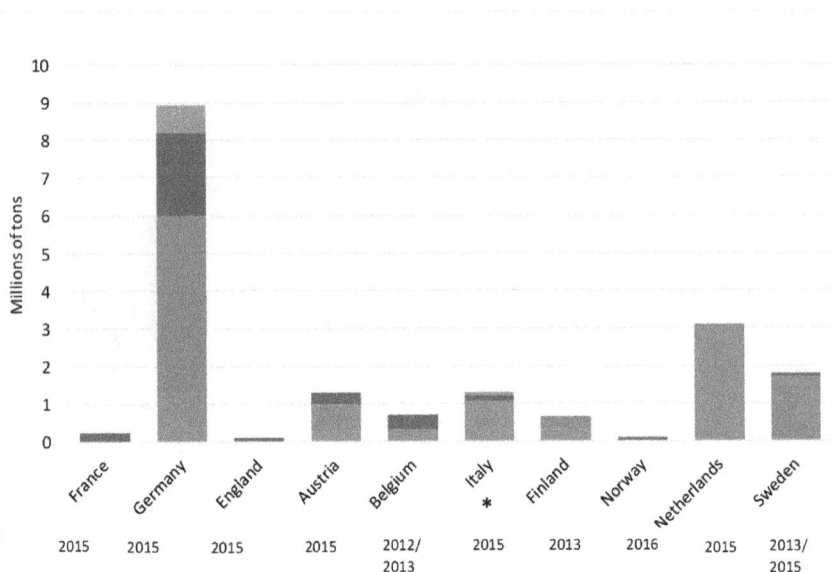

Sources : ADEME, ISPRA, UBA, BLFUW, Environmental Agency, RECOMBIO...
* Excluding RDF use from I&CW

- Non-dedicated thermal power stations (coal lignite)
- So-called dedicated installations (generally heat-oriented for industry or district heating)
- Cement works and lime kilns
- MSWI (elec and/or heat if heat generally for district heating)

Figure 3: Quantities of RDF used in the countries studied (DGE, RECORD, 2018)

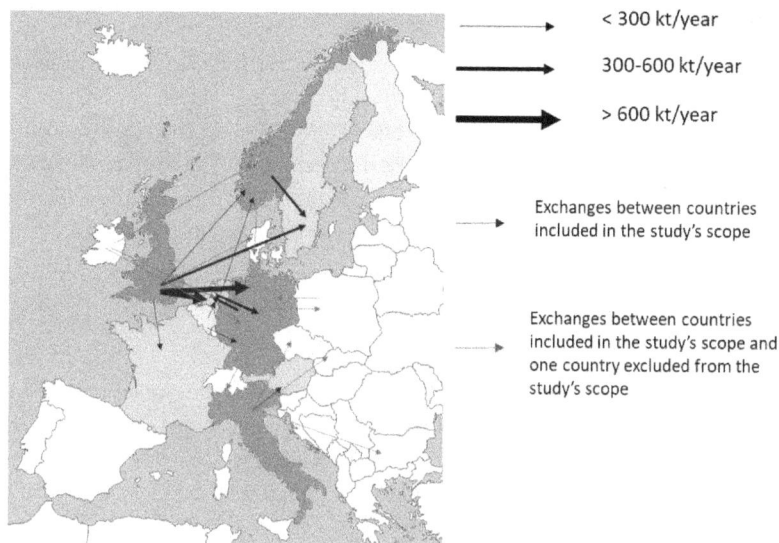

< 300 kt/year

300-600 kt/year

> 600 kt/year

Exchanges between countries
included in the study's scope

Exchanges between countries
included in the study's scope and
one country excluded from the
study's scope

NB: The colours have no significance in this graph, they simply serve to distinguish one country from another

Figure 4: RDF import-export flows (DGE, RECORD, 2018) (data from each country, reference year 2013-2014, 2015 or 2016)

2.2.2 Technical conditions of the facilities using RDF

The majority of the facilities in operation in Europe function by combustion, with gasification only occupying a very minor role (3% of facilities). Fluidised-bed furnaces and grate furnaces share the market, with fluidised beds mainly used when the facility does not only use RDF (biomass, sludge, bone meal, etc.) or uses RDF with a high LHV. **Cogeneration** has been adopted and recommended by the majority of the facilities visited.

Flue gas cleaning at these sites is similar to technologies in place in incineration plants. Placing the cyclone before the economiser allows the precipitation of hazardous substances to be reduced in the cyclone ashes and to class these ashes as non-hazardous waste if the regulations allow it, which reduces management cost.

The proportion of combustion residues (7-35%) seems all the greater when the waste has not been prepared or scarcely.

2.2.3 Economic conditions of the facilities using RDF

All the facilities visited indicated they were profitable.

Business model

There are almost as many viable models as sites visited (company and site sizes, part-nership with the supplier and/or with the energy consumer, public ownership…). In the majority of cases encountered, the (or one of the) RDF supplier(s) or heat user are shareholders in the facilities, sometimes on a shared basis. The more powerful sites are run by large groups. The complementarity of the waste and energy expertise and knowledge of the regional situation of the waste market are important when setting up the project.

Supply management

Integration of preparation and use within the same site, or the same group, allows the quality of the RDF to be more easily adapted to the needs of the recovery facility, and reduces quality control requirements. Integration within the same group that has several recovery facilities can allow each fuel to be taken to the most appropriate facility.

The supply radius is greater for large sites (150-300 km) than for average-sized sites (50-100 kt/year), which are usually supplied at less than 100 km. The sites visited that use imports are located in countries with an overall incineration overcapacity.

Contracts with RDF suppliers are generally medium term (1-5 years). Certain large sites mix the types of contract to reduce the risk of RDF price changes, ensures a constant quantity and quality and to manage the risk of a drop in demand among users.

RDF prices are highly variable from one site to another (-60 to +€30/t) depending on national market and RDF quality.

The evolution of the sources or of waste quality is not a source of concern among oper-ators regarding the future of their facility. Strategies have been developed to compen-sate for variations (supplier negotiations, imports, income diversification through recy-cling of reusable materials, etc.).

Management of the energy demand

The key success factor is the search for a long-term partnership with an energy con-sumer with a constant heat demand. Two means areused to establish the partnership: participation of users in the capital of energy supply facilities, and long-term contracting, usually lasting more than 10-15 years. Seeking several manufacturers requiring heat can also help to reduce the risk.

In the case of district heating, interconnection between networks allows the different connected groups to count on the cheapest power according to heat requirements and on the season. The search for a demand in cooling (hospitals, universities, etc.) allows facilities producing district heating to profitably supplement their income during the summer season.

2.2.4 Administrative conditions and acceptability

In all countries except for France, facilities using RDF are regulated based on the framework in force for incineration (dedicated facilities) or co-incineration (coal-fired plants, cement works using RDF, dedicated facilities) and not based on a dedicated framework. On the other hand, in other countries, facilities using RDF are authorised to use other fuels according to their permit (biomass, sludge, unprepared waste).

Relations with the neighbourhood may hamper the constitution and maintenance of the project, which can affect all facilities. Facilities owned by local authorities are particularly attentive to this for political reasons. The prevention of pollution and an open communication (site visits, brochures, meetings…) help improve relations.

3   Drivers behind the development of the production and use of SRF in other countries

Contrary to the approach implemented in France to build the sector, the facilities producing and those using RDF in a dedicated manner, do not form an organised 'sector' in the countries studied. There is not or has not been any real structuration of the value chain the public authorities (by means of e.g. regulatory framework or observation), with the exception of Italy.

In other countries, facilities producing and those using RDF are not the result of a public policy whose goal is to develop an RDF sector. They emerged through a combination of the following factors (in order of importance):

1. a climate of **high and increasing fossil fuel prices**, pushing energy consumers to look for alternatives;

2. combustible waste available at a competitive price, resulting from the implementation of **public policy instruments aimed at reducing landfill** in the form of bans and/or dissuasive taxation;

3. a European system of **CO_2 quotas** favouring alternative energy overall compared with the most carbonaceous fossil fuels;

4. **for some countries,** placing dedicated RDF facilities in the incineration category, thus **exonerating RDF heat from the CO_2 quotas, and in the absence of incineration taxes.**

Factors no. 2 and no. 4 were implemented nationally and explain why RDF occupies a different place depending on the country in Europe. Factors 1, 2 and 4 are not currently present in France: low energy prices, non-competitive SRF, SRF heat production facilities classified as co-incineration and subject to CO_2 quotas.

Furthermore, national energy investment or operational subsidies can apply to facilities using RDF and encourage the economic model.

Finally, the regulatory conditions surrounding RDF, the facilities producing RDF and user facilities differ from country to country and can be more or less favourable (see paragraph 2).

3.1   Energy context and role of heat consumers

The facilities visited that use SRF were first constructed with the aim of substituting fossil fuel in order to produce heat in the form of steam or hot water. Sites that use RDF, replace a variety of fuels: coal, gas, oil, biomass.

The rise in fossil fuel prices was quoted as a major motivation for users in a context of energy price rises. At all the sites with an industrial energy consumer, the latter has played a key role in the elaboration of the project, and is often **partial shareholder** (with the SRF supplier or operator) or **sole shareholder of the site**. Waste energy is said to be competitive compared with the fossil fuels previously used, both for district heating and industrial energy consumption. The improvement of the environmental impact, was also mentioned as a major motivation by energy users.

3.2   National waste management policies

In order to see RDF production

- landfill must be more expensive than RDF production and use, or forbidden for some streams. In all the countries studied, we found that the cost of sending non-hazardous waste to landfill exceeds €100/t and/or that there were landfill bans in force, except in France.

- direct incineration must be more expensive than RDF production and use, or capacities must be missing.

This context was found in Germany, Austria, England and Italy, where significant RDF production is seen. On the other hand, we found that in Sweden, Norway, Belgium and the Netherlands, direct incineration in municipal waste incinerators is relatively cheap, which explains the relative lack of RDF preparation.

RDF imports and exports are generated because of a delayed implementation of rising landfill prices and landfill bans among Member States. This created overcapacity in countries that started diversion from landfill relatively early (Netherlands, Sweden, Germany) and undercapacity in countries where the diversion is more recent (UK, Italy). Competitiveness of incineration facilities also plays a role and is affected by energy efficiency (connection to heat demand), taxes and subsidies.

3.3   National subsidies for RDF energy

Contrary to national waste policies, national energy public policies have only vaguely contributed to the development of facilities dedicated to energy recovery from RDF.

All the countries studied have indirect support systems that could be applied to RDF (heating networks, cogeneration, green certificates...) **No country, except for France, has developed a specific subsidy framework for RDF.**

Except for the Italian facilities, the facilities studied seem profitable without public support. Only six sites of the 13 visited have benefited from subsidies. For five of the six sites, support was not necessary to reach profitability. There was no information available for the sixth one. In Italy, green certificates had a major effect on the profitability of the two visited sites.

A higher waste treatment pricing and high oil prices at the time of investment explain this profitability. High oil prices is not a differentiating factor between the countries studied but it does give an advantage to amortized facilities.

3.4 CO_2 quotas and incineration / co-incineration classification

Directive 2003/87/CE establishes an exchange system for GHG emission quotas in the EU. It concerns combustion or co-incineration plants with a calorific power of combustion >20MW. The energy used by users subject to CO_2 quotas that was produced by incineration plants for hazardous or municipal waste, is exempt from quotas.

The CO_2 quota policy has led manufacturers using coal to look for alternative fuels because it has a significant influence on the steam cost (€3.4/MWh for a total cost of €12/MWh, i.e. 28 % of the cost). The changeover from coal to RDF is all the more interesting if the facility is considered as an incinerator, whereby its energy is exempt from quotas. CO_2 quotas have less of an influence on manufacturers using gas or oil, since the cost of producing steam is higher for these two types of fuel; the quotas represent approximately 6 % of the costs.

Directive 2010/75/CE on industrial emissions (IED) defines and regulates incineration and co-incineration. Despite a common definition in the IED, facilities dedicated to energy recovery from RDF, sized for heat requirements and whose main objective is energy production, are the subject of different incineration/co-incineration classification depending on the country in Europe.

Table 1: Status of facilities dedicated to energy recovery from RDF (RECORD, 2018)

Country	Incineration	Co-incineration
Germany	X	
England	No facilities concerned	
Austria	X	
Belgium		X
Finland		X
France		X
Italy	X	
Norway	No facilities concerned	
Netherlands	No facilities concerned	
Sweden		X

This difference in definition is associated with different regulatory regimes. This leads to distortions of competition, mostly concerning CO_2 quotas, but it also affects the design of the facilities (air entry and temperature measurement).

The fact that France classes dedicated facilities as co-incinerators increases the price of RDF energy by €0.7/MWh compared with facilities classified as incinerators (price of quotas €8/tCO$_2$). The impact would be €10 to 12/MWh with a quota price of €50/tCO$_2$, which could constitute a distortion of competition between French manufacturers and those in other countries. This is likely to favour industrial energy users located in countries where dedicated facilities are considered as incinerators (Germany, Austria, Italy); and, therefore, the export of RDF to these countries, to the detriment of local recovery.

4   Overall conclusion

The European RDF market has developed thanks to a demand for competitive renewable energy among manufacturers and entities (context of rising prices before crisis, CO_2 quotas); and an offer in fuel prepared from waste, the consequence of national policies aimed at reducing landfill.

The study has helped to improve the understanding of how the European RDF market works:

- influence of public policies on the national market and imports/exports;

- technical functioning of facilities using RDF;

- applicable regulations and subsidies;

- economic models.

Through the study, it has been possible to highlight the choices favourable to the development of the sector according to technical (cogeneration, ash management, etc.), regulatory (status of ash, status of the facilities using RDF, etc.) and economic (types of partners, etc.) points of view. The study showed that facilities using RDF do not receive subsidies specific to RDF in other countries, and that if they receive other subsidies, the latter are rarely determining factors in the economic model.

However, the economic model of the facilities is very sensitive to the price of waste and energy. While France would like the SRF sector to develop nationally, it is necessary to build an environment of public policies favourable to its development, by drawing inspiration from the conditions in force in other countries. Without these conditions, it will be necessary to subsidise the sector to enable it to emerge.

Mathilde LE BIHAN
RDC Environment
57, avenue Gustave Demey
1160 Brussels
0032 490 45 60 57
mathilde.lebihan@rdcenvironment.be

Small and medium size power plants for a decentralized treatment of RDF and industrial waste

Christian Eder

Wehrle Werk AG, Germany

Abstract

It is common practice to incinerate RDF and pre-treated waste in fluidized bed incinerators. Over the past 15 years, Wehrle and his partner TBU delivered engineering services of the technology for major improvements to 12 lines and for the construction of a further 12 fluidized bed combustion plants in Germany, Austria and Switzerland.
Now the market needs more local decentralized solutions with a lower CO_2-Footprint regarding also the transport of industrial waste and RDF. The direct or regional use of these fuels needs small and medium sized power plants which are capable to fulfil the legal requirements and a safe operation with long plant travel times.

Key words:

Fluidized bed, small fluidized bed, medium fluidized bed, RDF, waste fuels, decentralized plants, waste incineration

1 Technical concept

1.1 Plant description (rough concept)

The plants incinerate high-calorific waste fuels as RDF and industrial waste and recycle the produced energy.

The Fluidized bed system can treat different pre-treated waste fuels as:

- RDF / SRF
- Industrial waste
- Waste wood
- Sewage sludge
- Digestates
- Liquid manure
- ...

The fluidized bed is integrated into the steam boiler.

The fuel receiver is designed as a steel construction and designed for a storage time of approx. 8 hours. Filling can be carried out with a crane or a conveyor system. The bottom of the fuel feed is designed as a walking floor and supplies the dosing screw automatically depending on its weight. The housing of the dosing screw is mounted on load cells. Two rear screws are arranged at the end of the dosing area of the dosing screw. The boiler output is controlled by the speed of the dosing screw. The filling level of the dosing screw at the outlet of the dosing area is kept constant by two return screws. Therefore the the fuel heat output of the solid fuels is controlled by the speed of the dosing screw and the loss of weight.

The supplied fuel heat output is calculated indirectly from the boiler data. With the number of rotations of the dosing screw, a value for the supplied fuel heat output per rotation is calculated and subsequently also the total fuel heat output supplied. This signal is used both for pilot control of the combustion air quantity and the fuel quantity.

The dosing screw is operated at less than 5 rotations per minute. Within one rotation the discharge from the screw is not constant and therefore too uniform.

The processed waste mix tends to bake together due to the high proportion of sludge. The outlet of the screw is therefore designed in such a way that the fuels are advanced almost uniformly and with scraping stars into the chute of the injection chute.

The scraped off fuel chunks are conveyed with transport air through the injection tube and onto the fluidized bed distributed. The injector air is introduced via an annular gap sector at the bottom of the injection pipe. Thus the highest impulse is generated at the deflection of the falling fuels to the inclined injection pipe, whereby deposits are avoided as far as possible. The bottom of the injection pipe is designed as a wear part and is easy to replace.

In order to avoid backfire and to detect blockages the following parts are used parameter is monitored:

- Differential pressure of the injector air between inflow and ring sector

- Pressure and temperature at the end of the chute and

- Pressure and temperature at the inlet of the injection pipe

1.2 Open nozzle floor

In all three systems, the nozzle floor is designed as an open nozzle floor. The nozzle pockets are in the hopper is integrated and lined with refractory material. The nozzles are located on both sides of the nozzle pockets arranged and inclined downwards. The nozzle pockets are as smooth as possible, in order to provide as little support as possible for impurities.

All nozzles are arranged in a horizontal plane and have the same geometry and thus the same pressure loss. This allows an even distribution of the fluidisation gas with a low inlet pressure.

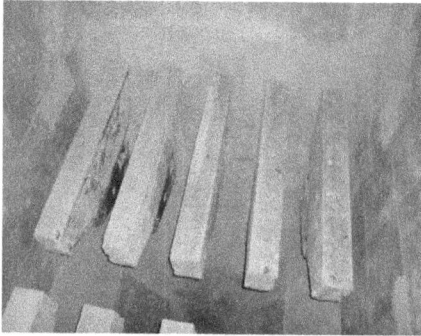

Open nozzle floor

1.3   Bed material discharge

In normal operation, the hoppers between the nozzle level and the discharge are filled with bed material. The bed material is discharged by a pusher out of the funnel. The volume released in the process is refilled from the bed.

Fluidisation gas flows into the fluidized bed between the nozzle pockets and cools the bed material. In addition, the fluidized gas acts as a separator. Therefore, the proportion of fine material in the amount of bed material removed from the bed is less than in the bed.

When the discharge is at a standstill, the bed material column is supported on the bottom of the discharge.

This means that the discharge acts as a shut-off during standstill and as a controlled discharge during operation.

The bed material discharged with the discharge enters directly into a cross-flow separator. Fine-grain and blown back into the furnace. Coarser material is fed via the double pendulum valves and temporarily stored in skip containers.

Due to the combination of open nozzle bottom, steep hopper, coarse mechanical discharge, Cross-flow separator with fine material return and coarse discharge continuously remove impurities and the need for sand is kept to a minimum.

Cut with open nozzle floor, bed material discharge system and cross current classifier; bed material discharge unit

1.4   Firing system

At the ABRG and Villas Austria plants, very high flexibility with regard to fuels is required. The range goes from a fuel mix of sewage sludge with a minimum of calorific waste to the sole incineration of high calorific waste.

These two plants are therefore equipped with staged incineration with recirculation gas.

Incineration takes place in three stages:

    1 Fluidized bed (sub-stoichiometric)

    2 Lower combustion zone (sub-stoichiometric)

    3 Post-combustion zone (over-stoichiometric)

The temperature is controlled in each zone. Therefore in normal operation there are no local temperature peaks.

The recirculation gas has two essential functions:

The heat released in the bed, and thus the bed temperature, is regulated by different Oxygen supply controlled, realized by mixing recirculation gas with low oxygen content and air.

The final combustion chamber temperature is controlled by the total quantity of recirculation gas.

The Mc Step Cimo firing system does not have to meet such high requirements and therefore a simpler Furnace guide installed without recirculation gas.

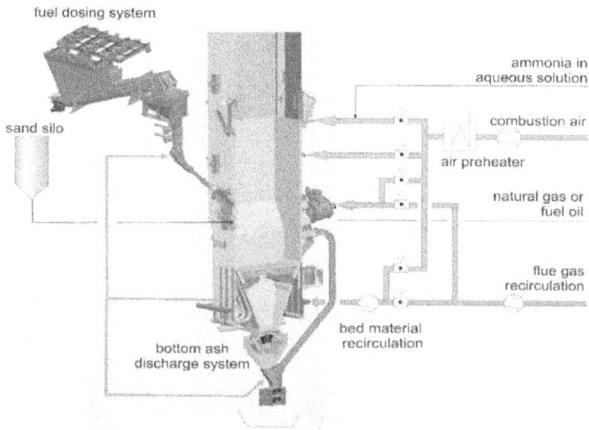

Staged combustion with recirculation gas

1.5 Reduction measures for NOx

For the existing installations, the NOx reduction measures are not limited to the SNCR installation. The following accompanying measures are implemented to minimise the formation of NOx:

- Design of the entire plant with regard to a low final combustion chamber temperature
- Operation of the entire furnace at a uniformly low temperature by controlling the temperature for the individual sections in the furnace
- Adjustment of a reducing atmosphere in the fluidized bed and in the lower freeboard zone.

A further reduction in nitrogen oxide emissions is achieved by injecting ammonia. Ammonia water (25% NH3 in H2O) is evaporated in a heated evaporator and added to the dilution air stream. The dilution air is blown in with ammonia above the upper combustion zone and intensively mixed with the flue gas. At temperatures between 800°C and 900°C, ammonia reacts with nitrogen oxides of the flue gas to form nitrogen and water vapour. A part is burned to NOx.

NH4OH (approx. 25%) is delivered in transport containers and is then transported over an intermediate storage of drip tray. The dosage depends on the clean gas value of NOx.

1.6 Boilers

In the case of waste fuels with a high calorific value, no measures are taken to reduce the co-firing of the boilers, such as adiabatic furnace linings or air preheaters in the boiler passes. The boiler generally consists of two empty passes and one convection pass. The firing system is integrated in the first pass.

The flue gases formed in the furnace are cooled to less than 550 °C in two radiation passes. In the third pass convection heating surfaces are arranged as evaporators.

In a downstream train with uncooled walls, the pipe bundles of the feed water preheater. The cleaning of the heating surfaces of the 3rd pass and the feed-water preheater is carried out with soot-blowers which are operated with steam or as compressed air lances, depending on the chosen steam parameters.

First pass of the Mc Step Cimo boiler in production at Wehrle Werke AG

1.7   Flue gas cleaning

Dry and semidry flue gas cleaning technologies are mostly applied for cleaning of following hazardous substances like HCl, HF, SO3, heavy metals, PCDDs and PCDFs, and dust.

Basically the plants consists of

- dry reactor or spray absorber,

- in addition baghouse filter and

- secondary plants to adsorption materials and for dust handling.

Baghouse filters are additionally to the efficient dust separation are helpful in building an adsorption reactor on the mesh. Depending on use, there are different adsorption materials available. For the adsorption of SO2 is often used Ca(OH)2 or NaHCO3.

For the semidry method is atomized water in a spray absorber for optimized reaction temperature. The adsorption material Ca(OH)2 is given powdery in the spray absorber and mixed with flue gas.

For dry treatment is placed NaHCO3 direct in flue gas piping before baghouse filter. The reactions are effected with the hazardous substances in the whole piping system beginning with the insertion of the adsorption materials to the baghouse filter. To reduce the consumption of Ca(OH)2 or Na-HCO3 a partial flow returns in the spray absorber. As a residue of the cleaning method is produced by the 100 % desulfurization only calcium sulfite, -sulfate and lime.

In application to an extensive flue gas cleaning plant are in residues also salts from isolated acidities like CaCl2 and CaF2 as soon as deposited dusts, heavy metals, organic harmful substances and adsorbent materials like activated carbon involved.

For an efficient flue gas cleaning application of Ca(OH)2 a temperature of less than 160°C is necessary. In opposition to Ca(OH)2, NaHCO3 using is also by higher temperatures (> 160°C) with a high efficiency possible. Adsorption of gaseous heavy metals and there chemical compounds and gaseous organic compounds on activated carbon is most efficiently to temperatures under 180°C.

A wet flue gas cleaning technology is also available on demand.

1.8   Plant types

The presented systems are niche products in terms of both size and equipment. Their technology was transferred from large-scale plants. Individual components first had to be developed for this plant types. With the experience gained from construction and operation, an essential part of the development is completed for fuel feed, nozzle bottom, bed material ejection, SNCR system, boiler and flue gas cleaning, a robust, cost-effective technology has been developed for the power range of 2 MW and above. standardized.

The criterion of transport essentially results in 3 plant types:

Between 1 and 5 MW, the boilers are designed in such a way that they can be used in one piece can be transported.

The individual boiler trains can be transported between 5 and 10 MW.

The classic construction site assembly is planned for more than 10 MW.

Standardization is intended to reduce planning, execution and erection costs without sacrificing quality and can be significantly reduced.

The aim is to develop decentralised plants for the disposal of waste streams and the provision of energy for customers from the field of industry and municipal waste management with high quality and low cost to be able to offer.

Firing capacity		$1,0 \text{ MW}_{th}$	$2,5 \text{ MW}_{th}$	$5,0 \text{ MW}_{th}$	$7,5 \text{ MW}_{th}$	$10,0 \text{ MW}_{th}$
(at 8.000 h/a operation)		High calorific value ~ Low calorific value	High calorific value ~ Low calorific value	High calorific value ~ Low calorific value	High calorific value ~ Low calorific value	High calorific value ~ Low calorific value
RDF (9 – 16 MJ/kg)	[t/a]	1.800 - 3.200	4.500 - 8.000	9.000 - 16.000	13.500 - 24.000	18.000 - 32.000
Sewage sludge (4.2 – 10 MJ/kg)	[t/a]	2.900 - 6.900	7.200 - 17.250	14.400 - 34.500	21.600 - 51.750	28.800 - 69.000
Sieving overflow, Digestate, Manure (4,2 – 10 MJ/kg)	[t/a]	2.900 - 6.900	7.200 - 17.250	14.400 - 34.500	21.600 - 51.750	28.800 - 69.000
Waste wood (11 – 14 MJ/kg)	[t/a]	2.000 - 2.600	5.000 - 6.500	10.000 - 13.000	15.000 - 19.500	20.000 - 26.000
Industrial waste (12 – 20 MJ/kg)	[t/a]	1.400 - 2.400	3.500 - 6.000	7.000 - 12.000	10.500 - 18.000	14.000 - 24.000
Hazardous waste (12 – 22 MJ/kg)	[t/a]	1.300 - 2.400	3.200 - 6.000	6.400 - 12.000	9.600 - 18.000	13.000 - 24.000

Throughput with different waste fuels and different standardizes fluidized bed plant sizes

2 CAPEX-OPEX, model calculation for RDF,

2.1 Key data of the plant

Fuel throughput per year	approx. 24.000	Mg/a
Fuel throughput per hour	< 3,0	Mg/h
Operating hours per year	8.000	h/a
Firing capacity of the fluidized bed system	5,0	MWth
Decouplable heat max. (district heating, process steam,...)	4,4	MWth
Bleed steam quantity after turbine for heat supply max.	4,0	MgD/h
Power generation in condensation mode max.	1,25	MWel
Power generation in maximum extraction mode	0,90	MWel

2.2 Investment costs (CAPEX) containing the following positions:

- Fluidized bed system without fuel supply - Steam turbine with steam circuit - Electrical measuring and control technology complete - Hall cladding of the fluid bed plant, industry standard - Turbine plant hall - Concrete foundations for the above plant components (from upper edge of building gravel layer)	11.180.000,-	€
Annual subsequent annuities in % of capital (Depreciation 15 a, interest 3%)	8,38%	%/a
Total annual subsequent annuities in € from capital	937.000,-	€/a
Annual subsequent annuities in €/Mg fuel	**39,02**	**€/Mg**

2.3 Substitution of externally generated electricity

Electricity:		
Gross electricity generation (in extraction mode)	7.296	MWh$_{el}$/a
minus(-) own electricity demand of the thermal plant	-1.200	MWh$_{el}$/a
Net electricity generation	6.096	MWh$_{el}$/a
Savings on electricity working price (approach 100% own electricity generation)	30	€/MWh$_{el}$
Savings on electricity supply costs Total	**183.000,-**	**€/a**
Savings on electricity supply costs per tonne of fuel	**7,62**	**€/Mg**
Savings in indirect electricity costs for own power generation (without renewable energy levy)	294.000,-	€/a
Savings on renewable energy levy per tonne of fuel	**12,25**	**€/Mg**

2.4 Substitution of primary energy sources (oil / gas / coal / ...)

Local heat / process heat		
saleable heat	25.600	MWh$_w$/a
Heat reimbursement, internal consumption approach	40	€/MWh$_{th}$
Heat income Sum	**1.024.000,-**	**€/a**
Heat revenue per tonne of fuel	**42,67**	**€/Mg**

2.5 Operating expenditure (OPEX) includes the following items:

Personnel costs (6 persons for 24/7 operation)	250.000,-	€/a
Maintenance costs	281.000,-	€/a

Operating costs - fluidized bed material - Chemicals - Sodium bicarbonate + activated carbon - Analytical costs - Recurring tests - Insurances	442.000,-	€/a
Disposal and landfill costs of residual materials	507.000,-	€/a
Operating expenditure (OPEX) total	**1.480.000,-**	**€/a**
Operating expenditure (OPEX) per tonne of fuel	**61,67**	**€/Mg**

2.6   Calculation of the disposal/fuel price:

Revenues / Savings		
Savings on electricity procurement costs per tonne of fuel	+ 7,62	€/Mg
Savings in ancillary grid costs through own generation	+ 12,25	€/Mg
Heat	+ 42,67	€/Mg
Total income / savings	**+ 62,54**	**€/Mg**
Expenses		
Debt service, annual subsequent annuity	- 39,09	€/Mg
Operating expenditure (OPEX)	- 61,67	€/Mg
Total expenses	**- 100,76**	**€/Mg**

Processing costs of the plant for waste / RDF	-38,22	€/Mg

3 Summary

The technologies used for firing, boiler and flue gas cleaning have proven themselves for decades in large-scale plants. The fact that their application to small plants is economically and operationally possible has been proven over several years by the small plants implemented.

Compliance with the specified emission limits was not a problem for any of the plants. The availability was significantly higher than that of large plants. For example, with good operational management, the annual revision only takes about 1-2 weeks. The availability over several years is more than 95 %. On the one hand, this is due to the robust and simple design of the system. On the other hand, it must be acknowledged that special operational measures were taken for this purpose. For example, wear-prone the equipment is kept in stock so that only the entire equipment can be replaced during the annual revision and the wear areas on these equipment can be repaired until the next overhaul.

The systems are operated according to private-sector criteria. The provision of cheap energy with the safe and inexpensive disposal of the resulting waste is an essential basis for securing these industrial sites.

4 Literature

Stubenvoll, J.   2015 Vortrag 16./17.09.2015 VDI Klärschlammbehandlung
                                         Paderborn

Author's address :

Dipl.Betriebswirt (FH) Christian Eder
Wehrle Werk AG
Bismarckstraße 1-11
D-79312 Emmendingen
Telefon +49 7641 585370
E-Mail eder@wehrle-werk.de

Waste materials effect the implementation of new exhaust gas purification techniques in plants of the Austrian cement industry

Gerd Mauschitz, Thomas Laminger

TU Wien,
Institute of Chemical, Environmental and Bioscience Engineering

Ersatzstoffeinsatz und Implementierung neuer Abgasbehandlungstechnologien in der österreichischen Zementindustrie

Abstract
The cement industry conserves natural resources through the use of substitute fuels and secondary raw materials. So they make an important contribution to environmental protection. The use of these materials makes the implementation of new clean gas techniques necessary. This article describes the implementation of the DeCONOx-process and the ExMercury-process for elimination of nitrogen oxides, organic pollutants, carbon monoxide and mercury from the exhaust gases generated during cement production.

Inhaltsangabe
Der Vortrag wird mit dem Ziel in Angriff genommen über die Ersatzstoffmengen der österreichischen Zementindustrie zu berichten und mit dem DeCONOx-Verfahren und dem ExMercury-Verfahren jene beiden Abgasnachbehandlungsmethoden verfahrenstechnisch vorzustellen, die in den letzten Jahren im Zementerzeugungsprozeß als *emerging technologies* Eingang gefunden haben.

Keywords
Ersatzbrennstoffe, sekundäre Rohstoffe, Abgasreinigungsverfahren, DeCONOx-Verfahren, ExMercury-Verfahren, Zementindustrie
Substitute fuels, secondary raw materials, clean gas technology, DeCONOx-process, ExMercury-process, cement industry

1 Input of secondary raw materials and alternative fuels in Austrian cement plants

In 2017 approximately 3.3 million tonnes of cement clinker were processed in nine Austrian cement plants to generate about 4.9 million tonnes of different cements.

The following process conditions make it possible for the cement industry to use various waste materials as secondary raw materials, alternative main cement constituents and alternative fuels (LOCHER, 2000):

➤ flame temperatures of more than 2000°C;

➢ exhaust gas temperatures at the rotary kiln inlet of more than 1000°C with exhaust gas residence times of more than 10 seconds;

➢ extensive sintering of the solids passing through the rotary kiln at temperatures of 1350°C to 1500°C;

➢ residence time of the kiln feed in the sintering zone of 10 to 20 minutes;

➢ neutralization of acid exhaust gas constituents by passing them in counter-current to the raw meal and

➢ destruction, respectively permanent immobilisation of harmful substances in the cement clinker matrix.

Consequently all Austrian cement plants utilized in 2017 about 700000 tonnes of secondary raw materials, approximately 1 million tonnes of alternative main cement constituents and about 510000 tonnes of alternative fuels. Figure 1 shows the development of the resource conservation factor over the last ten years. It can be seen that Austrian cement plants conserve natural material resources in a high extent and make in this way an important contribution to environmental protection.

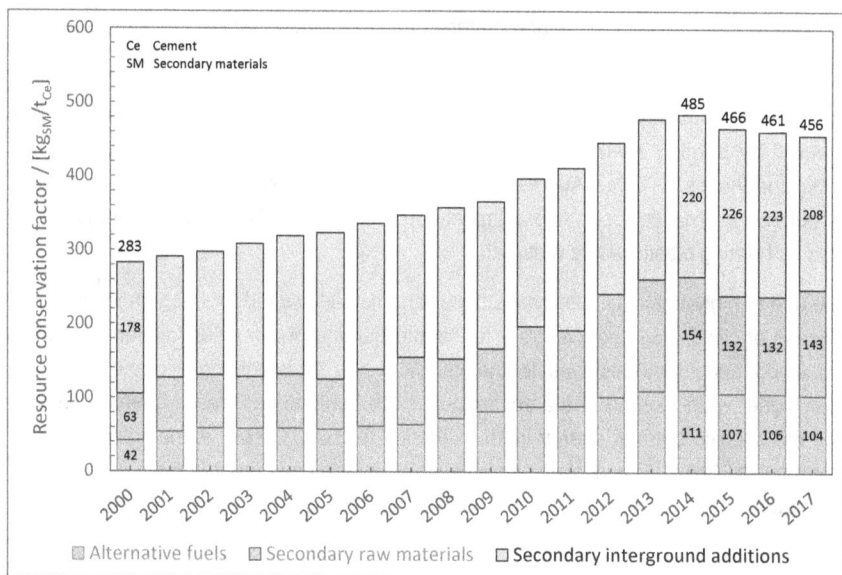

Figure 1 Development of the resource conservation factor at the Austrian cement industry since 2000 (MAUSCHITZ, 2018B)

The use of alternative fuels instead of conventional fossil fuels makes progress in the Austrian cement industry (see Figure 2).

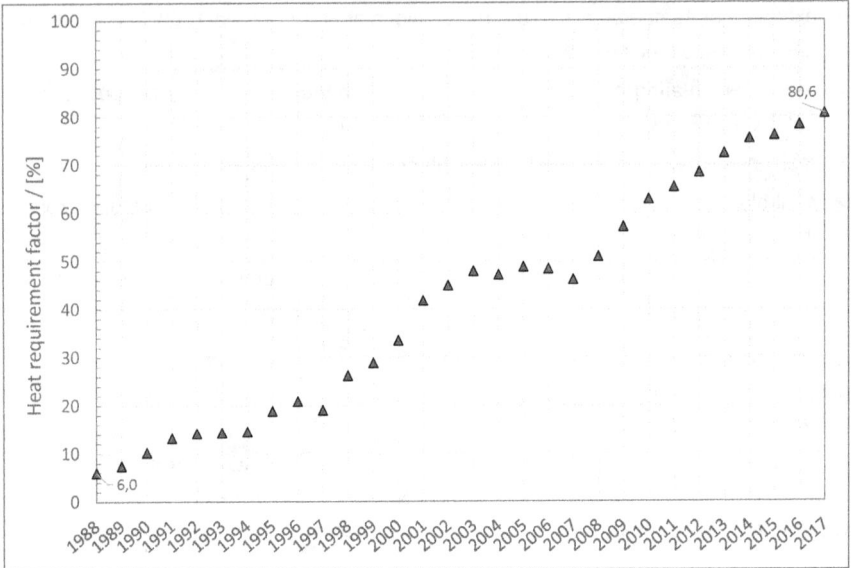

Figure 2 Progress of the heat requirement factor of the Austrian cement industry (MAUSCHITZ 2018B)

In 2017 approximately 81% of the heat requirement at the Austrian cement industry was covered by burning alternative fuels (e.g. old tyres, processed waste materials from trade and industry). Three Austrian Cement plants produced their whole process heat by burning alternative fuels. Approximately 50% of the total heat requirement is generated by burning plastic waste material.

The use of higher calorific alternative fuels is particularly suited for the high temperature process of clinker burning. Contrary to this the burning of low calorific alternative fuels e.g. wood, agricultural residues, let the specific exhaust gas volume increase caused by their higher water content. As a consequence the exhaust gas losses have climbed up and the specific thermal energy input increased to 3.839 GJ per tonne clinker in 2017 (MAUSCHITZ 2018B). This decline can be avoided by the use of more residual heat e.g. in raw material drying processes or district heating.

2 DeCONOx plant operation in the Austrian cement industry

The DeCONOx process provides simultaneous mitigation of NOx, CO and TOC emissions by combining SCR-DeNOx technology and regenerative post-combustion in the same plant.

With SCR-DeNOx technology the nitrogen oxides (NOx) are converted into nitrogen (N_2) and water vapour (H_2O), using ammonia water (NH_4OH) as a reducing agent in a temperature range from 240°C to 400°C. $V_2O_5/WO_3/TiO_2$ is used as ternary catalyst material. The chemical reactions for SCR-DeNOxing can be summarized as follows:

$$4NO + 4NH_3 + O_2 \rightarrow 4N_2 + 6H_2O \qquad \text{Equation I}$$

$$6NO_2 + 8NH_3 \rightarrow 7N_2 + 12H_2O \qquad \text{Equation II}$$

The reaction products are natural air constituents therefore environmental problems can be ruled out.

In the regenerative post-combustion part of the DeCONOx-plant carbon monoxide (CO) is oxidized to Carbon dioxide (CO_2) (Equation III) to avoid high CO concentrations in the exhaust gas due to environmental protection requirements.

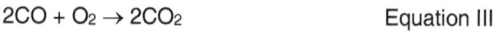

$$2CO + O_2 \rightarrow 2CO_2 \qquad \text{Equation III}$$

If there are organic pollutants in the exhaust gas then cleaning by thermal oxidation is possible in which the hydrocarbons are oxidized to harmless products. The following reaction equations show how the hydrocarbons first dissociate at high temperatures into combustible components carbon (C), hydrogen (H_2) and oxygen (O_2) (Equation IV) and consequently they are converted in a second reaction step to water (H_2O) (Equation VI) and CO (Equation V) or CO_2 (Equation VII):

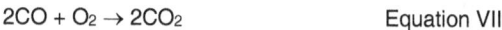

$$C_xH_yO_z \rightarrow xC + y/2H_2 + z/2O_2 \qquad \text{Equation IV}$$

$$2C + O_2 \rightarrow 2CO \qquad \text{Equation V}$$

$$2H_2 + O_2 \rightarrow 2H_2O \qquad \text{Equation VI}$$

$$2CO + O_2 \rightarrow 2CO_2 \qquad \text{Equation VII}$$

Often the complete oxidation of CO to CO_2 requires temperatures of more than 800°C. Persistent organic pollutants (POPs, e.g. dioxins, furans, biphenyls) can only be thermally destroyed by much higher temperatures of 1200°C in oxygen-rich atmospheres and sufficiently longer residence times.

The higher the calorific value of the raw gas the less supplementary fuel must be fired in the combustion chamber of the DeCONOx for destroying the organic air pollutants. The amount of supplementary fuel burnt is substantially lower if a regenerative post-combustion process is used with heat exchange between the hot clean gas and the cold raw gas.

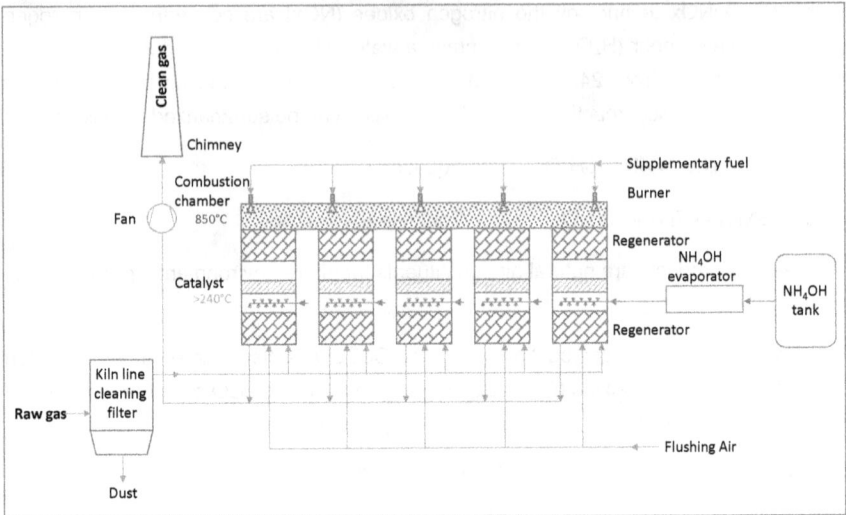

*Figure 3 Simplified process engineering flow chart of the five tower DeCONOx plant,
installed in an Austrian cement plant*

Dependent on the raw gas flow three to seven reactor towers are installed in the De-CONOx-plant (Figure 3). The raw gas first flows through the lower regenerator layer to reach the catalyst inlet temperature of at least 240°C. Then the raw gas flows through an intermediate space into which the reducing agent ammonia water is injected by a nozzle bar followed by the catalyst layer. The ternary catalyst material consists of a TiO_2 matrix with V_2O_5 and WO_3 as promoting agents. The upper regenerator layer follows.

The combustion chamber is located above the reactor towers and it works at a temperature of at least 850°C. The residence time of the exhaust gas in the combustion chamber is more than three seconds.

The clean gas leaves the DeCONOx plant with a temperature of approximately 160°C and is about 30°C hotter than the raw gas temperature on the entry of the DeCONOx plant. A heat recovery concept (e.g. district heating) helps to improve the energy efficiency.

The regenerative plant operation requires constant change-over operations to change the direction of the exhaust gas flow into the reactor towers during which the exhaust gas picks up heat from the regenerator elements during the upward flow and gives off heat to the regenerator elements again during the downward flow. The exhaust gas flaps are exposed to high stress because the change-overs take place about every 60 seconds.

The availability of the DeCONOx plant have successfully improved by the operators from 96% to about 99.9%. Excessive combustion chamber temperatures caused by excessive CO raw gas concentrations were primarily responsible for the switch to bypass operation.

After setting measures to improve the plant efficiency the NOx elimination efficiency could be raised to about 78.5%, so that an average NOx clean gas concentration of about 150 mg/m³(stp) with an average ammonia clean gas concentration of about 2.1 mg/m³(stp) could be maintained permanently (10.0 vol. % O_2).

The operators have successfully improved the CO elimination efficiency by more than 99% from more than 8000 mg/m³ (stp, dry, 10.0 vol. % O_2) to about 63 mg/m³ (stp, dry, 10.0 vol. % O_2). The TOC emissions decreased by about 99.7% from about 70 mg/m³ (stp, dry, 10.0 vol. % O_2) to about 0.14 mg/m³ (stp, dry, 10.0 vol. % O_2).

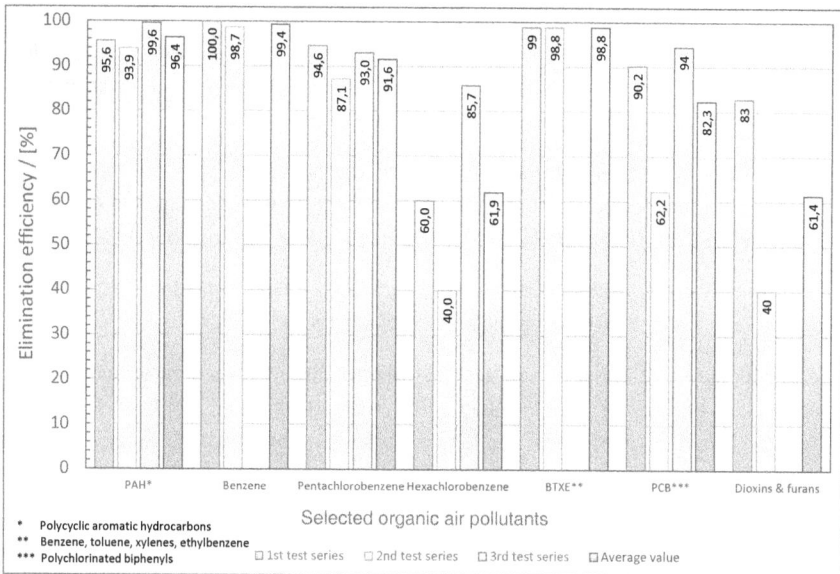

Figure 4 Elimination efficiencies for different organic air pollutants achieved by the DeCONOx plant (MAUSCHITZ 2018A)

Figure 4 shows that under the given operating conditions in the DeCONOx plant the elimination efficiency of benzene, the leading aromatic hydrocarbon, is extensively decomposed to 99%. The same applies for toluene, ethylbenzene and xylenes although the higher chlorinated aromatics, e.g. pentachloro- und hexachlorobenzene experience lower levels of elimination. The same also applies for PCDDs and PCDFs and their precursor substances the polychlorinated biphenyls, those are important representatives of the persistent organic pollutants (POPs). Finally it must be mentioned that the described DeCONOx plant with its permanent operating temperature in the combustion chamber of 850°C was not designed to eliminate totally POPs.

3   ExMercury plant operation in the Austrian cement industry

Due to their relatively high volatilities, mercury and its compounds are components of rotary kiln exhaust gases from the cement industry and are regarded as environmentally relevant trace metal emissions into the atmosphere, which are limited by strict pollution control laws.

The Hg input into the emissions balance cycle is predominantly influenced by burning solid fossil fuels (e.g. hard coal and lignite). Mercury may also enter the rotary kiln system using Hg contaminated raw materials (e.g. cinnabar in clay). In addition, mercury may enter the clinker burning process through secondary raw materials and alternative fuels. Depending on their origin, such substitutes may have a higher or lower mercury content than those natural feedstocks in their place.

In the exhaust gas stream of coal-fired industrial furnaces (e.g. cement rotary kilns), mercury preferably occurs due to kinetic inhibition not as mercuric chloride ($HgCl_2$) or mercuric oxide (HgO) but is present in its elemental form Hg^0 up to about 90%.

Due to their relatively high volatility elemental mercury and its compounds can be emitted gaseously and/or adsorbed on the surface of solid particles.

The vapour pressure curve of non-adsorbed metallic Hg^0 shows theoretically that the condensation should only be intensified at temperatures of about 100°C and would be completed to about three quarters at temperatures of about 50°C. However, there is a remarkable reduction in the volatility of mercury and its compounds by capillary condensation within the fine pores of the kiln meal. Therefore more volatile mercury compounds (such as $HgCl_2$) can be bound completely on adsorbents (e.g. activated carbon) even at relatively high temperatures of about 150°C as long as the maximum absorption capacity of the adsorbent is not exceeded.

The ExMercury process (Figure 5) is based on these principles. This secondary flue gas cleaning method can significantly reduce Hg emissions from cement plant exhaust gases. First the Hg-contaminated kiln dusts must be heated up to desorb the Hg from the dust particles. Then the mercury and its compounds are bound by adsorption on e.g. activated carbon. After that the Hg-loaded activated carbon is discharged from the plant and deposited as hazardous waste.

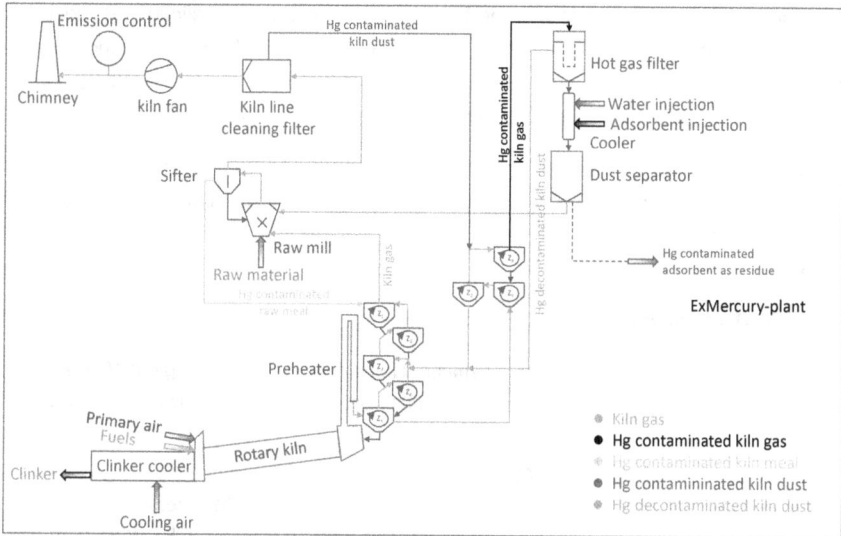

Figure 5 Simplified process engineering flow chart of an ExMercury plant, installed in an Austrian cement plant

In the ExMercury process, first a part of the separated Hg-enriched filter dust from the kiln line dust precipitator is heated up to approximately 400°C. For this purpose, a gas volume of about 3% to 5% of the total amount of furnace gas is sucked off at the level of the lowest cyclone stage of the preheater.

Due to the increase in temperature, the filter dust desorbs about 90% to 95% of its Hg load into the hot kiln gas. The Hg-enriched kiln gas loaded with particulate matter leaves the highest cyclone stage of the ExMercury plant. In the following hot gas dust precipitator the dust particles will be separated effectively at a temperature of 360°C, using ceramic filter candles.

Both, dust separated in the cyclone heat exchanger of the ExMercury system and dust from the hot gas dust precipitator are collected and returned together to the cyclone preheater of the rotary kiln. The Hg-loaded kiln gas is then cooled in a spraying tower to about 100°C to 120°C. Powdered adsorbent (activated brown coal coke) is injected into the cooled gas stream which adsorbed the vaporous mercury and its compounds. Subsequently, the Hg-contaminated sorbent particles are separated from the kiln gas by using a conventional pulse jet regenerated bag house filter, equipped with high-performance textile filter media. In order to achieve the fullest possible loading of the adsorbent with mercury, the adsorbent is recirculated until the maximum Hg-loading is reached. Finally the separated Hg-contaminated adsorbent must be deposited as hazardous waste.

After setting measures to improve the ExMercury process the Hg elimination efficiency could be raised to approximately 80%, so that an average Hg clean gas concentration of about 10 µg/m³ (stp, dry, 10.0 vol. % O_2) could be maintained permanently, using an annual adsorbent consumption of about 20 tons (brown coal coke).

At present, the use of other adsorbents, e.g. activated carbon, brominated activated carbon, inorganic sorbents are tested to further increase the Hg elimination efficiency of the ExMercury process.

4 References

Locher, F. W.	2000	Zement - Grundlagen der Herstellung und Verwendung. Verlag Bau+Technik GmbH, Düsseldorf, ISBN 978-3-7640-0400-2
Mauschitz, G.	2018b	Emissionen aus Anlagen der österreichischen Zementindustrie - Berichtsjahr 2017. Verlag Zement + Beton Handels- und Werbeges.m.b.H., Wien, 31p.
Mauschitz, G.; Hagn, S.; Philipp, G.; Secklehner, A.	2012	Pilot trials for catalytic reduction of nitrogen oxides in exhaust gases from Austrian cement plants. CEMENT INTERNATIONAL, **10**, (2012), pp. 40 - 51
Mauschitz, G.; Secklehner, A.; Hagn, S.	2018a	The DeCONOx process - an example of advanced exhaust gas cleaning technology in the Austrian cement industry. CEMENT INTERNATIONAL, **2**, (2018), pp. 34 - 53

Addresses of the authors:

Dipl.-Ing. Dr.techn. Gerd Mauschitz
TU Wien, Institute of Chemical, Environmental and Bioscience Engineering
Getreidemarkt 9/166
A-1060 Vienna
Phone: +43 1 58801 166150
E-mail: gerd.mauschitz@tuwien.ac.at

Dipl.-Ing. Dr.techn. Thomas Laminger
TU Wien, Institute of Chemical, Environmental and Bioscience Engineering
Getreidemarkt 9/166
A-1060 Vienna
Phone: +43 1 58801 166152
E-mail: thomas.laminger@tuwien.ac.at

Separation of the Fine Fraction to Improve the SRF Quality - Possibilities and Effects

Alexander Curtis, Roland Pomberger, Renato Sarc

Chair of Waste Processing Technology and Waste Management

Montanuniversitaet Leoben, Leoben, Austria

Abstract

This article deals with the quality assurance and improvement of solid recovered fuels (SRF) used for substitution of fossil fuels (e.g. coal) in the cement industry. In order to be able to increase the degree of energy substitution, on the one hand the quality of the SRF has to be increased and on the other hand all possibilities of fuel feeding ports (main burner (SRF PREMIUM Quality), calciner (SRF MEDIUM Quality), etc.) of the rotary kiln must be used (Sarc, 2015). In this study, the sieving of a fine fraction after pre-shredding and after post-shredding in a two-stage SRF production process was investigated. The fractions resulting from the sieving were examined for their combustion parameters. Extensive sampling and subsequent chemical-physical analysis of the sieve fractions revealed that sieving off the fine fraction leads to an improvement of the combustion characteristics of the SRF PREMIUM Quality, but that two heavy metals investigated accumulate slightly. The fine fraction (below 20 mm grain size) is very well suited for use in the calciner.

Keywords

SRF, sieving, fine-fraction, combustion characteristics, calciner, main burner, SRF medium quality, SRF premium quality, quality assurance

1   Introduction

The cement production process is quite energy-intensive with 3,839 GJ/t of clinker (Mauschitz, 2018), which is why the search for alternatives for the standard fossil fuels such as coal has been going on for quite some time now. The thermal substitution rate, that is to say the share of energy from solid recovered fuels (SRF) that are burned in place of primary standard fuels (e.g. coal) in relation to the demand for energy in a cement production facility, was at 80.6 % in Austria in the year 2017(Mauschitz, 2018). In order to attain this high substitution rate, the input of energy in the cement production process must occur in several feeding ports (primary firing, calciner firing, rotary kiln inlet) with various substitute fuels. The SRF being used at the several input feeding ports must be adjusted with regard to its chemical (heavy metals, chlorine) and physical parameters (calorific value, grain size distribution, density). In addition to contractual arrangements between SRF producers and cement industry, the legal and technical framework for the quality of the SRF (heavy metals) is set by the Austrian Incineration

Ordinance (AVV) (BGBl. II Nr. 476/2010) (Bundeskanzleramt, 2010). Table 1 shows limit values for the selected parameters, which were investigated in this work.

Table 1   Limit Values for Co-Incineration of SRF in the Cement Industry (AVV).

Parameter		Limit Values AVV [mg/MJ$_{TS}$]	
		Median	80-percentile
Antimony	Sb	7	10
Cadmium	Cd	0.45	0.7
Mercury	Hg	0.075	0.15

The input of SRF into the rotary kiln via the primary firing is carried out pneumatically via a channel on the burner gun and this must incinerate completely within a quite brief retention time, during the trajectory. With an optimal burn-out, only incombustible fuel components (i.e. ashes) get into the clinker and are then utilized materially (from a technical point of view). Any changes in the fuel properties have an impact on the conveying characteristics, the dosing, the burn-out behaviour, the flame form, the flame temperature, the clinker quality, etc., and is thus able to be determined by means of simulations and practical tests and to be carefully coordinated with the operator of the cement facility. The demands on a calciner fuel are not as high, however, this SRF also has to fulfil many criteria (e.g. calorific value, chlorine content, heavy metals). For the production of SRF for the cement industry, various combustible materials are put to use.

The quality of the alternative fuels produced depends on the one hand on the mix and the quality of the input materials used and on the other hand on the technology used. One possibility to influence the combustion characteristics of the SRF for the primary firing (main burner) is to separate the fine fraction (Sarc, 2014; Bohrn, 2016).

In a two-stage SRF production process, there are basically two options for separating a fine fraction:

• **after pre-shredding** (first shredding stage)

• **after post-shredding** (second shredding stage).

In this article, results from sieving tests for the separation of a fine fraction directly after the first shredding stage (**pre-shredding sieving tests**) and results from tests directly after the second shredding stage (**post-shredding sieving tests**) of a two-stage SRF production process are presented. It was investigated how a sieving process affects the quality parameters of the remaining main burner SRF like grain size (GS) distribution, pollutant distribution and the mass ratio between the coarse fraction and the fine fraction. In addition, the resulting fine fraction was examined for its suitability to be used as SRF for the calciner in the cement facility. Further fundamental and subject-specific aspects of the production of SRF for energy supply in the cement industry as well as on the topic of quality assurance and the usual market qualities have been described and discussed by Lorber et al. (2012), Sarc et al. (2013), Sarc et al. (2014) and Aldrian et al. (2016).

2 Materials and Methods

The tests with two industrial sieves (small and large) were conducted between June and November 2016 (post-shredding tests, small sieve) and between March and May 2018 (pre-shredding tests, large sieve) on the premises of a waste treatment plant in Austria.

The subsequent sieving tests (grain size distribution) and chemical analyses (Calorific Value, Cl, Sb, Cd, Hg, etc.) were conducted during the same time period for each test in the laboratory of the Chair of Waste Processing Technology and Waste Management of the Montanuniversitaet Leoben.

2.1 Test Material

Different input materials important for SRF production were tested, one of these was residues from **packaging waste** pre-processing and the other materials were from pre-processed **mixed commercial waste**. Pre-shredding tests used both types of waste. In total approx. 30 t processed in the SRF facility with the entire amount used for the sieving tests with the large industrial sieve (65 mm sieve cut). In the post-shredding tests only residues from packaging waste pre-processing were investigated. In total, approx. 1,000 t processed in the SRF facility of which several hundred kilograms were taken for the sieving tests with the small industrial sieve (5 mm mesh size). Quality assured, high-calorific (SRF PREMIUM Quality) and medium-calorific (SRF MEDIUM Quality) SRF ready for incineration in the cement facilities are able to be produced from these input materials by means of suitable mechanical treatment steps.

2.2 Post-Shredding Sieving Tests - Industrial Sieve (Small)

The experiments to separate the fine fraction (GS <5 mm) after the **second** shredding stage (post-shredding) at the end of a two-stage production line, were carried out with an industrial sieve (small) with a mesh size of 5 mm. The sample material was recovered by means of an automatic sampling flap after the post-shredding step. Two scenarios have been discussed and evaluated. Scenario A (small scale tests) considers sieving of the SRF after post-shredding in a coarse and a fine fraction. In the scenario B (large scale tests) the coarse fraction (GS > 5 mm) produced is processed in a further fine shredder which increases the combustion properties of the material (Bohrn, 2016).

2.2.1 Small Scale Tests Scenario A

The tests (sieving at 5 mm) were carried out with a small industrial sieve with small (approx. 100 kg) amounts of sample material.

To obtain a better overview, the following Figure 1 illustrates the process of the entire small scale tests conducted.

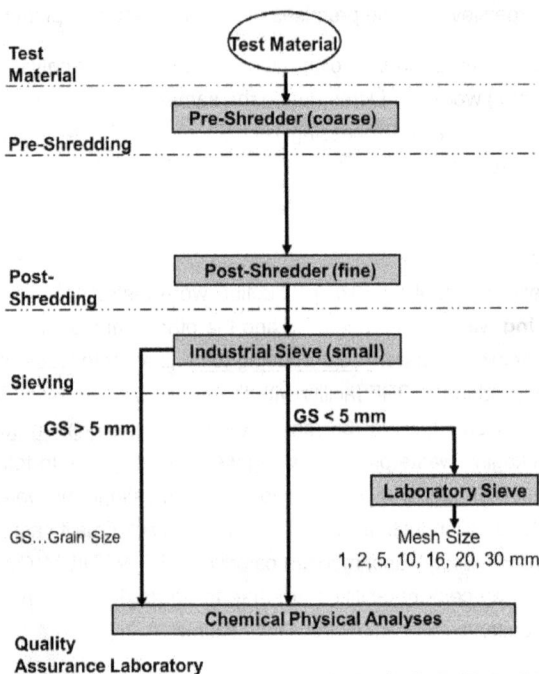

Figure 1   Small scale tests with the small sieve, scenario A (Bohrn, 2016, modified).

The test material (packaging waste) passes through the entire SRF production line and is sampled after post-shredding. The samples were taken from a sampling flap and sieved at 5 mm. The fine fraction GS <5 mm was subject to both a sieve analysis and a chemical analysis. The coarse fraction GS > 5 mm was only subject to a chemical analysis.

2.2.2 Large Scale Tests Scenario B

The tests (sieving at 5 mm) were carried out with a small industrial sieve with large (approx. 400 kg) amounts of sample material. To obtain a better overview, the following Figure 2 illustrates the process of the entire small scale tests conducted.

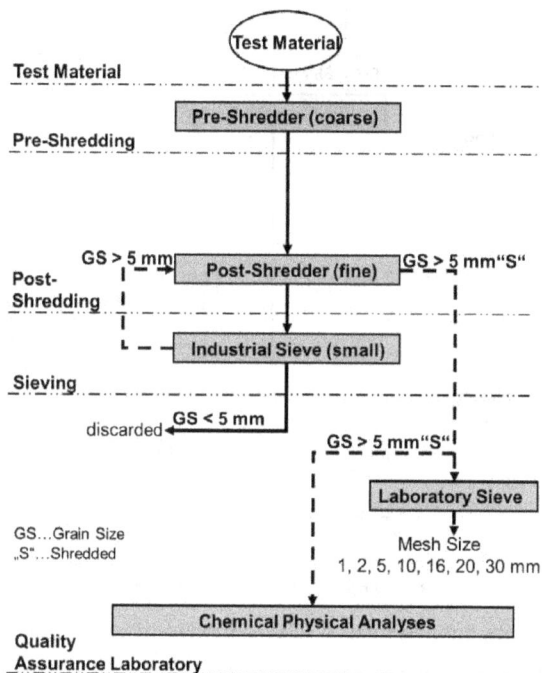

Figure 2   Large scale tests with the small sieve, scenario B (Bohm, 2016, modified).

The test material (packaging waste) passes through the entire SRF production line and is sampled after post-shredding. The samples were taken from a sampling flap and sieved at 5 mm. The resulting fine fraction GS < 5 mm was discarded. The coarse fraction GS > 5 mm was crushed again by the post-shredder and the resulting fraction **GS > 5 mm"S"** was subject to both a sieve analysis and a chemical analysis.

2.3   Pre-Shredding Sieving Tests - Industrial Sieve (Large)

To obtain a better overview, the following Figure 3 illustrates the process of the entire tests with the large industrial sieve conducted.

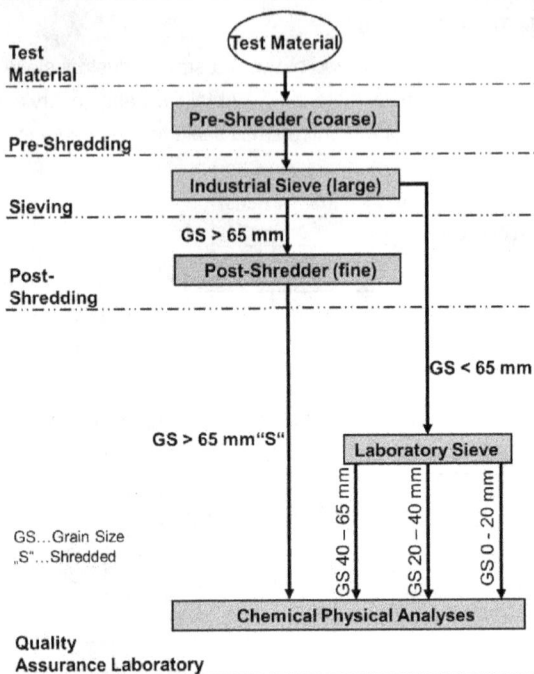

Figure 3　Large scale tests with the large sieve.

The tests to separate the fine fraction (GS < 65 mm) after the **first** shredding stage (pre-shredding) were carried out with an industrial sieve (large) at a sieve cut of 65 mm with large amounts of test-material (several tests with a total of 30 tons of pre-shredded test-material (packaging and mixed commercial waste). The width of the sieving holes was 83 mm. The inclination of the sieve deck caused smaller effective sieving holes (projected surface) which in this case corresponds to a 65 mm sieve cut. Thus and because of the complex geometry of the holes of the sieve, in this article, the sieve cut and not the hole diameter is used as a term for the sieve deck of the large sieve. The laboratory sieving machine is specified with the actual sieving hole diameter.

In order to determine the optimum (combustion properties of SRF PREMIUM Quality and SRF MEDIUM Quality) sieve cut, representative samples of the fine fraction (GS < 65 mm) were further examined in the laboratory. The samples were collected by suitable sample vessels from the falling mass flow after the large industrial sieve. The coarse fraction GS > 65 mm was shredded with the post-shredder to improve the sampling. The resulting fraction GS > 65 mm"S" was sampled via a sampling flap and chemically analyzed in the laboratory.

2.4 Laboratory Analyses of SRF-Quality

2.4.1 Analyses Post-Shredding Sieving Tests

The samples from the **post-shredding sieving tests** were subject to a chemical analysis (CV, Cl, Sb, Cd, Hg) and a particle size analysis (1, 2, 5, 10, 20, 30 mm mesh size).

2.4.2 Analyses Pre-Shredding Sieving Tests

The samples of the fine fraction GS < 65 mm from the **pre-shredding sieving tests** were subdivided by means of a laboratory sieve into the grain sizes classes GS 0 - 20 mm, GS 20 - 40 mm and GS 40 - 65 mm. For each sieve class including the sieve fraction GS > 65mm"S" (from shredding of the coarse fraction GS > 65 mm (Figure 3)) several chemical and physical parameters (e.g. calorific value, Cl, Sb, Cd, Hg) were determined in the laboratory.

3 Results and Discussion

3.1 Results Post-Shredding Sieving Tests - Industrial Sieve (Small)

Next Figure 4 shows the results of the tests with the small industrial sieve. The sieving allowed a division into a fine fraction (GS < 5 mm) and a coarse fraction (GS > 5 mm).The distribution (transfer factors) to the coarse and fine fraction of mass, energy, chlorine and heavy metals is shown as bars. Several combustion-relevant parameters (specific energy, concentrations) are shown as numbers next to the bars.

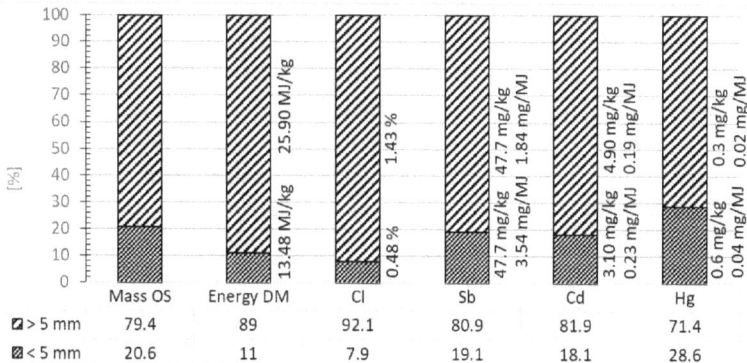

	Mass OS	Energy DM	Cl	Sb	Cd	Hg
> 5 mm	79.4	89	92.1	80.9	81.9	71.4
< 5 mm	20.6	11	7.9	19.1	18.1	28.6

Figure 4   Results of the large scale tests with the small sieve (5 mm mesh size) (Bohrn, 2016, modified).

The fine fraction constitutes on average 20.6% of the mass of the entire sample, the coarse fraction 79.4%. From this distribution, the transfer coefficients of the considered parameters were calculated. The distribution of heavy metals between the two fractions shows a fairly consistent picture. The proportion of antimony and cadmium in the fine fraction, each at 19.1 and 18.1%, accounts for around one-fifth of the total load of the SRF PREMIUM Quality. Mercury is represented at 28.6% in the fine fraction and 71.4% in the SRF PREMIUM Quality (Bohrn, 2016). The legal limit values (AVV) are complied with.

The chlorine load shows a strong one-sided distribution. Chlorine is mainly (92.1%) included in SRF PREMIUM Quality.

The separation of the fine fraction < 5 mm and subsequent shredding of the coarse fraction > 5 mm results in improvements in the combustion properties of the SRF PREMIUM Quality. The fine fraction is very well suited for use in calciner in a mix with other SRF MEDIUM Quality fuels with a slightly higher calorific value.

Post-shredding of the fraction > 5 mm (dashed line in Figure 2) resulted in a significantly finer fuel (GS > 5 mm"S") in the particle size range between 5 and 30 mm. The thus increased specific surface ensures a better burnout behavior. The bulk density, which decreases after separation of the fine fraction, can be restored by post-shredding.

Post-shredding leads to a drying of the material due to the mechanical energy input and thus to a higher calorific value.

3.2   Results Pre-Shredding Sieving Tests – Industrial Sieve (Large)

As already mentioned, the fine fraction GS < 65 mm was sampled and subdivided by means of a laboratory sieve into the grain sizes classes GS 0 - 20 mm, GS 20 - 40 mm and GS 40 - 65 mm. All grain size classes were analyzed. From mass distribution of these sieve classes, the mass flow ratio between the coarse and the fine fraction for the sieve cuts 40 mm and 20 mm, the transfer factors for the heavy metals, chlorine and energy could be calculated with the material flow analysis software STAN (Technische Universitaet Wien, 2018). The distribution (transfer factors) to the coarse and fine fractions of mass, energy, chlorine and heavy metals is shown as bars. Several combustion-relevant parameters (specific energy, concentrations) are shown as numbers next to the bars. The results of these investigations and calculations are described in the following.

Next Figure 5 shows the results of the tests with the large industrial sieve, laboratory sieve and chemical analyses. The sieving with the large industrial sieve allowed a division into a fine fraction (GS < 65 mm) and coarse fraction (GS > 65 mm).

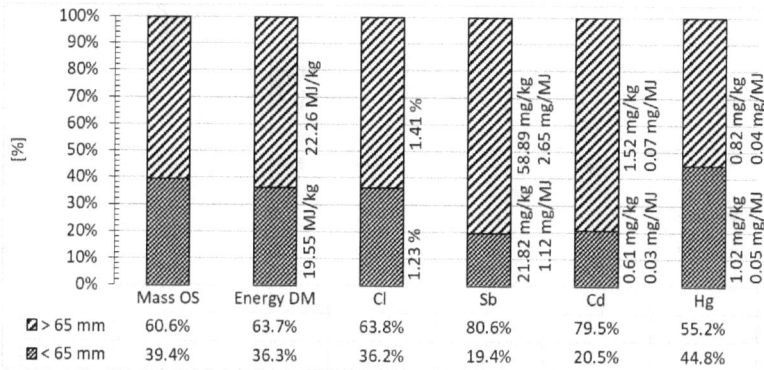

	Mass OS	Energy DM	Cl	Sb	Cd	Hg
> 65 mm	60.6%	63.7%	63.8%	80.6%	79.5%	55.2%
< 65 mm	39.4%	36.3%	36.2%	19.4%	20.5%	44.8%

Figure 5   Results of the large scale tests with the large sieve (65 mm sieve cut).

The fine fraction constitutes on average 39.4 % of the mass of the entire sample, the coarse fraction 60.6 %. The calorific value in the coarse fraction (SRF PREMIUM Quality) increases by separating the fine fraction. However, chlorine, antimony and cadmium accumulate in the coarse fraction (SRF PREMIUM), whereas Hg accumulates in the fine fraction. The legal limit values (AVV) are complied with. The fine fraction has a very high calorific value and is therefore not suitable for use in the calciner.

The calculations with STAN with a fictitious sieve cut of 40 mm show the following results for the fine fraction <40mm and the coarse fraction> 40mm (Figure 6).

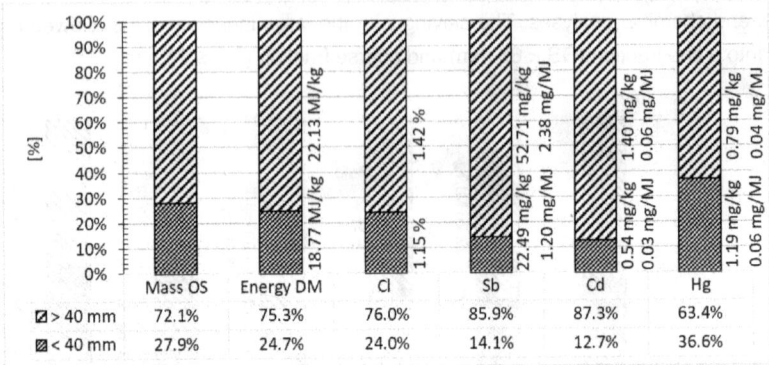

Figure 6   Results of the large scale tests with the large sieve (calculated sieve cut of 40 mm).

The fine fraction constitutes on average 27.9 % of the mass of the entire sample, the coarse fraction 72.1 %. The calorific value in the coarse fraction (SRF PREMIUM Quality) increases by separating the fine fraction. However, chlorine, antimony and cadmium accumulate in the coarse fraction (SRF PREMIUM), whereas Hg accumulates in the fine fraction. The legal limit values (AVV) are complied with. The fine fraction has a very high calorific value and is therefore not suitable for use in the calciner.

The calculations with STAN with a fictitious sieve cut of 20 mm show the following results for the fine fraction < 20 mm and the coarse fraction > 20 mm (Figure 7).

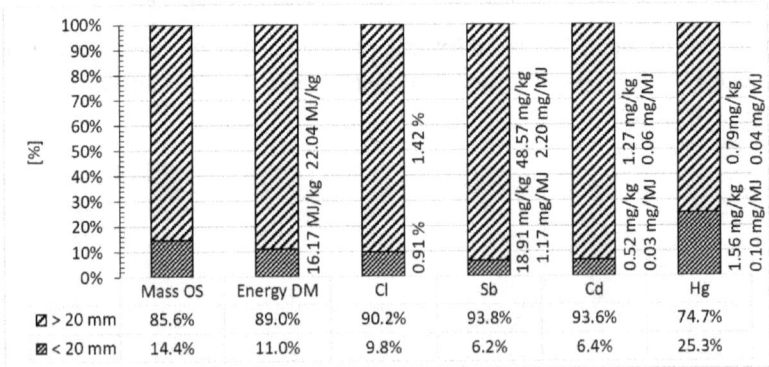

Figure 7   Results of the large scale tests with the large sieve. (calculated sieve cut of 20 mm).

The fine fraction constitutes on average 14.4 % of the mass of the entire sample, the coarse fraction 58.6 %. The calorific value in the coarse fraction (SRF PREMIUM Quality) increases by separating the fine fraction. However, chlorine, antimony and cadmium accumulate in the coarse fraction (SRF PREMIUM Quality) due to the higher plastic share, whereas Hg accumulates in the fine fraction. The limits of the AVV are met except for Hg which slightly exceeds the median value (0.07 mg/MJ). The fine fraction has a medium calorific value and is very well suitable for use in the calciner.

4   Literature

Aldrian, A., Sarc, R., Pomberger, R., Lorber, K.E., Sipple, E.-M.,   2016   Solid recovered fuels in the cement industry-semi-automated sample preparation unit as a means for facilitated practical application. Waste management & research: the journal of the International Solid Wastes and Public Cleansing Association, ISWA 34 (3), 254–264. 10.1177/0734242X15622816.

Bundeskanzleramt   2010   Austrian Incineration Ordinance. https://www.ris.bka.gv.at/GeltendeFassung.wxe ?Abfrage=Bundesnormen &Gesetzesnummer= 20002239 (Last access: 27. March 2019).

Bohrn G.,   2016   Praktische Untersuchung der Feinfraktion von „EBS PREMIUM Quality" und technisch-wirtschaftliche Planung einer großtechnischen Umsetzung. Master Thesis, Montanuniversität Leoben.

Lorber, K.E., Sarc, R., Aldrian, A.,   2012   Design and quality assurance for solid recovered fuel. Waste management & research: the journal of the International Solid Wastes and Public Cleansing Association, ISWA 30 (4), 370–380. 10.1177/0734242X12440484.

Mauschitz, G.,   2018   Emissionen aus Anlagen der Österreichischen Zementindustrie Berichtsjahr 2017 (Emissions from Austrian cement industry reporting year 2017). Available online at: https://www.zement.at/beton-2/nachhaltigkeit-und-umwelt/emissionen. (Last access: 01. March 2019).

Sarc, R., Lorber, K.E.,   2013   Production, quality and quality assurance of Refuse Derived Fuels (RDFs). Waste management (New York, N.Y.) 33 (9), 1825–1834. 10.1016/j.wasman.2013.05.004.

Sarc, R., Lorber, K.E., Pomberger, R., Rogetzer, M., Sipple, E.M.,   2014   Design, quality, and quality assurance of solid recovered fuels for the substitution of fossil feedstock in the cement industry. Waste management & research: the journal of the International Solid Wastes and Public Cleansing Association, ISWA 32 (7), 565–585.

10.1177/0734242X14536462.Abfallforschungstage
2004. Auf dem Weg in eine nachhaltige Abfallwirtschaft. Cuvillier Verlag, Göttingen, ISBN 3-86537-
121-3.

Sarc, R, Pomberger, R,   2015   „REUQ-Ausweis" für EBS Entwicklung eines Res-
Eferdinger, S.,   sourcen-, Energie-, Umwelt-, Qualitätsausweises
                                   (REUQ) für Ersatzbrennstoffe (EBS) ("REUQ certifi-
                                   cate" for SRF Development of a resource-, energy-,
                                   environmental-, quality certificate (REUQ) for solid
                                   recovered fuels (SRF), Association of Austrian Waste
                                   Management Companies (VÖEB)), Verband Öster-
                                   reichischer Entsorgungsbetriebe (VÖEB), Wien, Aus-
                                   tria.

Technische Universität   2018   STAN (short for subSTance flow ANalysis)
Wien   http://www.stan2web.net/ (Last access: March 2018)

Acknowledgements

The Center of Competence for Recycling and Recovery of Waste 4.0 (acronym Re-
Waste4.0) (contract number 860 884) under the scope of the COMET – Competence
Centers for Excellent Technologies – financially supported by BMVIT, BMWFW and the
federal state of Styria, managed by the FFG.

Author's address

Dipl.-Ing. Alexander Curtis
Lehrstuhl für Abfallverwertungstechnik und Abfallwirtschaft

Montanuniversitaet Leoben

A-8700 Leoben
Telefon +43 681 8425 6967
E-Mail: alexander.curtis@unileoben.ac.at

Co-processing as a tool for Sustainable Waste Management in Brazil

Christiane Pereira, Cora Buchenberger, Klaus Fricke, Olga Kasper, Hubert Baier; Alaim Silva de Paula

Technische Universität Braunschweig, Leichtweiß-Institut für Wasserbau, Abteilung Abfall- und Ressourcenwirtschaft, Braunschweig, Germany

Abstract

The implementation of the provisions of the National Waste policy in Brazil brought a new systematic approach for waste management, particularly in relation to the introduction of technology for the waste recovery, enforcing a series of new activities, that should be implemented in the short and medium term for adequacy of current practices in alignment with the terms of the Law. It is also necessary to articulate an environmental agenda with social inclusion, economic development, citizenship, construction of urban infrastructure and the introduction of technology, mobilizing efforts under a cross-sectional view, strongly related to sustainable development. The market development for alternative fuels is a consequence of the implementation of the legal mechanisms, demanding specific infrastructure, technology and effective waste management systems.

Keywords

Sustainable Waste management, public policy, climate protection, municipal solid waste, alternative fuels; AF; Greenhouse Gas emissions; cement sector; co-processing; sanitary landfill.

1 Introduction

The cement and waste sector have fostered discussions on minimizing impacts on the production process and understands that investing in innovation is a fundamental part of this effort, thus replacing the energy source with Alternative Fuels (AF) has become strategic for both sectors. Cement companies have invested constantly to achieve, by 2030, international standards in the use of raw materials and alternative non-fossil fuels through co-processing in their units in Brazil. The waste sector looks for alternatives where complementary income can be generated and, at the same time, to avoid air and water pollution.

The new condition of low economic growth in the country goes back to the search for alternatives, which means that the current challenge is not related to cement performance, production and the market, but related to the reduction of emissions of greenhouse gases (GHG) and the possibility of reconciling climate change and the need of the cement sector to seek sustainability through energy efficiency and fuel derived from alternative materials.

It is important to contextualize that in order to achieve an economic equilibrium of the AF business where financial goals are set for projects that meet the expectations of private investors, a multilateral effort will be necessary where the central public power will have the essential role of promoting quality both in the criteria of choice of cases to be prioritized, as well as in the correct and fair equation of the variables involved in the business so that the effort is well equalized among those involved, generating an environment of trust among those involved. This has been the biggest obstacle that has prevented development of energy recovery from MSW so far, and which needs to receive special attention from the central government to be developed.

This information aligned with the waste sector will demonstrate the potential of the sector, showing its importance for the coming years, serving as a reference for other developing countries where the substitution of fossil energy source is also present in public policy discussions.

In this way, this article opens a multisectoral dialogue that, in addition to ensuring the transversal approach of the theme, will allow its implementation in a continuous and lasting manner, representing an alternative of waste valorization that will support the guidelines of the National Policy on Solid Waste, both economically and environmentally.

2   Waste Management

The National Waste Policy published in 2010, establishes principles found in Europe and Germany, as well as the hierarchy of procedures concerning sustainable solutions for the problems associated with the waste management, focusing on resource preservation and climate protection and creates a positive agenda for encouraging the adaptation of landfills as electric power plants; closing and remediating the wild dumps and promoting waste valorization and social inclusion. Other practices as material recycling and energy recovery are promoted by the National Policy as well.

9 years after its publication we can observe a slow movement, however slow doesn't mean small when we consider the giant waste market that Brazil represents. ABRELPE, in its publication Waste Panorama 2017 presented the following figures: total MSW generation: 78,4 million tonnes per year; 337.000 employees, annual budget of 7,13 billion Euros; 60% of final disposal in sanitary landfill.

Regarding the waste recovery technologies in large scale, there are already implemented three Mechanical Treatment Plants to produce AF material from MSW , of which 2 have a nominal capacity of 300.000 t per year and one of 150.000 t per year. There are two Mechanical Plants to recover recyclables from selective collection in operation, both with a nominal capacity of 90.000 t per year.

Besides the above mentioned plants there are contracts signed for the implementation of one incineration plant (mass burning) with a capacity of 300.000 t per year and several composting plants with a capacity of 90.000 t per year.

The cities under 250.000 inhabitants are already focusing on the implementation of technologies in consortium models. More than 50% of the cities in Brazil have already developed their Integrated Waste Management Plan where Mechanical-Biological Treatment (MBT) solutions lead the discussion.

Despite the good news, Brazil still has 40% of final waste disposal being done at wild dump sites and their remediation represents a huge market for consultants and operators.

The waste management market has two important players: the public and the private sector. Observing the private sector, for example traditional industries, we can identify a quick movement for technologies. The reason for that is not really their environmental responsibility but the reverse logistic that has been implemented after 2010, meaning that they have to take responsibility with the waste and invest at the sustainable destination.

To change traditional practices, we need to open a multidisciplinary discussion which integrates multiple market segments to enable the design of tools for the implementation of sustainable management of municipal solid waste. The discussions range from technologies in the form of fermentation, composting, recycling and energy recovery to the supply of information, advice on introducing a sustainable waste management and measures to reduce the waste through environmental education programs, also the engineering and scientific content as well as relevant aspects for the implementation of projects, such as financing, environmental licensing, monitoring, and other aspects of the market.

Actually, the demand is concentrated in the search for infrastructure, technology and effective management systems, including technical-operational aspects, as well as the consolidation of appropriate technologies for the implementation and monitoring of future treatment plants.

The opportunity for new businesses in Brazil is focused on forming partnerships among national operators and foreign technology suppliers, but also increasing the interest of competitive segments of the economy, such as the cement industry, cellulose-methanol- other, and energy industries, that will divide the market with the traditional operators into the public cleaning sector.

3   Cement Market

After performing a series of investments to expand the production capacity from 60 million to 90 million tonnes between 2005 and 2014, due to income and employment growth, strong real estate lending, falling interest rates and inflation and investment in infrastructure programs, significant investments in expansion of installed capacity were motivated, currently reaching around 100 million tons per year.

Since 2015, the country has been facing a serious political-economic crisis, which has resulted in a reduction in investment in infrastructure and an increase in unemployment. The rise in interest rates and wage losses reflected heavily in the real estate market. As a result, construction activity suffered a sharp contraction, culminating in the worst crisis that the Brazilian cement industry has faced, with a cumulative drop in production of 25% in the last three years. The cement industry now deals with the reduction in demand in the country where the production in 2018 dropped down to 52,9 million tonnes in 2018. Due to the downturn in the economy in 2017, there is an excess capacity in the sector of approx. 50%.

This means that to overcome the recession period, it will be essential to introduce other measures to the operational efficiency through modernization of the industrial park and energy efficiency. Therefore, the longevity of companies is directly related to research on new sources to add in cement, technological evolution, energy efficiency and substitution of energy sources by alternative fuels (AF), either biomass or waste derived.

The Brazilian cement industry operates 99% of its kilns according to the high energy efficient dry process, characterized by pre-calciner and heat recuperation from the clinker cooler, online monitoring, mills with high efficiency separators and bag house filters. It is also a pioneer in a high use of biomass as an alternative fuel. Their modern cement plants consume thermal energy of 3,50 GJ/ t clinker and electricity of 113 kWh/ t for cement grinding.

In 2016, Brazil was the sixth largest producer and the eighth largest consumer of cement in the world. The industrial cement park currently consists of 100 plants, present in 88 municipalities and 24 states, 62 of which are integrated units and the rest are cement grindings. Most of these factories are located in the coastal region of the country, following the highest the consumer market and the limestone deposits.

According to the cement sector roadmap, currently about 60% of the integrated facto-ries are licensed to co-process waste derived alternative fuels and raw materials. The consumption of alternative fuels by the sector in Brazil has increased considerably. However, when comparing the current level of 15% of the thermal substitution rate (TSR) with other countries, there is a great potential for increasing the energy use of pre-processed residues and biomass, including urban solid waste. In 2014, the Brazilian cement industry used 1.5 million tonnes of waste derived fuels and biomass, represent-ing 15% of the total energy consumption, 8% of which from waste and 7% from bio-mass. Currently, about 70% of the used tires in the country are utilized by co-processing.

In this context, co-processed volumes fluctuate slightly above 1.0 million tonnes per year, distributed among mineral Alternative Raw Materials, non-residual tires and Alter-native Fuels. This corresponds to a rate of 10 to 12,5% of TSR for the total clinker ca-pacity in Brazil. Against this background, the roadmap for technological development in the cement sector aims to substitute 35% of thermal energy from fossils sources to al-ternative sources, where 10% of the AF shall be derived from non-hazardous solid commercial, industrial and treated municipal solid waste.

For the year 2050, the same study indicates a TSR of more than 50%, with AFs accord-ing to a level of around 20% derived from waste. This roadmap denotes the strategic importance of the issue of AFs for the cement sector in Brazil, in comparison with the developments in more developed countries the waste management market is more ma-ture in terms of waste treatment technologies, usage and a final and safe disposal.

4   Results and Discussion

Co-processing can be presented as a suitable practice that integrates reuse and final destination, where these two purposes occur in a single operation of burning properly pre-processed municipal solid waste mandatorily compatible to the process of clinker production, in rotary kilns of the cement industry. While an efficient, safe and economi-cal waste treatment and recycling process is occurring, through the use of these as al-ternative fuels and raw materials (AFR), an important economic product is being pro-duced, cement - providing a more comprehensive global environmental balance.

Among the main advantages are:

- avoidance of CO_2-emissions and substitution of the fossil energy source through an environmentally and technological adequate source

- preservation of non-renewable energy resources; the possibility of large-scale processing of alternative fuels, mitigation of environmental impacts of landfills, and

- indirectly contributing to public health and the creation of new jobs.

As proven impacts can be pointed out:

- instability of the kiln run by poorly pre-processed AF qualities
- high risk of ring formation and clogging
- additional investment for infrastructure such as reception, storage, feeding, and quality surveillance,
- need of process adaption,
- and above all the costly risk of increased consumption of electricity, refractory lining and other resources.

It is worth mentioning the existence of appropriate countermeasures for all these points. By implementation of a sound waste management and the use of alternative fuels in thermal processes reduces the amount of fossil fuels needed, contributing to the reduction of the greenhouse gases due to a lower CO_2-eq.-emission factor. This means, that AFs will reduce methane production on the dump side and will emit less fossil carbon on its use.

In this context, the Brazilian cement sector has fostered discussions regarding the minimization of impacts on the productive process and understands that investment in innovation is a fundamental part of this effort, thus replacing the traditional energy source became strategic for the sector which is mainly dominated by the coke of petroleum production ("petcoke") . Cement companies have invested steadily in sustainability initiatives in order to achieve, in the medium term, international standards in the use of raw materials and alternative non-fossil fuels through co-processing in their units.

Still, the new condition of low economic growth in the country goes back to the search for alternatives that generate greater competitiveness, which means that the current challenge is not related to cement performance, production and the market, but related to the reduction of greenhouse gas (GHG) emissions and the possibility of reconciling climate change and the need of the cement sector. That is, seeking sustainability through energy efficiency and fuel derived from alternative resources.

During the period of 2004 to 2018, the price of imported petcoke increased by 425%, which incentivized the use of AF. The link to the oil price and the large fluctuation forces the cement companies to make intensive efforts to reduce their manufacturing costs. Energy costs account about 30% of cement production costs.

Focusing on this new market context, the AF processed from municipal, commercial and industrial solid waste presents enormous potential due to increased environmental pressure for converting wild dumping into a safe and sanitary landfill in the near future.

However, there are some limiting factors to this alternative: the need for a mature system of separate collection and sorting of materials; distance from the facility in relation to large urban centres; difficulty in establishing contracts that guarantee the supply of waste; technical issues related to excess chlorine; and need for processing the MSW and C+IW to secondary products as fuel.

A combination of factors, such as lack of industrial parks facing the attending to the demands of technologies for waste recovery, lack of technical knowledge, economic valuation of primary resources and as a result of secondary, high energy value and dependence of the Brazilian market on hydroelectric power, and especially the obligation to provide waste recovery as consequence of the new law, sets new horizons for the management of waste, generating opportunities for new business.

Besides all these mentioned opportunities and challenges, the domiciliary waste management competence is public, that means that it will be necessary to seek for guarantees from the public administration to guarantee the continuously demand, so efforts for implementing a differentiated and sustainable waste management will not be compromised.

The market for co-processing of tires in Brazil has changed completely in the last years. During its implementation, the cement sector would receive a remuneration for co-processing these tires. Currently the cement factories need to assume 1/3 of the costs of logistics and co-processing of tires in order to gain access to these sources. This market change reinforces the idea that, despite not being its core business, the cement industry understands that the MSW has potential to be a source as renewable energy.

Regarding the potential to integrate AFs from MSW in the cement market, a recent diagnostic study called PLANSAB – Energetic report, 25 potential clusters for the evaluation of a AF project were identified. Within each cluster, municipalities that play a central role in terms of availability of MSW volume have also been identified. In the most ambitious scenario, it could be possible to process 10,5 million tonnes of MSW per year against the realistic scenario of 3,3 million tonnes of MSW per year, to generate 1,1 million tonnes of AF per year, when its quality suits to the production of the sales market.

Considering the investment of the new management, it will be necessary to invest in need-based mechanical biological treatment (MBT) plants. The investment into plants that process 10,5 million tonnes of MSW per year, to produce almost 4 million tonnes of AFs per year is about 1,2 billion of Euros. The investment for the cement sector to substitute fossil against alternative fuels will be about 17,5 million Euro per kiln unit. Exclusive the adaptations of extending the calciner or to retrofit ID-fan, burner or chlorine bypass systems, the following investments for laboratory, air pollution control and monitoring, reception, storage, dosing and feeding are mandatory.

The PLANSAB study results in the following affirmation:

- There is a REALISTIC potential for AF consumption between 1,2 and 2,5 Million tonnes/year spread all over the country given by the AF Study Boundary Conditions – this scenario represents 25 AF Business Opportunities Clusters

- The Technical Limit (Without AF Boundary Conditions) is about 4,0 Million tonnes/year, this means more than 10 Million tonnes/ year can be diverted from landfilling (around 15% of Total MSW in Brazil)

- 50% of 25 AF Clusters are located in Waste Dumping Areas. It is possible to divert more than 1,5 Million tonnes of waste to be used for the production of renewable fuel to the cement industry in the same region.

4.1 Implementation challenges

In the following, some peculiar aspects related to the using of AF from municipal solid waste are shown:

- **(Pre-processing) technic**: Low calorific value, high content of sulfur, chlorine or moisture or the presence of other substances harmful to the combustion process, the product or the emissions.

- **Economic**: Low cost of landfills, proximity to cement plants with urban centres and/ or biomass and waste generation sources, competition with other energy or material recovery alternatives, difficulties and logistics costs. Process control and emission standards for the use of waste derived fuels in thermal units such as power plants, lime and cement kilns, or are even more restrictive than for those using conventional fuels.

- **Legislation**: Long and bureaucratic environmental licensing processes, laws that hinder or prohibit energy recovery of solid waste, the possibility of having state wide resolutions that prevent or limit the deposit, custody and processing of hazardous waste generated outside the state, non-compliance with local and federal laws restricting landfill disposal

- **Contractual**: Difficulty in establishing long-term contracts for the supply of Municipal Solid Waste with public agents.

- **Social**: Relationship with local communities. Conflict between recycling of materials and energy recovery.

4.2 Climate protection

While energy efficiency measures reduce (fossil) fuel consumption in clinker production, the use of renewable-carbon energy sources represent an additional option to further reduce emissions from fuel combustion. In 2014, most of the sector's thermal demand was met by petcoke (85%), with biomass and residues in the remaining 15%. In 2050, biomass could assume a share of 11% share of the total energy demand and waste could account for 44%. Thus, together, they could account for about 55% of the total thermal energy demand in the sector. This would reduce the fuel emission factor from 91 $kgCO_2$/ GJ in 2014 to 77 $kgCO_2$/ GJ by 2050.

Over time, the substitution of petcoke in Brazil's "cement industry" (or "clinker kilns") with alternative fuels will represent an important contribution to the sector's emissions reduction, with the potential to mitigate around 13% of cumulative emissions by 2050.

Additionally, the introduction of a modern waste management system will lead to sufficient resource management inclusive treatment technologies and sanitary landfilling, and will put an end to the uncontrolled dumping and its methane emissions, which are several times more effective than CO_2.

5 Conclusions

The discussion about environmental and climate protection has changed intensely in the last few years. Climate protection has become the icon of sustainable development and is present on the global agendas, strategic plans and technical conferences. The promotion of multicultural and multidisciplinary discussion ensures an expansion beyond the geographical and political boundaries of these approaches and seek symbioses between the most diverse markets, as supplying equipment, monitoring services, management and even of consumption and public policies.

Regarding the waste and the cement market symbiosis, it is necessary to reach a balance between supply and demand of alternative energy, combining the strengths of the cement market, like qualified management, specialization in energy efficient technologies and quality control, with the knowledge from waste market like dealing with public administration and managing a large scale of residues and their complexity as seasonality, diversified characterization, among others peculiarities.

It is assumed that the synergy between the waste sector and the cement sector can be the first step to guarantee the sound disposal of a significant part of the AFs processed from waste, and the remaining quantity can be sent to other economic sectors. Indeed, cement plants where kilns are already installed are located close to their cement sales market, where residents build, live and also generate waste as well. In nearly all states of Brazil the geographic location of the cement plants is also close to the population centres which also coincide with the large waste generating centres.

In order to foster the commitment of the cement industry to the safety of the development of the topic, each cement plant needs to implement individual adaptations at its process and infrastructure to replace its fossil energy sources by AFs. The operators have to adapt their processes to guarantee best retention time and combustion conditions, to install by-pass systems as well as AF-reception and storage, materials handling, dryer, dosing and feeding systems and equipment for monitoring the quality as well as the exhaust gas emission in accordance to the legal application. Partly post-treatment equipment is required, too.

In addition to the economic question, the cement sector has voluntarily taken a growing environmental attitude. The cement industry worldwide accounts for about 8% of the CO_2-emissions of the industrial sector, where the most of these emissions (55%) are process–related by decarbonization of the limestone respectively calcination of the raw meal, and is not linked to the combustion of fuels.

On the other hand, with the view on the existing WBCSD commitment and developing initiatives to achieve the voluntary emission reduction targets, the substitution of fossil fuels has a relevant contribution potential. Its implementation path is already known and tested, but in comparison to more cost-effective measures this issue still not levers on the required priority level.

The Roadmap 2019 establishes the substitution of petcoke in the cement kilns in Brazil. Alternative fuels will contribute significantly to the reduction of emissions in the sector, with the potential to mitigate around 13% of cumulative emissions by 2050.

As a function of the defined boundary conditions, the potential volumes of AF consumption were calculated with the four views mentioned. There is an estimated potential for AF consumption of over 1,1 million tonnes in a realistic scenario, which means that around 3,3 million tonnes of MSW would be treated in AF-production plants for this purpose. This amount can be doubled in a challenging scenario where the required CAPEX amount is two-times larger.

The volumes of the potential ambition denote the volume that could technically be consumed, but with projects of an economic attractiveness lower than the national average. Within this potential, more than 10,5 million tonnes of solid non-hazardous waste could be treated which would mean something around 15% of the total volume produced in Brazil in 2017, resulting in the production of 3-4 million tonnes of AF to substitute the fossil regular energy.

It is important to understand that to implement a MBT plant makes only sense in regions, where kilns or power plants are operated. In regions without any thermal processes, new WtE plants or MWI make more sense!

Notably, there is a trend that shows the movement of incoming multinationals in the cement market that has the potential to provoke changes in the sector, breaking with the status quo in the face of the growing competition environment. Another relevant aspect concerns the technology that should predict the production of AF through MBT plants with reuse of the organic fraction, since it is known that in the face of the 3R´s policies, the dry residues will tend to reduce in the medium and long term.

Due to the growing competition, the Cement Companies´ Goal shall aim at reducing the clinker production cost. This can be achieved only by implementing the proper MBT technology for MSW treatment enabling to produce high quality AF, as Commodity!

Therefore, it can be concluded that the use of AF represents a renewable energy source which is able to play an important role not only in the sustainable management of waste, but also in efficient energy management.

Regarding GHG emission: not only the substitution of fossil energy sources will reduce the CO_2-emission, it is also the implementation of a reliable waste management system, which will reduce the methane emission and its huge CO_2-equivalent as well.

In addition, AFs contribute to the fight against climate change by reducing greenhouse gas emissions in front of and during its use and preserving natural resources. This initiative will allow Brazil to take an important step in the development of solid waste management, placing it in the same direction as developed countries do.

6   Literature

ABRELPE - Associação Brasileira de Empresas de Limpeza Pública e Resíduos Especiais	2018	Panorama dos Resíduos Sólidos no Brasil – 2017. São Paulo http://abrelpe.org.br/download-panorama-2017
K. Fricke, C. Pereira, A. Leite, M. Bagnati (Coord.)	2015	Gestão sustentável de resíduos sólidos urbanos: transferência de experiência entre a Alemanha e o Brasil. Braunschweig: Technische Universität Braunschweig
Roadmap Tecnológico do Cimento	2019	http://www.abcp.org.br/cms/wp-content/uploads/2019/04/Roadmap_Tecnologico_Cimento_Brasil_Book-1.pdf

PLANSAB	2019	http://www.cidades.gov.br/images/stories/ArquivosSNSA/Arquivos_PDF/plansab/Versaoatualizada07mar2019_consultapublica.pdf

Author's addresses:

Technische Universität Braunschweig
Dipl-Ing. RA Christiane Pereira
Dipl.-Reg. Wiss. Latam. Cora Buchenberger
Prof. Dr.-Ing. Klaus Fricke
Leichtweiß-Institut für Wasserbau, Abt. Abfall- und Ressourcenwirtschaft
Beethovenstraße 51a
D-38106 Braunschweig
Telefon +49 1753426247
E-Mail chrdiasp@tu-braunschweig.de
https://www.tu-braunschweig.de/lwi/abwi/mitarbeiter/pereira/index.html

Terra Melhor Ltda.
MSC. Olga Kasper
olga@terramelhor.com.br
http://terramelhor.com.br

WhiteLabel-TandemProjects e.U.
Dr. Hubert Baier
hubert.baier@wltp.eu
www.wltp.eu

Kappa Gears Consulting
Dipl.-Ing. Alaim de Paula

alaim.paula@kappa-gears.com.br

Comparison of temperature driven and oxygen concentration driven aeration method during biodrying of mixed municipal solid waste

Vojtěch Pilnáček

-Institute for Environmental Studies, Faculty of Science, Charles University in Prague, Czech Republic

Abstract

Using a model biodrying reactor, an optimal aeration method for biodrying of mixed municipal solid waste was investigated. Two aeration methods were employed: one method was controlled by oxygen concentration (where it was maintained between 16% and 20%) and the other method was controlled by temperature (where a temperature between 42°C and 45°C and an oxygen concentration between 16% and 20% were maintained). Change of moisture content, change of calorific value and total energy balance were investigated. A moisture increase resp. decrease between 0.94% resp. 27.54%, and a lower calorific value decrease resp. increase between 9.23% resp. 41.12% were reached. The energy balance was positive in case of method controlled by temperature. In case of method controlled by oxygen concentration energy balance of one of two performed runs was negative, second positive. Aeration method controlled by oxygen concentration is strongly influenced by organic matter content. The method is thus suitable only for drying wastes with known and stable composition. The outcomes of the different methods were influenced by ambient relative air humidity. Moisture gradients in dried substrates often described in the literature were not replicated in all cases.

Keywords

Biodyring; Municipal solid waste; Mechanical-biological treatment; Aeration regime; Ambient relative air humidity; Organic matter content

Abbreviations

LCV - Lower calorific value
MMSW – mixed municipal solid waste

1   Introduction

Biodrying is a process which utilises heat generated by aerobic decomposition of organic matter for substrate drying. The principal mechanism of the entire process is air which is artificially blown into the substrate. The air serves as a source of oxygen for aerobic decomposition processes and as a medium for the transport of the evaporated moisture. It, however, simultaneously acts as a cooling medium where the energy consumption negatively impacts the energetic balance of the process. The main purpose of the process is to obtain alternative fuel with a sufficiently high calorific value (VELIS ET AL., 2009). It is thus necessary to investigate optimal conditions in the reactor chamber to achieve the highest rate of moisture evaporation, the lowest energy consumption and the lowest decomposition of organic part of the dried waste. One of main factors affecting the above-mentioned output parameters is aeration rate.

Aeration rate was surveyed by ADANI ET AL. (2002). Their work indicates that optimal drying efficiency is achieved by using higher aeration rates resulting in lower temperatures in the reactor chamber. Similar results were shown by ZHAO ET AL. (2010) and CAI ET AL. (2013). When the highest aeration rate was used, the highest evaporation, the lowest degradation of organic matter and the highest calorific value were achieved. The higher the aeration rate used, the higher the moisture amount evaporated and transported. The substrate is then so dry that the microorganisms which decompose organic matter suffer from lack of water and their activity ceases (WALKER ET AL., 1999; ADANI ET AL., 2002; AVALOS RAMIREZ ET AL., 2012). HUILIÑIR AND VILLEGAS (2014) showed that when the aeration rate is too high the dried substrate is cooled and the decomposition of organic matter decreases. Moisture volume then still decreases, but the convection drying effect is stronger than the biodrying effect. VANDERGHEYNST ET AL. (1997) show that when too low aeration rate is used, oxygen deficit in the upper parts of substrate occurs. Microorganism activity then decreases.

The aim of this study was to compare two aeration methods in order to determine an optimal aeration rate. Two aeration methods were tested: one method was controlled by oxygen concentration and the other by temperature in upper part of the reactor. Drying efficiency (moisture content decrease, calorific value increase) and energy balance of both methods was compared.

2   Material and methods

2.1   MMSW samples

1 m^3 samples were taken from regular MMSW collection routes in the town of Mníšek pod Brdy in the Czech Republic, the site of a proposed mechanical-biological waste treatment facility. Metals, glass and other rigid materials that could damage the equipment used in the subsequent grinding step were removed. The sample was crushed using a grinder knife with a grain size of approximately 30 mm. The quartering method was employed to take approx. 30 kg subsample for the study. The subsample was transported in a plastic barrel directly to the laboratory. A further 1 kg secondary sub-

sample was taken, also using the quartering method, in order to perform further analysis (moisture content, calorific value and volatile matter content). Composition of waste after the sorting of metals, glass and rigid materials during different runs is described below in table 1. Four different drying runs in four different different time periods were performed.

Table 1   Composition of waste in during the experiment

Run	Plastics (%)	Biodegradable + paper(%)	Textile (%)
Temperature 1	26.4	59.7	13.9
Temperature 2	21.4	65.2	13.4
Oxygen 1	24.2	63.7	12.1
Oxygen 2	24.6	70.2	4.5

2.2   Model reactor

The model reactor system, designed specifically for this study, consisted of a reactor chamber, a Secoh SLL 50 blower and a biofilter. See figure 1. The reactor chamber was a PVC cylinder 120 cm high and 40 cm in internal diameter. A 3 mm mesh size sieve was placed above the bottom of the reactor chamber for waste stabilisation and air supply dispersion. Below the sieve was an approximately 10 cm high layer of glass beads 5 mm in diameter to facilitate the removal of leachate that could poten-tially be formed and which would drain out of the bottom of the reactor chamber into an air-tight Erlenmeyer flask connected to it. The air inlet was above the bottom, submerged in the drainage layer, and the exhaust outlet was in the reactor lid. The entire reactor was wrapped in a 10 cm thick layer of insulating material. Probes for dried waste sampling and probes for measuring experimental conditions during the process were constructed within the reactor walls. Papouch THT2 temperature and air relative humidity probes were placed in the upper and lower air passageways. An ASEKO GTE oxygen sensor was placed under the reactor lid. To adjust the flow rate, an electronically-controlled blower was used. The reactor and other parts of the sys-tem were connected with plastic tubing. To evaluate the influence of ambient tem-perature and relative air humidity on the system, an external temperature and air relative humidity probe Papouch THT2 were attached to the reactor assembly about 0.5 m from the air inlet. Ambient air oxygen concentration was not measured. It was considered as standard and stable. These sensors were connected to the computer via signal converters. The blower was operated by a USB I/O module Papouch Quido 2/2 and relays. A multi-range BK G4 BO diaphragm gas meter was connected be-tween the blower and the reactor chamber in order to measure the total volume of air that passed through the chamber.

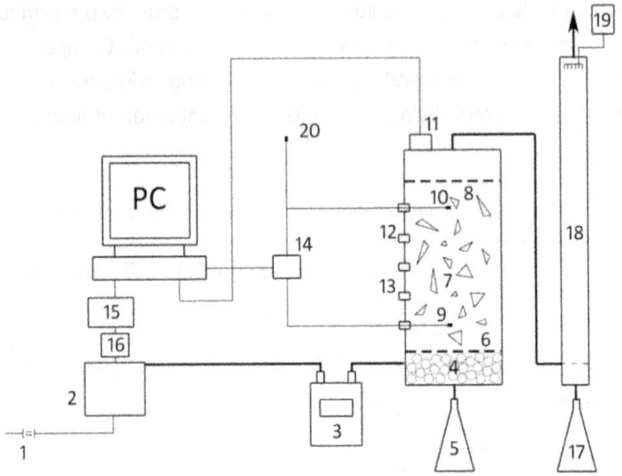

1 Voltage regulator
2 Air source
3 Gas meter
4 Drainage layer
5 Erlenmeyer flask to drain leachate
6 Perforated sheet
7 Dried waste
8 Perforated sheet
9 Temperature and air relative humidity meter
10 Temperature and air relative humidity meter
11 Oxygen sensor

12 Dried waste sampling port
13 Dried waste sampling port
14 PC converter
15 USB I/O module for switching an air source
16 Relay for switching the air source
17 Erlenmeyer flask for condensate
18 Biofilter medium (compost, peat, bark, expanded clay)
19 The feed solution to the biofilter
20 Ambient air temperature and air relative humidity meter

Figure 1   Reactor scheme

2.3   Aeration methods

The waste was dried using two aeration methods: one method was controlled by oxygen concentration and the other method was controlled by temperature in upper part of the reactor.

During the method controlled by oxygen concentration, the blower was operated in order to achieve an oxygen concentration between 16% and 20% in the upper part of the reactor. When oxygen concentration fell below 16% or rose above 20%, the blower was turned on and off, respectively. According to AVALOS RAMIREZ ET AL. (2012) an optimum oxygen concentration for aerobic degradation processes is between 15% and 20%. Dur-ing drying, temperature and relative humidity were recorded in the upper and lower lay-ers and oxygen concentration in the upper part of the reactor, as were the ambient tem-perature and relative humidity outside of the reactor. The oxygen concen-tration in the upper layer, the relative humidity and the temperature were recorded at five second in-tervals and the activity of the blower was recorded at one second inter-vals. The meas-urements were taken with a Papouch THT2 sensor and ASEKO GTE oxygen sensor using Wix software. Data processing was done using Microsoft Excel and R.

During the second aeration method based on temperature in the upper part of the reac-tor, the blower was turned on when the temperature in the upper layer reached 45°C and switched off when it decreased to 42°C. The upper temperature limit is based upon work done by ADANI ET AL. (2002). In addition to this, as mentioned above, the oxygen concentration in the upper layer was maintained in the 16% to 20% range.

2.4   Product collection and processing

Four drying runs were conducted - two controlled by temperature and two controlled by oxygen concentration. The drying runs took from 253 to 259 hours. See table 2. After the runs were completed, using openings in the reactor, samples were taken from the upper and lower layers to analyse drying efficiency. The weight of each sample was approximately 1 kg and each sample was tested for moisture content, volatile matter content and calorific value.

Table 2   Reactor conditions

Run	Waste mass (kg)	Time (h)	Air consump-tion (m³)	Air flow before (l/s)	Air flow after (l/s)
Temperature 1	30.0	257.50	538.707	78	78
Temperature 2	32.9	259.37	408.599	72	72
Oxygen 1	30.0	253.33	76.079	72	71
Oxygen 2	30.0	258.98	119.248	72	70

Moisture content was analysed according to CEN/TS 15414-1 Solid recovered fuels – Determination of moisture content using the oven dry method – Part 1: Determination of total moisture by a reference method.

Volatile matter content, three 500 g subsamples were predried at room temperature for 48 hours and were then ground to 1 mm grain size using a Retsch mill. The percentage of volatile matter in the resulting analytical sample was determined. Three 1 g test sam-ples obtained after the material was annealed in an oven at 850°C in a closed porcelain crucible for 7 minutes were gravimetrically analysed.

Calorific value was analysed according to EN 15400 Solid recovered fuels – Determina-tion of calorific value using calorimeter IKA Werke C2000.

2.5 Energy balance

The energy balance of the process was determined by the following:

$$E = \left(H_{after} \cdot m_{after}\right) - \left(H_{before} \cdot m_{before}\right) - P \cdot t$$

Where:

E – Energy balance (MJ)

H after – LCV after drying (MJ·kg-1)

H before – LCV before drying (MJ·kg-1)

m before – Mass of waste before drying run (kg)

m after – Mass of waste after drying run (kg)

P – Air supply wattage (65 W)

t – Air supply function time (s)

3 Results and discussion

3.1 Removal of water

Moisture contents were in good agreement with the relative air humidity in the individual layers in the two methods. The most significant decline was recorded in the aeration method controlled by the temperature in run no. 1. See table 3. The moisture in the lower layer decreased by 27.54%. See table 3.

The lowest drying efficiency was noted in the method controlled by oxygen. In the top layer, there was even a rise in moisture. This was a result of water condensation on the reactor ceiling. Condensation occurred in all processes due to the low intensity of aeration, but in this case it was particularly prominent (Table 3). These results were consistent with the work of DE GUARDIA ET AL. (2012).

Table 3 - Values of moisture before and after the process

Run	Moisture before (%)	Moisture after upper layer (%)	Moisture after lower layer (%)	Desiccation upper layer (%)	Desiccation lower layer (%)
Temperature 1	41.8 ± 1.88	23.87 ± 3.20	13.74 ± 2.79	17.41	27.54
Temperature 2	43.64 ± 1.16	25.08 ± 0.65	22.87 ± 1.60	18.56	20.77
Oxygen 1	36.88 ± 1.61	37.82 ± 2.12	21.76 ± 0.65	-0.94	15.12
Oxygen 2	38.52 ± 1.26	15.81 ± 0.43	16.09 ± 0.80	22.71	22.43

Mean of three samples, α=95%

3.2   Calorific value

The highest change in calorific value was observed in the process controlled by temperature run no. 1 as a result of intense aeration, low decomposition of organic matter and low moisture in the laboratory. See Tables 4 and 6 and Figure 2.

The lowest calorific value change was observed in the process controlled by oxygen concentration run no. 1. In the top layer there was even a negative change. This was due to, as mentioned above, significant water condensation on the reactor ceiling. A low calorific value change was also observed in the lower layer due to the low lower volume of air passed through (see table 2) and thus, low moisture transport. Similar results were reported by ADANI ET AL. (2002) and SUGNI ET AL. (2005).

Table 4 - Calorific value before and after the process

Run	LCV before (MJ/t)	LCV after upper layer (MJ/t)	LCV after lower layer (MJ/t)	LCV change upper layer (%)	LCV change lower layer (%)
Tempera-ture 1	9,077.12 ± 73.78	13,818.18 ± 344.23	15,415.33 ± 253.22	34.31	41.12
Tempera-ture 2	9,835.66 ± 21.21	13,610.96 ± 772.26	14,817.38 ± 668.71	27.74	33.62
Oxygen 1	12,358.60 ± 219.74	11,314.05 ± 51.51	14,548.64 ± 95.82	-9.23	15.05
Oxygen 2	11,102.23 ± 123.71	14,982.44 ± 197.09	17,001.06 ± 180.99	25.90	34.70

Means of three samples, α=95%

3.3   Energetic balance

The process controlled by temperature run no. 1 had the best energetic balance. The worse energetic balance was noted in the process controlled by temperature run no. 2 because of lower drying efficiency and higher energy consumption. The worst energetic balance was observed in the process controlled by oxygen concentration, where the impact of strong condensation was significant. See Table 5.

Three of the processes yielded positive net energy. This means that despite the energy used for drying, when burned, the dried waste produced more energy that if the waste was incinerated not having been untreated.

Table 5 - Energetic balance of the process

Run	Mass before (kg)	Mass after (kg)	Air supply operational time (s)	Energy consumed (MJ)	LCV before (MJ/kg)	LCV after (MJ/kg)	Energetic balance (MJ)
Temperature 1	30	23.26	439338	28.56	9.077	14.617	39.05
Temperature 2	32.9	26.43	342441	22.26	9.836	14.214	29.82
Oxygen 1	30	27.87	64231	4.18	12.359	12.931	-14.52
Oxygen 2	30	23.23	101854	6.62	11.102	15.992	31.80

3.4   Ambient air relative humidity effect

In the method controlled by temperature, the drying efficiency for runs 1 and 2 was different. For run 1, moisture decreased from 41.28 ± 1.88% to 23.87 ± 3.20% in the upper layer and to 13.74 ± 2.79% in lower layer. In case of run 2, moisture decreased from 43.64 ± 1.16% to 25.08 ± 0.65% in upper layer and to 22.87 ± 1.60% in lower layer. The difference in drying efficiency may be explained by different ambient air relative humidity during the execution of these runs (See Figure 2). Similar results show COLOMER-MENDOZA ET AL. (2012), who state that ambient air humidity, particularly air humidity of the reactor air input, could impact drying efficiency.

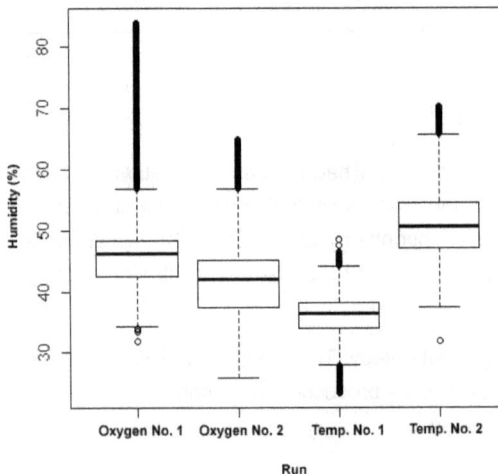

Figure 2   Ambient air relative humidity during different runs

3.5   Gradients

Figures 3, 4, 5 and 6 show a strong temperature gradient between the lower and upper layers of the reactor. This gradient was well described by ADANI ET AL. (2002) and SUGNI ET AL. (2005). They show that the gradient is caused by the cooling of lower part of the reactor by inlet air. The gradient is distinctly noticeable in all runs.

Another gradient described in the literature is the moisture gradient. In Table 3, the different drying efficiency differences between the lower and upper layers of the re-actor can be seen. The same phenomenon was described by ADANI ET AL. (2002), SUGNI ET AL. (2005) and ZHAO ET AL. (2010). This effect can be explained by the fact that in the upper layer of the reactor the air does not absorb moisture from as efficiently as the lower layer because of its higher relative humidity. The data in figures 3, 5 and 6 are in good agreement with this explanation. In the lower layer, a lower relative humidity than in the upper layer was achieved. The exemption is the data in Figure 4, (method controlled by oxygen concentration, run no. 2), which show balanced relative humidity values in the lower and upper layers approximately in the middle of the run. Subse-quently, the relative humidity in upper layer began to decrease. Relatively balanced moisture values displayed in Table 3 are in good agreement with the figure. A possible explanation could be higher temperature in the upper layer maintained for a relatively long time expressed as high average temperature and cumulative temperature in upper layer and high cumulative temperature difference between lower and upper layer (Ta-bles 6, 7 and 8). A lower temperature in the lower layer and a high temperature differ-ence between the layers probably resulted in air that was not so moistened when it passed through the lower layer. In combination with higher temperature in the upper layer, the air had higher capacity for moisture transport, so more moisture was evapo-rated and desiccation of upper layer could have been higher. Higher temperature in the upper layer was caused by higher organic matter content in the dried waste (see table 2) and therefore, in combination with optimal oxygen concentration, intense decomposi-tion of organic matter occurred. Higher MMSW organic matter content in this part of year is in general confirmed by work of DOLEŽALOVÁ ET AL. (2013). Increased decompo-sition confirms decreased volatile organic matter values in the upper layer in the method controlled by oxygen concentration, run no. 2. See Table 6.

Moisture and calorific value gradients are in good agreement in the method controlled by temperature and in the method controlled by oxygen concentration no. 1. The lower moisture was achieved the higher was calorific value. The exception again was the method controlled by oxygen concentration no. 2. Despite the same moisture content in the lower and upper layers, a calorific value gradient appeared again. This can be ex-plained by a decrease of organic matter content expressed as volatile matter change caused by intense decomposition. See Table 6. ZAWADZKA ET AL. (2010) also describe a fade-out of the calorific value gradient while keeping the temperature gradient.

Table 6 - Volatile matter change

Run	Volatile matt. before(%mass)	Volatile matt. after upper layer (%mass)	Volatile matt. after lower layer (%mass)	Vol. matt. change upper layer (%mass)	Vol. matt. change lower layer (%mass)
Temperature 1	63.24 ± 0.80	65.04*	65.44 ± 1.13	1.80	2.20
Temperature 2	69.04 ± 0.45	68.31 ± 0.42	68.25 ± 0.37	-0.73	-0.79
Oxygen 1	63.85 ± 0.42	70.73 ± 0.13	69.75 ± 0.44	6.88	5.90
Oxygen 2	68.64 ± 1.06	66.32 ± 0.80	70.78 ± 0.90	-2.32	2.14

* only one measurement performed, α=95%

Table 7 - Average temperatures

Run	Average temperature upper layer (°C)	Average temperature lower layer (°C)	Average temperature difference (°C)
Temperature 1	41.30	21.28	20.02
Temperature 2	39.46	23.40	16.06
Oxygen 1	49.69	33.51	16.18
Oxygen 2	46.43	27.00	19.44

Table 8 - Cumulative temperatures

Run	Cumulative temperature upper layer (°C·s)	Cumulative temperature lower layer (°C·s)	Cumulative temperature difference (°C·s)
Temperature 1	38 374 228	19 776 645	18 598 009
Temperature 2	36 335 760	21 545 243	14 790 517
Oxygen 1	45 319 331	30 564 476	14 754 856
Oxygen 2	43 289 776	25 169 723	18 120 053

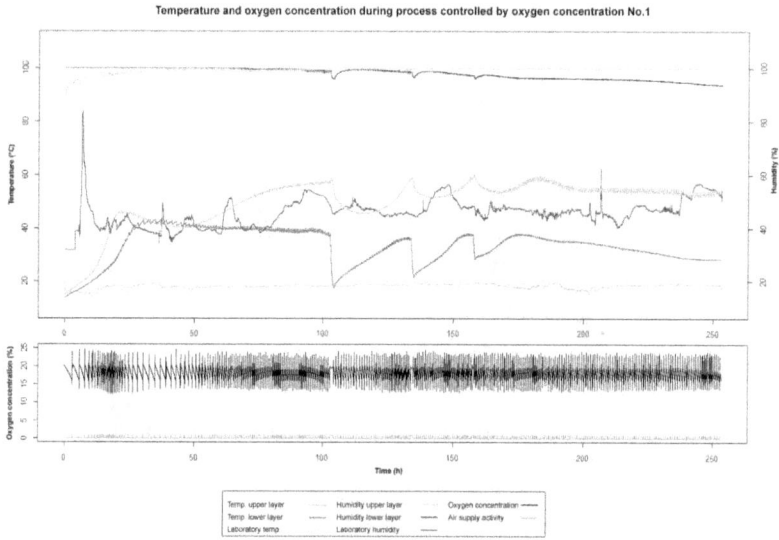

Figure 3   Temperature and O_2 concentration during process controlled by oxygen concentration No. 1

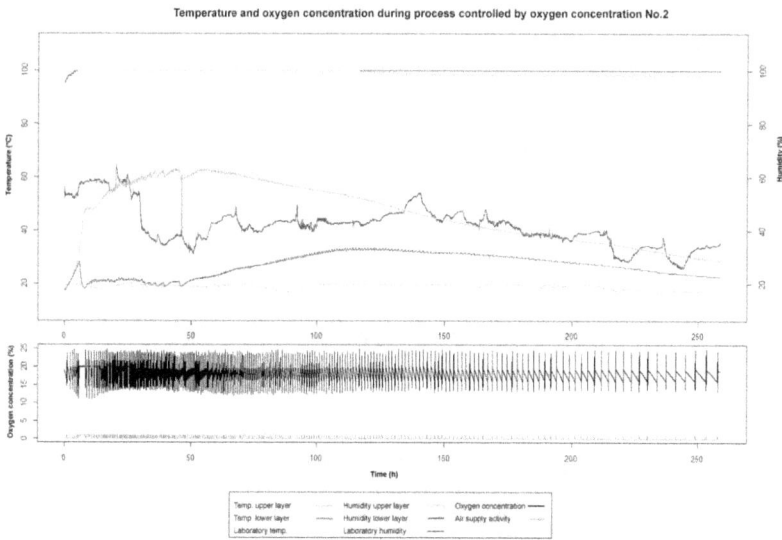

Figure 4   Temp. and O_2 conc. during process controlled by oxygen concentration No. 2

Temperature and oxygen concentration during process controlled by temperature concentration No.1

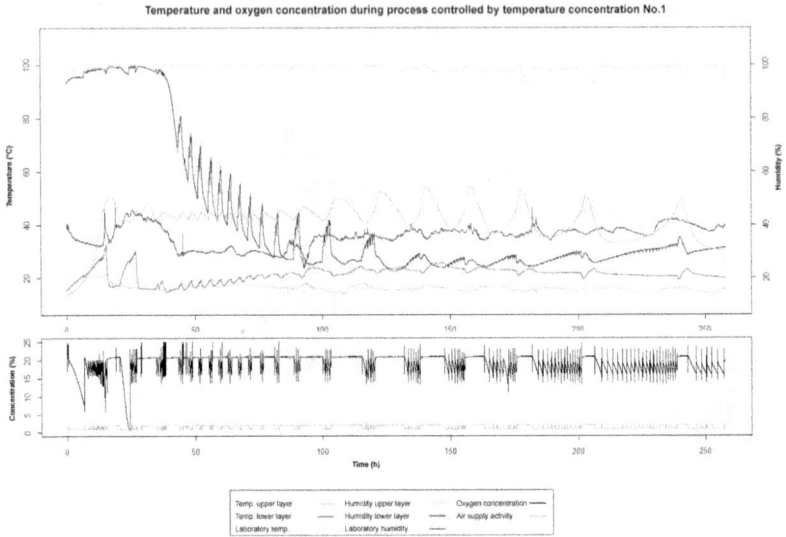

Figure 5   Temp. and O_2 conc. during process controlled by temperature No. 1

Temperature and oxygen concentration during process controlled by temperature concentration No.2

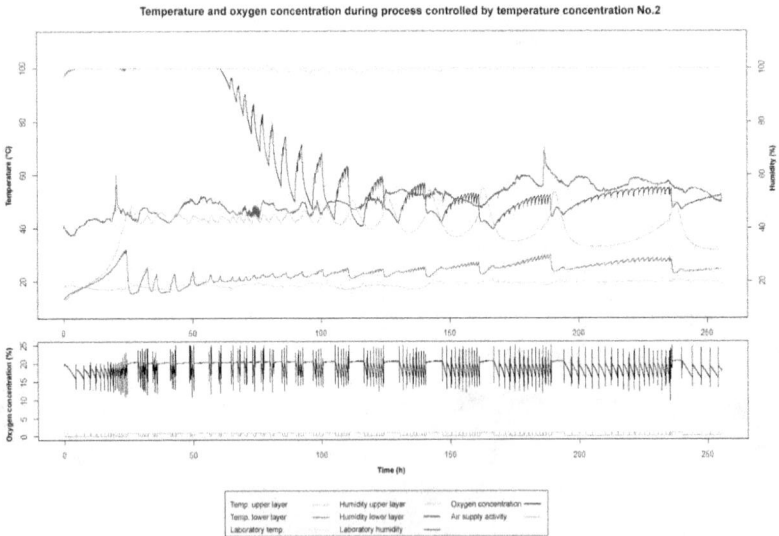

Figure 6   Temp.and O_2 conc. during process controlled by temperature No. 2

4   Conclusions

Two aeration methods were investigated: one method was controlled by oxygen concentration and the other method was controlled by temperature in upper part of the reactor. A better performance in moisture removal and calorific value increase were achieved in the method controlled by temperature. The energetic balance of the biodrying process was positive in both methods.

Temperature gradients, and moisture and calorific values during the runs were similar to those quoted by other researchers in the literature. The only exception was during the method controlled by oxygen concentration, run no. 2. In this case, the gradient moved in the opposite direction. In the upper layer, under the reactor ceiling, there was lower moisture content and relative humidity than in the lower layer. This effect was probably caused by a higher organic matter content in combination with optimal oxygen concentration in the upper layer of the reactor and consequently, a higher intensity of organic matter decomposition and a higher temperature maintained for relatively long time. An important outcome of this study is that higher organic matter content and aeration regime settings could involve sense of moisture gradient.

Drying efficiency differences between runs no. 1 and no. 2 under the method controlled by temperature were observed. These were caused by a difference in ambient air relative humidity. Characteristics of ambient air (mainly air humidity) could thus importantly involve drying efficiency. The impact of this factor was not considered in most of the experiments described in the literature.

5   Literature

Adani, Fabrizio, Diego Baido, Enrico Calcaterra a Pierluigi Genevini   2002   Bioresource technology 2002. The influence of biomass temperature on biostabilization-biodrying of municipal solid waste

Avalos ramirez, A., Godbout, S., Léveillée, F., Zegan, D., Larouche, J.   2012   Chemical Technology 2012, Effect of temperature and air flow rate on carbon and nitrogen compounds changes during the biodrying of swine manure in order to produce combustible biomasses

Cai, L., Chen, T., Gao, D., Zheng, G., Liu, H., Pan, T.   2013   Water Research 2013, Influence of forced air volume on water evaporation during sewage sludge bio-drying.

Cimpan, C., Wenzel, H..   2013   Waste management 2013, Energy implications of mechanical and mechanical-biological treatment compared to direct waste-to-energy.

Colomer-Mendoza, F. J., robles-martinez, F., Herrera-Prats, L., Gallardo-Izquierdo, Bovea, M. D.   2012   Environment, Development and Sustainability 2012, Biodrying as a biological process to diminish moisture in gardening and harvest wastes.

De Guardia, A., Petiot, C., Benoist, J.C., Druilhe, C.   2012   Waste management 2012, Characterization and modelling of the heat transfers in a pilot-scale reactor during composting under forced aeration.

Doležalová, M., Benešová, L., Závodská, A.   2013   Waste management 2013, The changing character of household waste in the Czech Republic between 1999 and 2009 as a function of home heating methods.

Economopoulos, A. P.   2010   Waste management 2010, Technoeconomic aspects of alternative municipal solid wastes treatment methods.

Huiliñir, C., Villegas, M.   2014   Bioresource technology 2014, Biodrying of pulp and paper secondary sludge: Kinetics of volatile solids biodegradation.

Sugni, M., Calcaterra, E., ADANI, F..   2005   Bioresource technology 2005, Biostabilization-biodrying of municipal solid waste by inverting air-flow.

Vandergheynst, J. S., Gossett, J. M., Walker, L. P.   1997   Process biochemistry 1997, High-solids aerobic decomposition: pilot-scale reactor development and experimentation.

Velis, C. A., Longhurst ,P. J., Drew, G. H., Smith, R., Pollard, S. J. T.   2009   Bioresource technology 2009, Biodrying for mechanical-biological treatment of wastes: a review of process science and engineering.

Walker, L.P., Nock, T.D., Gossett, J.M., Vandergheynst, J.S.   1999   Process biochemistry 1999. The role of periodic agitation and water addition in managing moisture limitations during high-solids aerobic decomposition.

Zhao, L, GU, W., HE, P., SHAO, L.   Water Research 2010, Effect of air-flow rate and turning frequency on bio-drying of dewatered sludge.

Zawadzka, A., Liliana Krzystek, L., Ledakowicz, S.   2010   Chemical Papers 2010, Autothermal biodrying of municipal solid waste with high moisture.

Author's address

Mgr. Vojtěch Pilnáček
Institute for Environmental Studies,
Faculty of Science,
Charles University in Prague
Benátská 2

128 01, Prague 2
Czech Republic

Telefon +420 604 324 605
E-Mail Vojtech.Pilnacek@gmail.com

Optimizing secondary fuel combustion in cement production by NIR spectroscopy

Leo Fit, Jost Lemke, Reinhard Teutenberg, Samuel Zühlsdorf

thyssenkrupp Industrial Solutions AG, Beckum, Germany

Carsten Steckert

LLA Instruments GmbH & Co. KG, Berlin

Abstract

The cement industry nowadays uses mostly secondary fuels to supply the necessary heat for the so-called clinker burning process. To allow the utilization of less processed secondary fuels, the prepol® step combustor was developed in order to burn coarse alternative fuels with particle sizes of up to 300 mm. Due to the low level of fuel pre-treatment, such fuels have significant fluctuation of heating value and moisture content. To further improve the throughput of the step combustor, near-infrared-based (NIR) online monitoring was tested in a cement plant in order to gain real-time data of the fuel's heating value and moisture content. An NIR analyser from LLA Instruments GmbH was successfully tested at the cement plant in Lägerdorf, Germany, and was permanently installed after the field test.

Keywords

secondary fuel, cement, near-infrared spectroscopy, prepol SC

1 Introduction

For decades, secondary fuels have played a major role in the cement industry due to the high heat demand of the clinker burning process and the relatively low price of the final product cement. Starting with secondary fuels such as used solvents and waste tyres, the cement industry continuously increased the utilization of secondary fuels and also shifted from rather valuable and scarce alternative fuels to cheaper fuels based on industrial and municipal waste.

While modern cement plants already operate at secondary fuel substitution rates of be-tween 60% and 100% of the total thermal input, these secondary fuel streams need to be thoroughly refined before firing them in the process. For the main burner, usually secondary fuel particle sizes of less than 30 mm, preferably less than 25 mm, are re-quired in order to ensure a sufficient retention time of the particles inside the burner flame and effectively prevent unburnt particles from falling into the material bed.

Modern gas suspension calciners consume about 60% of the total thermal input; they can achieve retention times of 3 s to 8 s. Therefore, the calciner fuel needs less pre-treatment, and the particle size of the fuel can reach up to 80 mm. Nevertheless, this process still needs a two-stage pre-treatment usually combining at least a primary shredding, a screening and air classification process as well as a secondary shredding process.

1.1   Development of the prepol® step combustor

As the cement industry demanded a solution for burning even coarser secondary fuel fractions with particle sizes up to 300 mm at least, thyssenkrupp Industrial Solutions AG developed the so-called prepol® step combustor (WO200602967, 2006). The step combustor is a refractory-lined combustion chamber which is attached to the gas suspension calciner in the clinker burning process.

The step combustor fuel is fed at the rear section of the combustion chamber using feeding screws. Based on the principle of underfeed firing, the fuel is laterally pushed on a so-called table. This concept causes the fuel to pile up in front of the screws, where the created pile effectively protects the screws from intense radiation from the combustion process. The conveyor shaft is water-cooled in order to protect it from overheating during start-up and shutdown of the combustion chamber.

Combustion air for the step combustor is taken from the so-called tertiary air. The tertiary air is taken from the product cooler at approximately 800 °C and used as additional combustion air in the gas suspension calciner. For the operation of the step combustor, a share of the tertiary air is sent to the combustion chamber in order to supply sufficient heat for drying and igniting the secondary fuel, as well as to supply sufficient oxygen for the combustion.

The main part of the step combustor consists of refractory-lined steps which give the combustor its name. After the fuel is pulled over the edge of the feeding table, the fuel is distributed on the first steps of the chamber. The steps are continuously cleaned by integrated nozzles which are supplied with pressurized air. Each air blast throws the material on a step further to the next few steps, at the same time all fine particles are entrained into the combustion air stream and directly taken to the gas suspension calciner. This principle causes a continuous shearing of the fuel; larger fuel lumps can be kept on the steps for up to 20 minutes. This concept therefore allows the efficient combustion of heterogeneous secondary fuels.

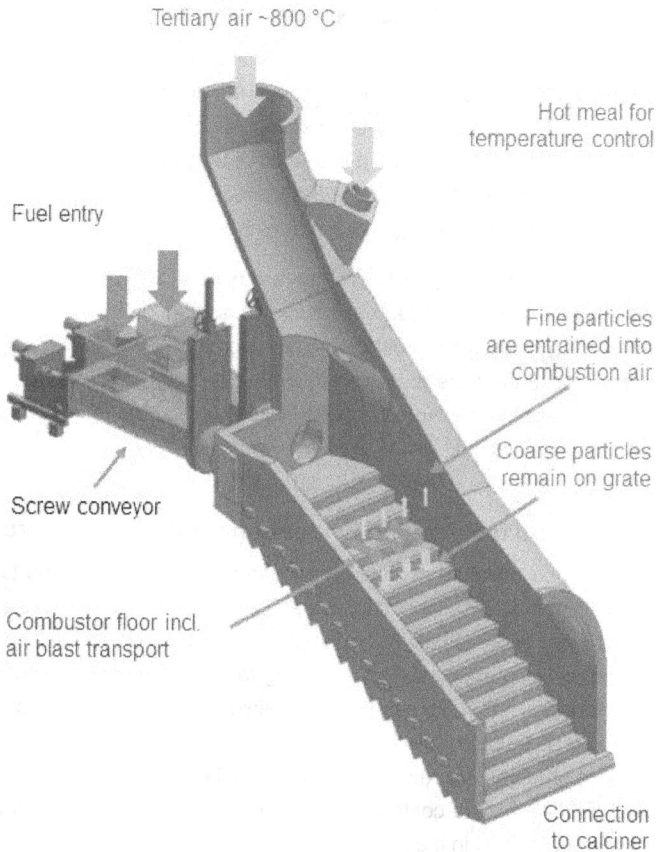

Tertiary air ~800 °C

Hot meal for
temperature control

Fuel entry

Fine particles
are entrained into
combustion air

Coarse particles
remain on grate

Screw conveyor

Combustor floor incl.
air blast transport

Connection
to calciner

Figure 1: Schematic illustration of the prepol® step combustor (prepol® SC) and its working principle

The state-of-the-art dosing systems for coarse alternative fuels usually consist of one or more fuel buffering devices such as walking floor bunkers or truck unloaders equipped with drag or apron conveyors. Subsequently, a screen is used to remove large particles that could cause issues in the fuel feeding system, as well as a metal separator. The dosing itself is then carried out by gravimetric or volumetric dosing, i.e. with belt scales.

2   Optimization of secondary fuel feeding

The step combustor itself has proven to be highly reliable and tolerant to variation of the fuel feeding rate and the fuel composition. However, the overall clinker burning process rather relies on a constant heat input as overheating can cause operational issues such as clogging inside the calciner due to the formation of sticky phases of alkali sulphates at excessively high temperatures. Also, significant product quality issues can arise with "underburnt" clinker if the calciner heat input and temperature are too low.

Intensively pre-treated secondary fuels such as are used in the main burner and conventional gas suspension calciner can already cause issues due to their fluctuations in moisture and heating value. This problem becomes greater if the secondary fuels are treated less, such as is the case for the step combustor fuel. Therefore, a dosing system should not necessarily achieve a constant mass flow feed but rather ensure a constant heat input into the system.

2.1   Monitoring the fuel feeding by near-infrared (NIR) spectroscopy

In order to monitor the moisture content and heating value of the fuel fed to the step combustor, near-infrared (NIR) spectroscopy can be used. The NIR spectra of the fuel particles are recorded and evaluated. By comparing the recorded spectra with calibration libraries, the particles are classified according to their material (i.e. polypropylene, polyvinyl chloride, paper etc.). For materials containing capillary moisture, such as paper and wood, the NIR spectra also supplies information about the moisture content. Consequently, the properly calibrated system can monitor the heating value of the fuel stream as well as the moisture content. Furthermore, the chlorine content is estimated based on the amount of PVC in the fuel.

3   Field test of an NIR system in a cement plant

To test the NIR spectroscopy, the belt scales of two dosing lines in an existing installation in Lägerdorf, Germany were equipped with an NIR analyser from LLA Instruments GmbH. Both dosing lines are designed for the supply of up to 12 tph of secondary fuels. After the screening, the material is distributed evenly on the flat scale belts; therefore this location was selected as being highly suitable for the installation of the NIR sensors. Furthermore, the fuel particles are well mixed on the belt scale; this is important, as the NIR sensor is only capable of analysing the surface of the material bed. Therefore, an even and well-mixed fuel layer is advantageous for recording a representative data set.

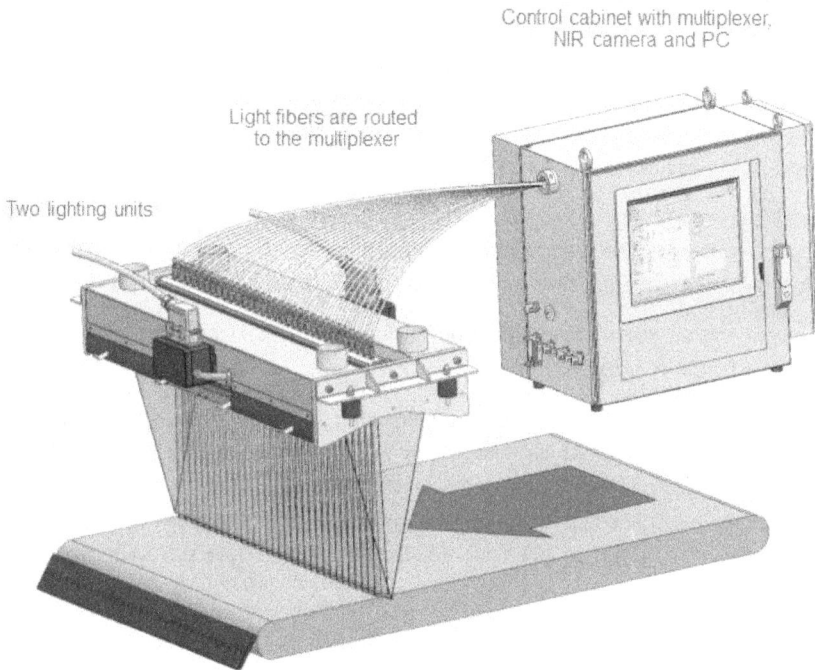

Figure 2: Schematic illustration of the NIR measuring and analysis system

Because the two belt scales are located close to each other, a system consisting of two IR lighting units with fibre optics and a multiplexer unit was used in the field test. For each belt, 32 fibres were distributed over the belt width, with the opening of each fibre covering a circular area with a diameter of approx. 25 mm. In combination with a 75 Hz sampling frequency and a belt speed of 1.5 m/s, a resolution of 20 mm by 37 mm was achieved. This resolution is considered to be sufficient as most particles have sizes between 30 mm and 300 mm.

The recorded NIR spectra were directly computed in the local control unit, and the resulting data (heating value, moisture, chlorine content) was then transferred to the central process control system. The continuous monitoring revealed a mostly constant gross calorific value of the secondary fuel, while moisture content and heating value fluctuated in a range from 10% to 35%, resulting in a variation of the heating value of more than 30%.

4   Conclusion and outlook

As at February 2019, the system has been operated in the field for more than one year. During the operating time, recordings have been stable and reliable. The system showed a high tolerance against the dusty environment, and no outages due to dust or other types of pollution were recorded.

The system contains no wear parts except the light sources in the lighting unit. Due to the high thermal load, the sensor needs to be recalibrated using a standardized reflector from time to time, and defective light sources need be replaced as required.

The cement plant management was very satisfied with the field test and decided to continue using the NIR analyser.

While the main objective has been the monitoring of heat and moisture input to the clinker kiln, the NIR system can also supply information about the chlorine content of the material. This provides further optimization potential, as the chlorine input into the clinker burning process is another crucial process parameter, and the operation of the so-called chlorine bypass could be adjusted based on the measured chlorine input.

5   Literature

WO200602967   2006   Patent publication: Method and device for incinerat-
                                           ing combustible material.

Author's addresses:

Dipl. Ing. Jost Lemke
thyssenkrupp Industrial Solutions AG
Graf-Galen-Str. 17
59269 Beckum, Germany
Phone: +49 2525 99-3915
E-mail: jost.lemke@thyssenkrupp.com

Leo Fit, M.Sc., MBA
thyssenkrupp Industrial Solutions AG
Graf-Galen-Str. 17
59269 Beckum, Germany
Phone: +49 2525 99-2085
E-mail: leo.fit@thyssenkrupp.com

Mould and Dust at Workplaces in Waste Management – State of the Art in 2019

Thomas Missel

Laboratory for Occupational and Environmental Hygiene, Isernhagen / Hanover

Abstract

Mould clinging to feedstock is inevitably released into the air during waste collection and recovery. The workplace atmosphere may be highly polluted with mould during operations. With today's state-of-the-art technology, highly elevated mould contamination should be avoidable in permanent workplaces, such as machine cabs and in cabins, provided that technical precautionary measures are always serviced and maintained properly. However, refuse collectors picking up waste still appear to face high mould contamination time and again. Safe and practical measures to protect refuse collectors against mould do not currently exist during the waste collection phase. Time and again, hygiene issues are found in refuse collection vehicle cabs. Recent inspections have also repeatedly detected significantly elevated mould contamination in break rooms and control rooms used as offices. In isolated cases, these levels were even hygienically unacceptable.

Keywords: Mould, workplace atmosphere, exposure, waste industry, jobs, protective measure, measuring exposure

1 Introduction

The collection and processing of organic waste in the waste industry inevitably release microorganisms into the air. Very high levels of microbial loads can occasionally be detected in work areas. If exposed, employees may, therefore, encounter adverse health effects. As a rule, mould is clearly dominant in bioaerosols generated during waste treatment. Its dominance can be explained by the fact that mould is better suited to substrate with relatively low humidity and can enter the air much more easily than bacteria. For its part, bacteria need more humidity to propagate and prefer to breed based on substrate. Since the 1990s, data on waste industry workers' exposure to mould has been increasingly gathered using standard microbiological sampling and detection methods. During this period, it became clear pretty quickly that people working in businesses that process waste and recyclable materials may still face persistently high mould levels and that there is a significant need to optimise technical and organisational protection equipment in these workplaces. Technical rules on protecting workers against biological agents at work have now been drafted and published in Germany. These rules detail tried-and-tested technical and organisational measures to reduce air pollution in workplaces. For waste treatment plants, this document is TRBA 214 'Waste treatment plants including sorting facilities in the waste industry'. In 2007, TRBA 214 set forth a technical control value (TCV) for reviewing the effectiveness of protective measures in workplaces. The TCV is not a health-based value but was derived from large pools of data measured during workplace assessments. The TCV was set at 5.0×10^4 CFU of mould $/m^3$. This presentation seeks to illustrate the current state of the art for avoiding or minimising exposure based on anticipated exposure data gathered during routine measurements and a comparison with the TCV contained in TRBA 214.

2   Materials and methods

The author of this article gathered the readings contained in this draft over the past 25 years during his expert activities as an auditor and advisor specialising in occupational health and safety and in several publicly funded research projects. Generally speaking, a dust sampling system in accordance with IFA was primarily used to collect mould from the air. The indirect filtration method under procedural instructions contained in IFA work portfolio, code 9420, serves as the basis for identifying mould in air dust samples. Alongside these 'traditional' mould readings, progression curves showing mould concentrations were recorded using the 'correlated particle counts according to Missel' methodology. Sensitive stray light particle counting devices are used to take the required dust measurements. These dust measurement devices measure the number of particles in different particle fractions between 0.3 and a size of 20 µm particle diameter at one-minute intervals.

'Correlated particle counts' are based on the statistic relationship between mould concentrations measured in accordance with IFA and dust particles of mould-relevant particle sizes. This statistical relationship is determined through linear regression of findings of repeated parallel mould and dust measurements, each integrated over a 30 to 60-minute period. Following a statistical data analysis of particle distribution and the elimination of anomalies, mould concentrations are shown in real time in graph form as progression curves. These calculations are based on particle concentrations measured in one-minute intervals and using the growth equation of regression lines as the conversion factor.

3   Assessing exposure to mould

3.1   Worker exposure during waste collection

The findings from workplace measurements taken in the course of waste collection activities prove that, even today, refuse workers working at the loaders of refuse collection vehicles may experience significant mould contamination. Time-weighted averages between 10^4 and 10^5 colony-forming units (CBU) $/m^3$ in the winter and between 10^5 and

10^6 CBU/m^3 in warmer times of the year were measured most frequently. Exposure can manifest itself in the form of lasting mould contamination, even though activities are conducted outdoors. It is not unusual for peak mould contamination to reach 10^7 CBU/m^3. Mould released during tipping can become concentrated and form real dust clouds on relatively short journeys between different container sites. These dust clouds can lie over the refuse collection vehicle and be relatively stable. Significant mould emissions can also enter the cab as a result. Even though it can be expected that the air-conditioning systems' filters should hold back mould spores, more recent studies have also found mould contamination of 10^4 to 10^5 CBU/m^3 in cabs, even if its windows are always kept closed. The findings show that refuse collection vehicles' air-conditioning systems apparently do not offer any really effective protection against bio-aerosols in very highly contaminated areas, such as a composting plant's reception hall.

3.2   Temporary storage and processing halls

Average mould contamination of 10^5 to several 10^6 CBU/m^3 can generally be expected in enclosed halls where waste and recyclable materials are delivered, processed and sorted. This level of contamination occurs largely regardless of the chosen ventilation method (passive and/or active ventilation, air change rates of one or two per hour). Experience shows that – even with more effective air supply and extraction - air contamination caused by microorganisms cannot be reduced to levels in the area of or even below the TCV contained in TRBA 214 (5.0×10^4 CBU/m^3), which is defined for permanent workplaces. This is explained by the limited dust collection capacity of extraction systems, which are generally located in the ceilings of waste treatment plants and thus far away from the source. Another factor is the large number of separate microorganism emission sources with a large total area and force. With current hall extraction systems, released microorganisms have to be spread out over a large area before they can be captured by the extraction system in the first place ($\leftrightarrow$ extraction near the source).

Time-weighted mould contamination can be kept below 1.0×10^5 CBU/m^{3,} even if all hall gates and doors are completely closed, in mechanical processing halls with dust avoidance and collection systems that have been well conceived and implemented carefully and consistently. However, these readings were taken during standardised, stationary sampling. Most of these samples were gathered in areas far away from where maintenance or servicing work had just been performed. In practice, more or less high dust and mould emissions will always inevitably occur when technicians service open machinery and transport systems and when plants are cleaned. Therefore, wearing breathing protection (filter class P2) to prevent excessive mould exposure is advisable for these activities, regardless of the effectiveness of the technology's dust collection systems.

3.3   Mobile machinery in plant halls

The TCV contained in TRBA 214 can normally be met without any problems in the cabs of machinery, such as wheel loaders and grab dredgers. Compliance with the TCV should be ensured, even if machinery is operated in waste storage and processing halls that are known to have high mould and bacteria contamination and if all gates and doors are closed. As a rule, the TCV is only found to be exceeded if the cab is opened in areas where microbial loading is high, although this cannot always be avoided as desired for operational organisation reasons.

In a few cases, problems have emerged because the air exchange rate in the vehicle's cab was not adjusted to reflect the machine's exposure and/or conditions of use (frequency of the door opening in contaminated areas, quality of mould aerosols, see below). The majority of the mould currently found in the workplace atmosphere of cabs definitely does not enter the vehicle through the ventilation system. Instead, it drifts into the cab when the door is opened or is released through excessively dusty interior surfaces. Adverse contamination of the inside of the vehicle may make itself felt in the fact that cab contamination with mould increases dramatically after the vehicle vibrates. Mould entering from outside can be readily proven by contrasting progression curves showing mould concentrations and the climatic environmental parameters.

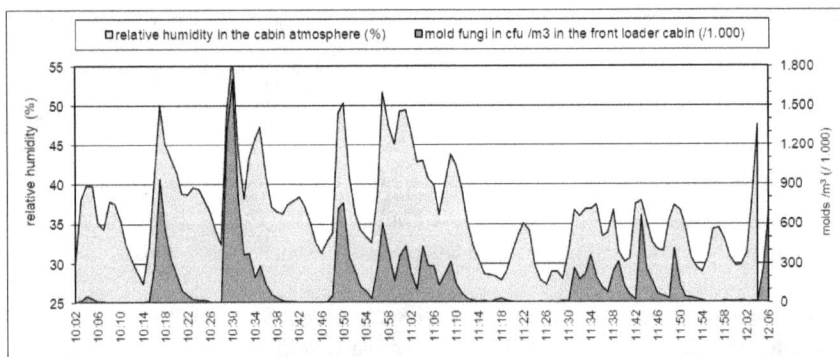

Figure 1   Progression curves showing mould contamination and relative humidity in the cab of a wheel loader. Emissions are only found when the cabin door is opened in a contaminated area (increase in relative humidity).

Under certain conditions, vehicle cabs can have a harder time complying with the TCV, for instance, if very high contamination in the ambient area (absolute) coincides with corresponding bioaerosol quality and work being organised in such a way that the cab has to be opened in these areas. Bioaerosols in enclosed work areas with high levels of Aspergillus fumigatus have proven fairly problematic. A. fumigatus propagates on a massive scale in heated waste and is perfectly positioned to form suspended mould spores that enter the air extremely easily and rapidly disperse over large areas. Large quantities of spores can enter the cab when the cab door is opened in areas contaminated with A. fumigatus. Not all vehicles' ventilation systems are adequately able to remove these spores, meaning that permanently elevated contamination levels can be documented over time and the TCV is not met. An individual risk assessment can thus be recommended for vehicle cabs operating in areas processing self-heated waste where disproportionately high air contamination well in excess of 10^6 CBU/m^3 can be expected because of the technology used.

3.4 Work cabins

Two fundamentally different ventilation principles are used in work cabins (e.g. vehicle cabs, sorting cabins and small control centres). The first is dilution ventilation, in which dust-laden cabin air is constantly mixed with fresh air, bringing about a permanent reduction in ambient air concentrations. The second is displacement ventilation. These systems are designed so that dust-laden air is constantly displaced by laminar fresh air flow and directly discharged outside or sent to exhaust systems. The principle of diluting dust-laden air by introducing fresh air into work cabins cannot prevent microorganism particles released within or drifting into the cabin from being kept away from workers' breathing areas. With this ventilation option, a more or less pronounced accumulation of particles is always detected in the cabin air (plateau in microbial air pollution).

Work cabins equipped with dilution ventilation systems will always have an equilibrium between mould spore emissions on the one hand and particle sedimentation and dilution of aerosols on the other. 'Zero load' can generally not be achieved in the waste management industry using dilution ventilation. By contrast, cabins with more effective displacement ventilation are found to have deeply dissected progression curves for air pollution that are characterised by short-lived, at times high, concentration peaks. However, displacement ventilation can easily make 'zero load' possible – at least when measured stationary using tripods - between individual peak loads.

It can be assumed that the TCV will be met in a control room stand equipped with an air-shower door system with an extraction device if the door opening intervals during operations are not too high. This is true even if the compartment is located in a highly contaminated hall near significant emission sources and only (less efficient) dilution ventilation is in place. However, it is essential that interior surfaces are clean, especially the ground. Significant need for optimisation was often found here at companies.

The TCV can always be reliably met in sorting cabins for manual material extraction, as well, using today's latest technology. The best experiences in terms of supply air were recorded with flat perforated sheet outlet boxes lined with filter mats, here in dimensions of 1,300 x 1,300 mm. These boxes introduce a laminar and stabilised fresh air flow into the cabin – in other words, with the lowest possible turbulence. The quantity of supply air per work station must reach at least around 1,000 m^3 per hour in order to avoid any adverse air turbulence, mainly at transverse collecting belts that are often overloaded with piles of sorting material. Extraction systems like those envisaged in TRBA 214 recommendations on ventilation in sorting cabins have proven superfluous in practice - or even counterproductive in a few cases. An appropriate surplus of supply air can ensure positive pressure in the sorting cabin and effectively counteract undesired dust slip from the hall through belts and discharge chutes and the well-known stack effect. Extraction systems can stand up against the very significant positive pressure mechanism.

Sorting cabins with maximum-efficiency ventilation technology also measure more or less highly fragmented progression curves in air pollution during the course of shifts. These curves also feature brief, sometimes high, peak concentrations. During peak loads, mould concentrations of up to two times the time-weighted average can be achieved during the day in question. Carrying out a measurement-based assessment of a ventilation system's effectiveness is thus especially challenging in sorting cabins with a high level of technical protection.

3.5   Communal rooms

Waste treatment facilities' communal rooms generally have much lower[1] fungi contamination than areas where waste is processed and people come into direct contact with biological agents. Nowadays, almost all waste management businesses have effectively introduced separate clean and contaminated areas. Much more stringent rules apply to hygiene in break rooms than in communal areas, such as changing rooms, where people do not eat, drink, store or prepare food. The obligation to minimise contamination should be adhered to in break rooms. When it comes to air hygiene, break rooms should follow rules defined for permanently used indoor rooms (see Chapter 4). The reference variable for hygiene inspections is current background exposure in natural outside air at the site.

These hygiene stipulations relating to ambient air quality are not always met during inspection measurements in break rooms, and are sometimes even missed by wide mar-

---

[1] Mould loads about 2 to 4 to the power of ten higher can be recorded at workplaces

gins. Spores frequently drift into break rooms through connecting doors leading to entrance areas that already have much higher loads. Contaminated dust carried on workers' clothing bearing microbial contamination has proven to be another key factor in ambient air in break rooms being contaminated with fungi spores, which are detected at times coinciding with breaks. As a rule, though, the source can only be ascertained on a case-by-case basis using a detailed measurement inventory.

Mould emissions that can be rated as 'hygienically unsafe' can certainly be measured time and again in break rooms and in control rooms used and set up like offices. It is not unusual for mould concentrations in the region of 10^4 to 10^5 CBU/m^3 to be recorded in rooms with high employee traffic. Aspergillus fumigatus has turned out to be a significant microbial component in hygiene issues in communal rooms. Spores of this mould species are known to float especially well and spread over a large area (c.f. Chapter 3.3). Aspergillus fumigatus is an infectious microorganism in risk category 2, which should be rated as 'hygienically unacceptable' where it occurs frequently in break rooms.

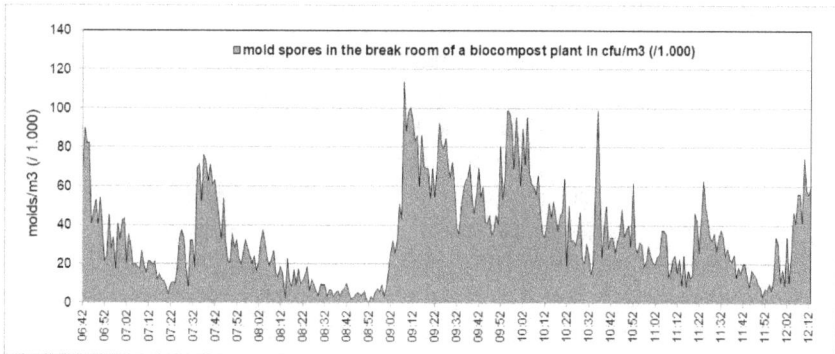

Figure 2   Progression curve showing mould spores in the break room of a composting plant

In principle, there are ways to reduce emissions of spores from more contaminated portions of the site that are directly adjacent to control rooms and/or communal areas through organisational or sealing-off measures and to constantly remove mould entering the building using technical ventilation systems. The relatively slow abatement curves for mould contamination coming down from peak concentration levels prove, though, that ventilation systems in communal areas, which generally use the dilution principle, are often incapable of dissipating mould that has drifted and/or been released into spaces requiring protection in an adequate amount of time (see Chapter 3.4).

4   Analysing exposure to diesel motor emissions and dust

Inhalable dust (I-dust) and alveolar dust (A-dust) levels are limited to 10 mg/m^3 and 1.25 mg/m^3 respectively in Germany under the Technical Rules for Hazardous Substances. Diesel motor emissions (DME), which are classified as carcinogenic, are subject to an obligation to minimise them in accordance with the latest technology. A technical standard concentration (TSC) of 0.1 mg/m^3 of elemental carbon in fine dust applied for DME up until 2004.

DMEs are much less important in workplaces in the waste management industry. Today, the TSC that used to apply should only be reached or exceeded in a small number of exceptional cases. Efficient capture of exhaust pollutants from heavy goods vehicles is clearly making itself felt here. Generally speaking, though, there had been hardly any problems with DME in waste treatment and recycling facilities in the past, either. Higher vehicle traffic generally only occurs in reception areas where significant dust sources are normally found. Coarser dust particles collect DME in the air and form sediment relatively quickly, leading to rapid decreases in concentrations of fine dust that floats easily, which today includes the majority of DME.

Alveolar dust does not represent a major problem in permanent workplaces in the waste management industry, either. As a general rule, finer dust particles in airborne dust dominate in workplaces with ventilation technology, while much coarser particles than fine dust are found in the air in unventilated or not effectively ventilated areas. As a rule of thumb, it can also be said that excessive dust and DME contamination can only be found in the waste management industry in (unprotected) areas where mould contamination is so high that the TCV in TRBA 214 is far exceeded. Respiratory protection already has to be worn in these areas due to high mould contamination (half masks, filter class P2). Potential risks associated with I-dust and, if applicable, A-dust and temporarily elevated DME - for instance in times of higher vehicle traffic and/or in the case of unfavourable ventilation - should be effectively counteracted by requiring breathing protection to be worn.

5   Discussion and recommendations

Nowadays, worker exposure to mould at permanent workplaces in waste management can be easily controlled with state-of-the-art technology. Breathing protection should generally be worn in halls since high mould contamination is unavoidable. Technology to ventilate work cabins has advanced enough for the TCV contained in TRBA 214 to be consistently and reliably met, even in the case of complicated fluid-mechanical parameters like those found in sorting cabins, for instance.

In connection with the TCV, it must be noted that this value is defined for mould alone in the absence of a standardised method of this kind to measure bacteria. The TCV is not an adequate instrument to assess work hygiene in areas with especially high levels of bacteria in the air. When screening compost in biological waste treatment plants, for instance, typical soil bacteria (Actinomycetes) is predominantly found in bioaerosols. Bacteria concentrations exceed mould levels in the air by 10 times or more.

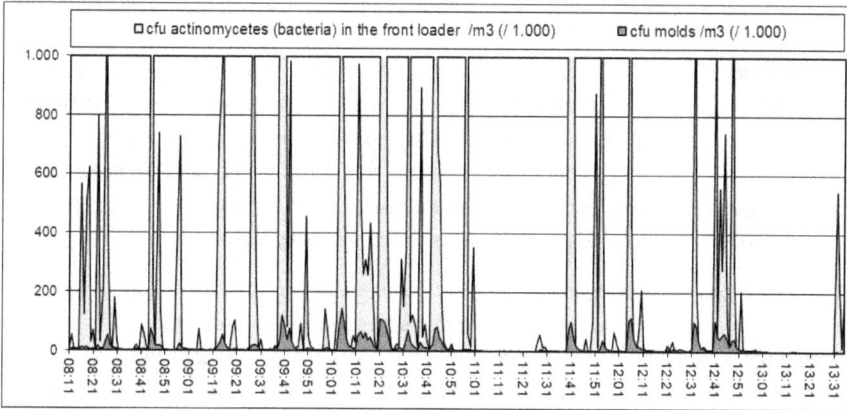

Figure 3   Bacteria and mould contamination in the cab of an older wheel loader without protective ventilation system during compost screening. High bacteria emissions are found when the screening unit is being filled.

Problems associated with elevated worker exposure may occur in halls with exceptionally high levels where mould contamination far exceeds 10^6 CBU/m^3 and/or where especially high levels of Aspergillus mould (A. fumigatus) are detected in the air. Halls where rotting organic waste is shredded and prepared for composting, for example, can generate high emissions, even if cab doors are opened for a short amount of time. A risk assessment can, therefore, be recommended for these businesses.

Refuse collectors being exposed during waste collection remains a problematic issue to which there appears to be no practical technical solution for the time being. Undesirably high mould contamination can also be measured in communal areas if these spaces are not located in completely separate buildings and are only separated from highly contaminated halls by a few corridors. Corridors only separated by a connecting door from a hall, such as a compost reception facility, must always be expected to have contamination of 10^5 mould per m^3. Dispersion of mould into neighbouring communal areas appears unavoidable. Ventilation systems located in communal areas are not able to rapidly break down emissions, meaning that significant increases in mould contamination are recorded during high-traffic break times. One significant factor repeatedly found to influence indoor hygiene in these areas has proven to be that mould is carried into break rooms on work clothing.

A technical recommended value for maximum permissible microorganism concentrations in keeping with the TCV has not been established for communal areas. The TCV only applies to workplaces with ventilation technology where contact with biological agents can occur because of the processes used. Much stricter requirements apply to hygiene in break rooms. The obligation to minimise levels must be respected in rooms where workers eat, drink and store food.

There is no analogue TCV or a 'target' for break rooms and other communal areas, such as changing rooms or office-like control rooms used for this purpose. In these areas, businesses should work towards the rules defined for indoor areas. In living spaces, both maximum permitted mould concentrations and rules for the detailed mould spectrum are geared towards current background exposure in outdoor air. The target reference value for indoor mould measurements is consequently highly variable.

The Laboratory for Occupational and Environmental Hygiene has defined the 'target value' for this kind of reporting as being that spore levels in a break room's ambient air should not exceed the (expected) annual mean local outdoor air concentration. Annual mean mould concentrations in outdoor air of 1.0×10^3 and 2.0×10^3 CBU/m^3 can be expected in rural areas, for instances. In control rooms used for office-like purposes and in uncontaminated areas, the ambient air should not have a spore content that is higher than the power of ten above the expected annual mean mould concentrations in local outside air.

Today, these 'target reference values' can often not be met in communal areas and control rooms.

Measuring progression curves showing mould concentrations using correlated particle counting has proven to be an optimal instrument for analysing technical precautions. One example is listed as an inspection of a wheel loader in a composting plant when 'traditional' mould measurements (cultivation) showed readings much higher than 5.0×10^4 CBU/m^3, thus exceeding the TCV. Based on progression curves showing trends in mould concentrations, however, it was revealed that peak concentrations only occur for a short time in the cabin and that cabin ventilation effectively counteracts a contamination plateau from forming during the course of the shift. Contamination spikes in the cab coincided with the driver's door being opened in contaminated work areas (Figure 1). The ventilation system was rated "very effective" based on ventilation parameters. The measures required to reduce mould emissions because the TCV was significantly exceeded in the cab were merely organisational in nature.

6   Literature

BIA (Hrsg.)	2002	Nr. 9420: Verfahren zur Bestimmung der Schimmelpilz-konzentration in der Luft am Arbeitsplatz. (Kennzahl 9420). In: BIA-Arbeitsmappe - Messung von Gefahrstoffen Bielefeld: Erich Schmidt Verlag. Ausgabe 2002, 18. Lfg. 4/97
Bekanntmachung des BMA	2007	Ausgabe September 2013: Die TRBA 214. Abfallbehandlungsanlagen einschließlich Sortieranlagen in der Abfallwirtschaft. Bundesarbeitsblatt
Felten, C., Küppers, M., Lösing, R., Missel, T. und Willer, E.	2004	Schutzwirkung partikelfiltrierender Atemschutzgeräte gegen Mikroorganismen – ein Feldversuch in der Abfallwirtschaft. Ergo-Med 3/2004, S. 70-76
Felten, C., Albrecht, A., Missel, T. und Willer, E.	2006	Schimmelpilzkonzentrationen an Arbeitsplätzen in Kompostierungsanlagen im Vergleich zum technischen Kontrollwert der TRBA 211. Schriftenreihe der Bundesanstalt für Arbeitsschutz und Arbeitsmedizin, FB 1081, 2007, ISBN-10: 3-86509-593-3, ISSN 1433-2086
Missel, T. und Felten, C.	2006	Wirksamkeitsüberprüfung Technischer Schutzmaßnahmen in der Abfallwirtschaft mit der Korrelierten Partikelzählung. Ergo-Med Nr. 3 06/2006, S. 84-89
Missel, T. und Hartung, J.	2002	Die Korrelierte Partikelzählung als indirektes Messverfahren für luftgetragene Mikroorganismen an Arbeitsplätzen. In: Tagungsband der Frühjahrstagung der Bundesanstalt für Arbeitsschutz und Arbeitsmedizin, Dortmund / Berlin 06/2002 Tb130, ISBN 3-89701-921-3, ISSN 1433-2132
Missel, T. und Hartung, J.	2005	Partikelzählung zur Erfassung von Schimmelpilzen in der Arbeitsplatzatmosphäre. (Projekt Fb 1043). Schriftenreihe der Bundesanstalt für Arbeitsschutz und Arbeitsmedizin – Forschung 2005, ISSN 1433-2086, ISBN 3-86509-298-5

Author's address:

Dr Thomas Missel
Labor für Arbeits- und Umwelthygiene
Tiefe Trift 6
D-30916 Isernhagen
Germany

Tel. +49 5139 9824 501
E-Mail:
info@schimmelpilz-messungen.de

Food Waste Statistics in Slovenia

Tanja Vidic, Master of Sanitary Engineering

Statistical Office of the Republic of Slovenia, Ljubljana, Slovenia

Abstract

SURS set up a working definition of food waste and established the methodology for monitoring the data on food waste generation and treatment at the national level. As a data source already edited final data from regular annual statistical surveys on waste in Slovenia were used. In addition to regular surveys, a special ad-hoc questionnaire for collecting additional information (method of food waste collection, method of recording and reporting of food waste, estimation of the share of edible and inedible part of food waste, estimation of the share of food waste among biological waste and mixed municipal waste) from public waste collection services was used. High-quality and comparable data on food waste are seen as crucial for the proper future measurement and establishment of appropriate food waste prevention and reduction policies and programs.

Keywords *food waste, statistics, generation, source, treatment*

1 Introduction

The Statistical Office of the Republic of Slovenia (SURS) set up a working definition of food waste and established the methodology for monitoring the data on food waste generation and treatment at the national level. Within the first pilot project SURS, based on available statistical data on waste, calculated the amount of food waste by source of food waste along with food waste treatment by type of treatment for 2013–2015. With a purpose to upgrade the methodology on food waste statistics at the national level, especially in terms of improving the quality of data on the share of food waste collected among mixed fractions of waste (among biological and mixed municipal waste) and data concerning the share of edible and inedible parts of food waste, SURS applied for the second pilot project. The upgraded methodology is the basis for calculating the amount of food waste (generation and treatment) at the national level in the following years.

1.1 Definition of food waste

In cooperation with members of the SURS working group (SURS-WG) in the first step of pilot projects the definition of food waste was set up. SURS invited several stakeholders (governmental and non-governmental organizations, institutions, companies, etc.) that are directly or indirectly connected with food waste generation and treatment to participate in the working group to help establish and then also upgrade and improve national methodology for monitoring food waste generation and treatment.

The definition is based on existing definitions of food (defined in Regulation (EC) No. 178/2002) and of waste (in accordance with the Environment Protection Act (OJ RS, No. 39/2006 with all amendments). Based on those definitions, within the national methodology for monitoring the amount of food waste, the following definition of food waste was determined:

"Food waste includes raw or processed food and remains of this food lost before, during or after food preparation or during food consumption, including food discarded during production, distribution, sale and implementation of food-related services and in households. Food waste does not cover:

- *Remains of food intended for processing into animal fodder in line with EU regulations*
- *Food for humanitarian purposes*
- *Paper tissues, napkins and towels collected as kitchen waste together with biological waste*
- *Packaging discarded together with food waste*

Food waste covers the edible and inedible part of individual foodstuffs. The edible part of an individual foodstuff is that part of the foodstuff that could at some point, under normal circumstances, be used for human consumption, but was due to various reasons (e.g. expired "use by" date, too large meals, inappropriate storage, etc.) discarded during production, distribution or sale or during food preparation or consumption. The inedible part of an individual foodstuff is that part of the foodstuff that is not suitable for human consumption or that, under normal circumstances, is not deemed suitable for human consumption but is generated as waste during production, distribution or sale or during food preparation or consumption. Such waste is for example peelings, bones, pits, eggshells, etc."

In accordance with the methodology developed at the national level all food waste collected by public services and special collectors of biological waste (i.e. all food waste that ended in the waste treatment system) was included in the final amount of generated food waste. On the other hand, all food that was home composted, used as animal feed or left on farmland (e.g. on fields and gardens) as a surplus or because of insufficient quality was excluded; only the amounts of food waste that end up in the waste management system are included in the calculations.

2 Methodological approach

Proceeding from the definition of food waste, the list of waste codes under which food waste is collected, recorded and reported, was prepared. Waste codes were selected from the List of Waste (LoW) as defined in Commission Decision of 18 December 2014 amending Decision 2000/532/EC on the list of waste pursuant to Directive 2008/98/EC of the European Parliament and of the Council. Some other sources considered (while defining the waste codes under which food waste is collected) are:

- Data acquired with a special (one-time) questionnaire for collecting additional information regarding recording and reporting of food waste by public services;
- Data acquired with a special (one-time) questionnaire for collecting additional information regarding recording and reporting of food waste by food selling companies (carried out only within the scope of the first pilot project);
- List of waste codes that could contain food waste defined by Eurostat in 2014;

Food waste in Slovenia is collected by collectors of biological waste (special collectors) and public services. Special collectors collect food waste separately from other waste mostly from business entities dealing with food production, distribution, processing and serving. Public services collect food waste as municipal waste within the system of public waste removal; hence food waste is mostly collected in mixed fractions as a part of biological waste. Even with separate collection of biological waste in Slovenia, part of food waste still ends up in mixed municipal waste. For this reason the amount of waste reported under codes 20 02 01 (Biodegradable waste), 20 01 08 (Biodegradable kitchen and canteen waste) and 20 03 01 (Mixed municipal waste) collected by public services within public waste collection was not fully included in the final amount of food waste. In this case rather only a certain share was taken into account, defined with the analysis of data obtained with mentioned special questionnaire.

2.1 Data sources

For calculating the amounts of food waste at the national level, already edited final data from **regular annual statistical surveys on waste generation and treatment** in Slovenia were used. In addition to regular surveys an **additional data source – a special ad-hoc questionnaire for collecting additional information from public waste collection services** (public services) – was used to calculate the data on food waste.

2.1.1 Regular annual statistical surveys on waste generation and treatment

Within regular annual surveys on waste generation and treatment SURS obtains data from the administrative source – Slovenian Environment Agency (ARSO) – which is the institution responsible for collecting data on waste generation and treatment in Slovenia within the IS-Odpadki information system. Data on generated waste (ODP-generation survey), data on collected waste (ODP-collection survey) and data on waste treatment (ODP-recovery/disposal survey), collected by ARSO are then taken over and statistically processed by SURS.

The ODP-collection survey is the main data source for food waste amounts collected within municipal waste by public services and also by special waste collectors. Public services collect waste mostly from households and in small amounts also from production and service activities (when that waste is left to the system of public waste collection). Special waste collectors collect food waste (coded mostly as 20 01 08 Biodegradable kitchen and canteen waste, 20 01 25 Edible oil and fat, 20 03 02 Waste from markets) only from production and service activities, separately from other types of waste.

The ODP-generation survey is the main data source on the amounts of food waste generated in production and service activities. These amounts of food waste are mostly collected separately by special waste collectors (collectors of biological waste), and recorded and reported under the following LoW waste codes:

- 02 02 02 Animal-tissue waste
- 02 02 03 Materials unsuitable for consumption or processing
- 02 03 04 Materials unsuitable for consumption or processing
- 02 05 01 Materials unsuitable for consumption or processing
- 02 06 01 Materials unsuitable for consumption or processing
- 02 07 04 Materials unsuitable for consumption or processing
- 16 03 06 Organic wastes other than those mentioned in 16 03 05
- 19 08 09 Grease and oil mixture from oil/water separation containing only edible oil and fats

The data source for food waste amounts that are separately collected from production and service activities by special waste collector was the ODP-generation survey because of more reliable and accurate data regarding generated amounts of food waste and also regarding the activities of origin and adequacy of the use of the waste code can be gathered directly from original waste producers.

Data on food waste treatment are obtained from the ODP-recovery/disposal survey, where food waste treatment is defined and divided by individual processes.

2.1.2 Additional data source - questionnaire for public services

In cooperation with the Chamber of Local Public Economy (ZKG) SURS conducted the ad-hoc survey with a special questionnaire for public services to get more reliable information on food waste collected within the public waste removal scheme. With questionnaire we obtained additional information on:

- The method of food waste collection by public services;
- Waste codes used to record and report food waste collected in the process of public waste collection from households and also from production and service activities;

- Estimation of food waste share among biological waste and mixed municipal waste collected in the process of public waste collection;
- Estimation of edible and inedible part of food waste collected in the process of public waste collection.

Data obtained with questionnaire for public services showed that public services report food waste mostly under the following waste codes:

- 20 02 01 Biodegradable waste
- 20 01 08 Biodegradable kitchen and canteen waste
- 20 01 25 Edible oil and fat
- 20 03 01 Mixed municipal waste
- 20 03 02 Waste from markets

However, within some of these waste codes not just food waste, but also other biological waste, such as green cut from gardens and parks is reported. With the aforementioned special questionnaire for public services also information about the actual share of food waste within the used LoW codes was obtained.

2.2 Methodological approaches to quantify food waste within mixed fractions of waste

Within the first phase of data analysis the share of food waste within mixed fractions of waste was estimated.

Public waste collectors within the municipal waste collection system mostly collect food waste within mixed fractions of waste. The amount of food waste collected together with other biological waste represents a problem. This amount is mostly recorded and reported under LoW codes 20 02 01 and 20 01 08. The share of food waste within the mentioned LoW codes was estimated on the basis of data obtained with special questionnaires for public services. A certain share of food waste from households and also in smaller part from production and service activities still ends within mixed municipal waste (20 03 01). The share of food waste within mixed municipal waste was also estimated on the basis of information obtained with special questionnaires for public services. The estimated shares of food waste within mixed fractions of waste are not a result of sorting analyses, but are only estimates based on the observations of public services.

Calculated shares (for Slovenia, 2016) of food waste within:

- Biological waste, reported under waste codes 20 02 01 was 35%
- Biological waste, reported under waste codes 20 01 08 was 63%
- Mixed municipal waste (waste code 20 03 01) was 10%

Food waste recorded and reported within LoW group 02, LoW codes 16 03 06 and 19 08 09 is collected separately and is pure food waste. The analysis included only amounts that were generated within NACE Rev. 2 activities that are in some way connected with food and consequently with food waste. Pure food waste is also collected by special waste collectors (collectors of biological waste), which recorded and reported this waste under LoW codes 20 01 08, 20 01 25 and 20 03 02.

Waste code 16 03 06 is also used to report other types of organic waste that does not fall within the scope of food waste. Based on the data obtained with the ODP-generation survey the precise data on the amounts of waste under LoW code 16 03 06 from the activities of food distribution and trade, which fall within the scope of food waste, was calculated.

2.3 Methodological approach for defining food waste generation by source

For distribution of food waste generation by souce, data from the ODP-generation survey were used. In defining food waste generation by source four main groups were defined:

1. Food production (incl. primary)
In this activity the majority of food waste is generated by the process of food production. This activity includes section 10 (Manufacture of food products) and section 11 (Manufacture of beverages) by NACE Rev. 2. Within this activity waste producers record and report generated food waste mostly under the following LoW waste codes: 02 02 02, 02 02 03, 02 03 04, 02 05 01, 02 06 01, 02 07 04, 19 08 09. A small part of food waste from production activities is recorded and reported under codes of waste falling into the framework of the municipal waste (groups 15 01 and 20 by LoW). Under LoW codes which are used by food production activities only food waste is covered. All possible incorrect uses of waste codes for recording and reporting of food waste from production activities were corrected already in the data analysis phase.

2. Food distribution and food trade
This activity includes mostly food that is wasted because the "use by" date expired or because the quality of food changed during sale. Food waste generated during distribution and in food stores (group 46 Wholesale trade, except of motor vehicles and motorcycles and group 47 Retail trade, except of motor vehicles and motorcycles by NACE Rev. 2) is recorded and reported mostly under code 16 03 06 and is regularly collected separately by special waste collectors. The main data source for data on food waste from food distribution and food trade is the ODP-generation survey. A small part of food

waste from food distribution and food trade is also collected as part of municipal waste by public services and by special waste collectors.

3. Restaurants and food service activities

In addition to the generation of food waste in the accommodation and food service activities (activity I Accommodation and food service activities by NACE Rev. 2), the scope was also extended to the generation of food waste in other institutions where food is served, such as schools, kindergartens, hospitals, nursing homes, etc. Within these activities food waste is mostly recorded and reported under codes 20 01 08 and 20 01 25 and separately collected by special waste collectors. For this reason, data on food waste generation in this activity were obtained from the ODP-generation survey and from the ODP-collection survey. A small part of food waste from restaurants and food service activities is also collected by public services. Data on food waste that was recorded and reported as municipal waste was obtained from the ODP-collection survey; its allocation by activities was prepared with the help of ODP-generation survey. Food waste that was recorded under other LoW codes, which are not municipal waste, was obtained from the ODP-generation survey, where waste generators report their data.

4. Households

Within households food waste collected by public services within the public waste removal scheme is covered. According to the information obtained with the special questionnaire, public services recorded and reported food waste under the following waste codes: 20 01 08, 20 01 25, 20 02 01, 20 03 01, 20 03 02.

Most food waste collected within the public waste removal scheme is collected together with other types of waste. Public services use different waste codes for recording and reporting food waste; most of them use waste code 20 02 01. Despite the separate collection of biological waste in the territory of the Republic of Slovenia, some food waste still ends up in mixed municipal waste bins.

2.4 Methodology for estimating edible and inedible part of food waste

According to the national definition food waste is divided into the edible part, which could be reduced with proper awareness and attitude towards food, and the inedible part. The inedible part of food waste, which generally cannot be reduced, includes fruit peelings, egg and other shells, bones, citrus pits, etc.

The estimation of the share of edible and inedible parts of food waste collected by public services (from households and in small part also from production and service activities) was estimated by public services. All data were acquired with the special questionnaire for public services in which the share of edible and inedible part of food waste under specified food waste codes was estimated. The share was estimated based on experiences and on-site observations of public services, and not on sorting analyses, since those are not conducted for biological waste.

3 Results and discussion

Data on food waste obtained with regular annual surveys on waste statistics were in the analysis merged with data on estimated shares of food waste (within mixed fractions of waste, estimation of edible and inedible part of food waste, etc.) obtained from the additional data source - questionnaire for public services. The analysis that was fully conducted by SURS, using the pivot table within MS Office Excel programme, was prepared on data for the reference period 2013–2017. Data on food waste in Slovenia for 2013–2017 were published for the first time on SURS's website (https://www.stat.si/StatWeb/en/News/Index/7826) and became a regular part of the first release of publication of data on waste statistics.

3.1 Food waste generation

According to the developed methodology, in 2013–2016 in Slovenia the amount of food waste increased, while in 2017 it slightly decreased compared to the previous year. In 2017, a resident of Slovenia discarded on average 64 kg of food, 4% less than in 2016, when they discarded on average 67 kg of food. This is however still 11% more than in 2013, when they discarded on average 57 kg of food.

Table 1   Food waste generation, Slovenia

	2013	2014	2015	2016	2017
TOTAL FOOD WASTE GENERATED (kg per capita)	57	61	65	67	64
TOTAL FOOD WASTE GENERATED (tons) ...	118,450	125,102	133,898	137,638	131,761
... in food production (incl. primary food production) (tons) [1]	7,950	9,516	10,001	10,726	10,485
... in distribution and food stores (tons)	9,165	9,478	12,933	14,492	13,115
... in food services (tons) [2]	38,313	41,348	44,824	43,899	40,568
... in households (tons) [3]	63,023	64,761	66,141	68,521	67,594
EDIBLE PART (%)	36	37	40	38	38
INEDIBLE PART (%)	64	63	60	62	62

Source: SURS

[1] Residues of organic origin derived from food production activities which are diverted to the production of animal feed are not included in the amounts of food waste.

[2] Includes food waste that is generated in restaurants and other institutions serving food (such as schools, kindergartens, hospitals, nursing homes, etc.).

[3] It does not include food waste composted by households on home composting system or disposed of in a sewage system.

Half of the total amount of food waste is generated in households. In 2017, households in Slovenia generated almost 67,600 tons of food waste; this amount was just over 1% lower than in 2016 and just over 7% higher than in 2013.

A third of the food waste originates from catering and other food-serving activities, e.g. in schools, kindergartens, hospitals, nursing homes. In these activities, less than 40,600 tons of food waste was generated in 2017, which is almost 8% less than in 2016 and almost 6% more than in 2013.

Slightly more than a tenth of food waste is generated in distribution and food stores due to transport damages, improper storage, expired "use by" date, etc. In 2017, about 13,100 tons of food waste was generated in this activity, or almost 10% less than in 2016, when around 14,500 tons of food waste was generated.

A little less than a tenth of the food waste originates from food production (including primary food production). In 2017, this activity generated almost 10,500 tons of food waste, which is around 2% less than in 2016, but at the same time almost 32% more than in 2013. The residues of organic origin that are deriving from the activity of food production and are diverted to animal feed production do not belong to food waste.

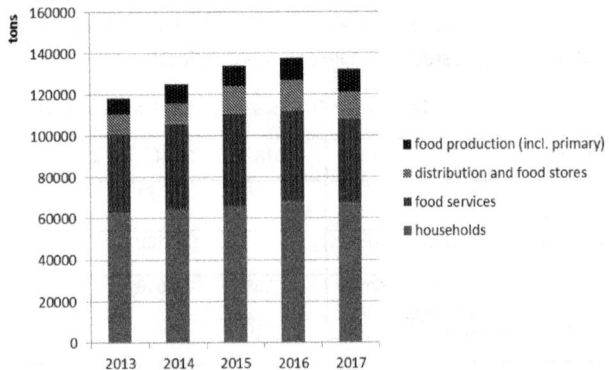

Figure 1 Food waste generation, Slovenia (SURS)

Out of 131,800 tons of food waste generated in Slovenia in 2017, it was estimated that 38% was the edible part, which could be reduced or avoided by raising awareness and proper attitude towards food, and 62% was the inedible part, e.g. bones, peels, egg shells, shells, hulls, etc., which normally cannot be avoided.

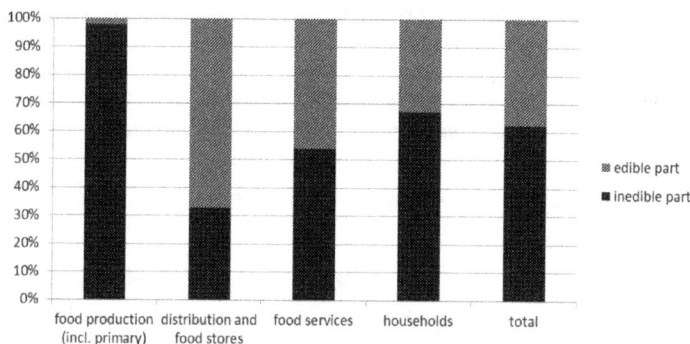

Figure 2 Estimation of edible and inedible parts of food waste by food waste origin, Slovenia, 2017 (SURS)

For food waste that was generated in food production activities it was assumed that mostly inedible part of food waste is generated, since normally all edible parts are processed and used. The share of edible part of food waste within the activity of food production is very small (2%). In food distribution and in food trade two-thirds (67%) of edible part and one-third (33%) of inedible part of food waste is generated. In accommodation and food service activities (in kindergartens, schools, hospitals, nursing homes, etc.) almost half (46%) of food waste is edible part and more than half (54%) is inedible part. We could say that food waste generated in restaurants and other institutions serving food is by its composition very similar to food waste generated in households, but in these activities also a lot of pre-prepared food is used. This is the reason that the share of inedible part of food waste generated in households is higher than in accommodation and food service activities. In households one third (33%) of food waste is edible part and two thirds (67%) is inedible part.

We would like to remark that the calculations of the shares of edible and inedible parts of food waste are only rough estimates and not a result of any kind of standardized measurement (e.g. sorting analysis). Division of food waste into edible and inedible parts is particularly demanding in activities of food production, food distribution and food trade.

3.2   Food waste treatment

Food waste that ends up in the waste management system is mostly processed in biogas plants, followed by processing in composting plants. The share of food waste processed in biogas plants increased during 2013–2016 (from 35% in 2013 to 48% in 2016), and it slightly decreased in 2017 (to 46%). The share of food waste processed in composting plants ranged between 29% and 34% between 2013 and 2017. In 2017, almost 39,600 tons of food waste was processed in composting plants, or 30% of food waste generated in 2017.

In accordance with the current waste legislation, after 2015 all food waste collected together with mixed municipal waste must be biologically stabilized in the plants for mechanical biological treatment of mixed municipal waste (MBT) before disposal. The share of food waste that was biologically stabilized before disposal increased in 2013–2016 and remained at 22% in 2017.

Other food waste treatment includes processes such as co-incineration and incineration, oil refinement, other biological recovery processes and disposal. In 2013–2017, the share of other food waste treatment dropped from 23% to 2%, mainly due to the reduction in the amounts of directly disposed food waste.

Table 2   Food waste treatment, Slovenia

	2013	2014	2015	2016	2017
recovery in biogas plants (tons)	41,616	45,116	52,418	66,336	60,083
recovery in composting systems (tons)	38,957	42,068	45,148	39,805	39,578
biological stabilisation (tons)[1]	10,890	26,444	28,190	29,131	28,976
other treatment (tons)[2]	26,987	11,474	8,143	2,366	3,124

Source: SURS

[1] Includes food waste that was collected as part of mixed waste fractions, and was biologically stabilized prior to disposal in the mechanical and biological treatment plants of municipal waste.
[2] Includes other recovery operations (co-incineration, oil refining, other biological recovery) and other disposal operations (incineration and landfilling).

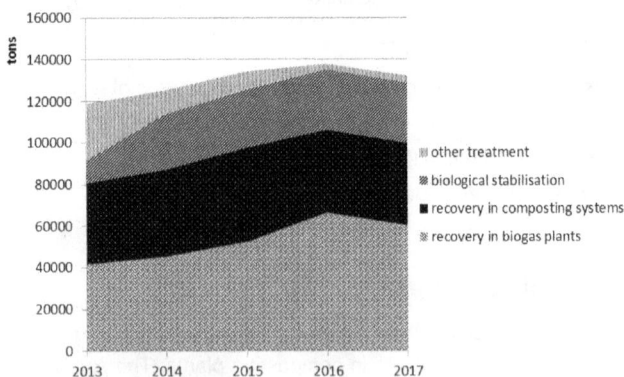

Figure 2 Food waste treatment, Slovenia (SURS)

4 Conclusions

Measurement of food waste is a necessary step towards furthering efforts to reduce food waste. Prevention or reduction of food waste is essential, not only from a moral and social but also from an environmental and economic point of view. Quality and comparable data on food waste generation and treatment are necessary to create an effective policy of food waste prevention, food waste management and evaluation of the effectiveness in achieving the set goals of reducing food waste at the national and international level.

With the establishment of the methodology for monitoring the amount of food waste at the national level and the preparation of data on food waste in Slovenia within pilot projects, SURS presents the facts in the field of food waste at the national level. The established methodology will be used for calculating the amounts of food waste (generated and treated) in the coming years. SURS will continue to search data sources to upgrade and improve the methodology for monitoring food waste generation and treatment at the national level. It will also comply with international requirements when they are set.
The described methodological approach has certain restrictions:

- The share of food waste within biological waste and mixed municipal waste collected within the public waste removal scheme by public services is the result of observations of the public services and not systematic and standardized measurements. It would be necessary to define more precisely the methodological approach of measuring the weight of this part of food waste.

- Division and measurement of edible and inedible parts of food waste by activities. The share of edible and inedible parts of food waste is a result of observations and assumptions not actual measurements, therefore these shares are rough estimates. Particularly difficult is the division of edible and inedible part of food waste from production and service activities, especially in food distribution and food stores. For calculating the shares (edible and inedible or avoidable and unavoidable parts of food waste) EU-wide criteria are needed. In the context of the complexity of dividing and measuring edible and inedible part of food waste, we are also questioning whether it makes any sense to divide food waste into edible and inedible part.

- Use of waste codes for recording and reporting food waste within regular annual reporting. Food waste collection in Slovenia is fragmented and varies from area to area. The same issue applies to use of waste codes for recording and reporting food waste by waste collectors. For these reasons, uniform guidelines at the national and international level for recording and reporting food waste are necessary for all collectors of food waste and also exercising control over the use of appropriate codes for recording and reporting food waste.

5 References

Programme of Statistical Surveys	2018	Annual Programme of Statistical Surveys for 2018 (OJ RS, No. 75/18)
National Statistics Act	2015	National Statistics Act (OJ RS, No. 37/2015, 69/2015)
Vidic, T.; Zitnik, M; Golobic, T	2018	METHODOLOGICAL EXPLANATION - FOOD WASTE, SURS, Ljubljana. Published on: https://www.stat.si/StatWeb/File/DocSysFile/10184/27-178-ME.pdf
Zitnik, M; Vidic, T	2018	Food Waste, Slovenia, 2017, SURS. Published on: https://www.stat.si/StatWeb/en/News/Index/7826

Author's address:

Tanja Vidic, Master of Sanitary Engineering, Advisor,

Statistical Office of the Republic of Slovenia

Environment and Energy Statistics Section

Litostrojska cesta 54, p.p. 3570, 1001 Ljubljana

T: +386 1 234 07 80

E: tanja.vidic@gov.si

Quantitative Survey of Organic Waste Generated In Oman and the Feasibility for Usage in Biogas Production Plants.

Al-Hosni, Suad

Oman Environmental Services Holding Company (be'ah)

Al Azaiba, Muscat, Sultanate of Oman

Abstract

There are three main sources of organic waste generation in Oman, which includes; Commercial & Industrial (C&I) waste, farms and the municipal solid waste (MSW). Around 600 tons of source segregated organic waste received daily in one of the largest landfills that be'ah is currently operating, which is Barka Landfill. In average almost 30% of the total waste received daily is a source segregated biowaste, where food waste alone is considered to be more than 150 tons/day. The green waste could reach 150 tons/day, and the remaining 40% is distributed between poultry waste, slaughterhouse waste, fish waste, waste water sludge and dead animals. As part of be'ah integrated solutions to treat this waste is to implement multiple biogas plants projects for whole of Oman for organic waste diversion. Be'ah is targeting to divert a total of 900 tons/day from whole of Oman and to produce up to 10 MWel by 2023 through development of 10 biogas projects in different scales.

Keywords

Bio-waste, Biogas, Food waste, Commercial & Industrial (C&I), landfill, MSW

1   Introduction

The Oman Environmental Services Holding Company S.A.O.C "be'ah", established in 2007 by a Royal decree, as the Omani government's arm in handling all forms of solid waste in the country. As one of the very first companies in the region assigned to devise and implement innovative sustainable waste management strategies, be'ah has strived towards a vision of conserving the environment and achieving harmony between the economy, society and the environment.

In Oman, the current method for organic waste management is to be disposed in engineered landfills. The disposal of organic waste in the landfills can be considered as a loss of resources in the long term, let alone the adverse environmental impact associated with such practices. The disposal of organic waste at landfills results in production of landfill gases and leachate which cost millions of omani rials which currently a burden on the government scared resources, these costs are to apply gas recovery and leachate treatment systems within the landfills. The food waste itself makes up 27% of total municipal solid waste reaching the landfill, which is considered as a significant quantity that result in intensive investment by the government to construct, operate and to close these landfills at later stage. Diverting the food waste from the landfills is the best solution to avoid any environmental problems, reduce landfilling cost and contribute in adding value to the economy of the country. Therefore, be'ah seeks to achieve its main objectives of diversion of 60% of the solid waste from landfills by 2020 and 80% by 2030 through an introduction of value recovery processes such as waste to energy plants and material recovery projects. The be'ah organic waste management plan incorporates both centralized and decentralized organic waste diversion approaches such as biogas as an alternative source of energy and compost as an economic value product. In order to develop a sound strategy, it is imperative to have a quantitative and qualitative understanding of the organic waste generated from each governorate in Oman.

1.1   Barka Landfill Organic Case Study and Assesment

A field study was conducted on Barka landfill site in 2016 to monitor the weighbridge data and trucks entering the landfill for disposing organic waste. The waste received at Barka landfill is coming from North Al Batinah, South Al Batinah and parts of Muscat governorates.

The total quantities of waste received currently at Barka landfill is about 2100 t/d. The organic fraction is around 30% (source segregated waste) which is exceeded 500-600 t/d. This fractions of organic waste are classified as follow in figure 1:

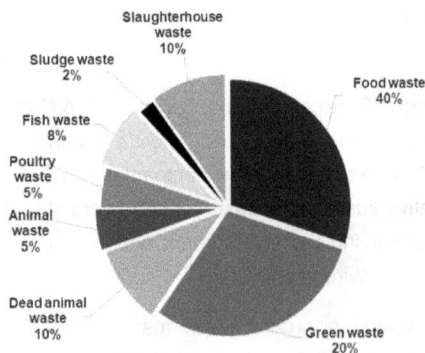

Figure 1   *classifications of organic waste received at Barka landfill*

1.2   Organic Waste Assessment of Oman

The overall result of the organic waste distribution in Oman is reflected in the below map. The darker color representhigh generation of organic waste. Dhofar governorate; is the largest area in Oman for generating of the organic waste, due to existance of significant quantities of green waste, poultry and animal manure and due to the moon-soosn season. Muscat and Al Batinah governorates together generate significant quantities of organic waste, mainly food waste, green waste, slaughterhouse waste, animal manure and many others.

Figure 2   Overall distribution of the organic waste in Oman

This assessment result in identifying the potential site to implement organic waste projects in term of capacity and periority. After identifying the potential areas with highest waste generation, its vulable to conduct a quantitative and qualitative assesmnt for more identification for the different types of organic waste available. This project is help us later in conducting the feasibility study for biogas proects implementation and especially for deciding the excat capacities for different scales biogas plants.

2 A quantitative field survey for three potential governorates of highest production

A quantitative survey was conducted for Muscat, North Al batinah and Siuth Al batinah governorates for indicating the exact quantities of each kind of organic waste generated and the different categories of organic waste to be considered for the purpose of organic waste diversion from engineered landfils at these areas. 1.3.1 Survey objectives

2.1 Survey objectives

This survey aims to determine the total tonnage of organic waste generated from each Governorate in Calendar year 2018, broken down by:

- Develop a sampling methodology to provide representative basis for the survey, that reflects each commercial & industrial activity and thus to consider the following:

  - Type of commercial and industrial activity (fish markets, fruits and vegetables markets, farms, slaughterhouses, hotels, restaurants and Hypermarkets)

  - Material type (food waste, green waste, wood waste, sludge, etc.)

  - Management/disposal method for each type of waste

  - Identify the local trading of organic materials (the rate of production and sales volume) from commercial and industrial activities and to identify the amount of waste generated for each.

- Review the questionnaire developed by be'ah to record the tonnage of organic waste from each commercial & industrial activity, material type and management method.

- Analyse data gathered, with appropriate application of estimators and conversion factors, to produce final report and comprehensive data analysis for each governorate and thus to decide the organic waste diversion mechanism by applying different project of biogas plants implementation and composting.

- Deciding the best strategy for biogas implementation for the whole country with different biogas plants scales and different capacities based on the feedstock availability.

2.2 Methodology

This survey was to target three main governorates in Oman with highest organic waste generation. There are Muscat, North Al Batinah and South Al Batiah (figure 3).

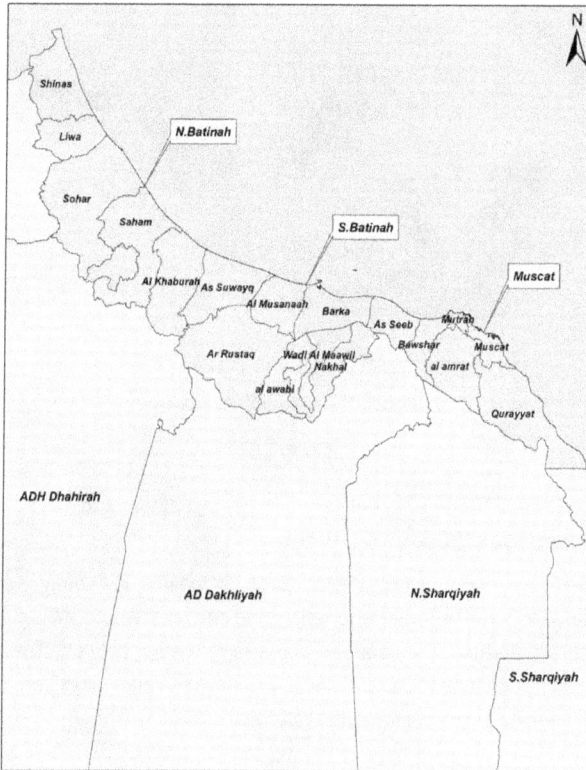

Figure 3   *the targeted governorates covered by the survey (Muscat, North and South Al Batinah)*

The scientific methodology should be applied for this study to insure the quality of data gathering which as per below:

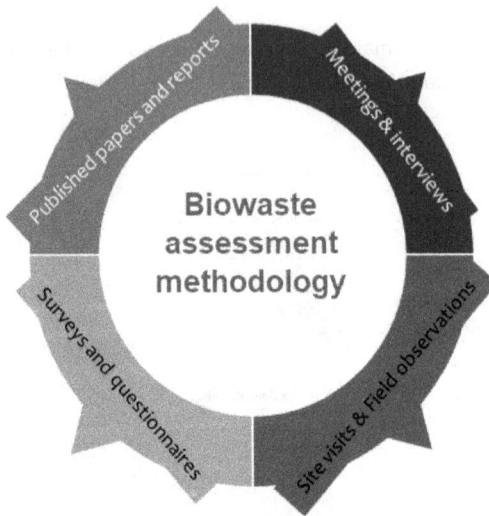

Figure 4   organic assessment methodology

2.3 Targeted categories of organic waste

The survey is to cover the main categories of organic wastes including; agricultural waste (green waste), food waste, fruits and vegetable markets waste, slaughterhouse waste and fish waste generated from commercial and industrial entities. The targeted commercial and industrial entities for organic waste generation was classified as to certain categories with specific elements for survey purpose.

Table 1:   Targeted categories of organic waste and survey elements

Organic waste category	Targeted elements
Food processing industries	• Food production capacity & food waste capacity • The disposal and the waste collection mechanism • Energy consumption (electricity, refrigeration, transportation fuel)
Hotels,Restaurants and Hypermarkets food waste	• Number and type of facilities of food production • Food production capacity & food waste capacity

	• The disposal and the waste collection mechanism • Energy consumption (electricity, refrigeration, transportation fuel) • Any food recycling processes taking place
Farms and agricultural waste	• Farm category (crops, animal, poultry, combined, etc.) • Type of other activities and facilities (animal slaughter facility, poultry slaughter facility, etc.) • production capacity & waste quantities • The disposal and the waste collection mechanism • Energy consumption (electricity, refrigeration, transportation fuel)
Municipal slaughter-houses	• Slaughterhouse category (cattle, poultry, combined) • production capacity & waste quantities • The disposal and the waste collection mechanism • Energy consumption (electricity, refrigeration, transportation fuel)
Municipal fish markets and fruit and vegetab-les markets	• production capacity & waste quantities • The disposal and the waste collection mechanism • Energy consumption (electricity, refrigeration, transportation fuel)

2.4 Sampling size

The minimum number of samples required for each category based on the available information are summarized in the below table.

Table 2:   Minimum number of samples required in each category

	Musca t	Al Bati-nah N	Al Ba-tinah S	Note
Fruits & vege-tables markets	All	All	All	there is one fruit and vegetable central mar-ket in each wilayah or more
Fish Markets	All	All	All	there is one fish central market or more in each wilayah

Slaughterhou-ses	All	All	All	The biggest municipal slaughterhouses i required for this section
crops farms	30	30	30	The biggest crop farms are required (governmental or private)
Animal farms	30	30	30	The biggest animal farms are required (governmental or private)
Poultry farms	15	15	15	The biggest poultry farms are required (governmental or private)
Hotels	15	10	10	5-3 stars hotels are required and any other big hotels/resorts (e.g.; Intercontinental H., Hormuz Grand H., Grand Millennium, Grand Hayat H., City season, Holiday Inn, etc.)
Restaurants	30	10	10	Biggest restaurants with more than one branch is required (e.g.; McDonalds, KFC, etc.)
Hypermarkets & Malls	10	10	10	Biggest hypermarkets with more than one branch are required (e.g. Carrefour, lulu, Mars, city center, avenues, etc.) includes the catering kitchens and restaurants.
Total	~145	~147	~141	

3   Quantitative survey results and discussion

3.1 Organic waste generation in Muscat governorate

As shown in Figure 4, the fruits and vegetables markets waste is the common waste produced in Muscat area which considered to be 72% of the total organic waste produced followed by hypermarkets with about 12% in total. The slaughterhouses waste to be the third common one that considered to be 9% of the total and finally the fish waste.

For Muscat governorate, farms waste is in low quantities as farmers usually recycling almost all wastes produced to a good fertilizer to be used in the farms again. Whereas, hotels and restaurants are working in commercial bases so they are trying to minimize the wastes as much as they could to reduce the cost and to increase the profit.

As mentioned above, the highest wastes is produced in the fruit and vegetable markets and the hypermarkets, however, the hypermarkets waste is a mixture of different wastes such as foods, papers, plastics and solid wastes (iron or steel). Therefore, it may be difficult to be utilized as a source of organic wastes before any kind of sorting or segregation to be applied from the source in order to separate the required quantities.

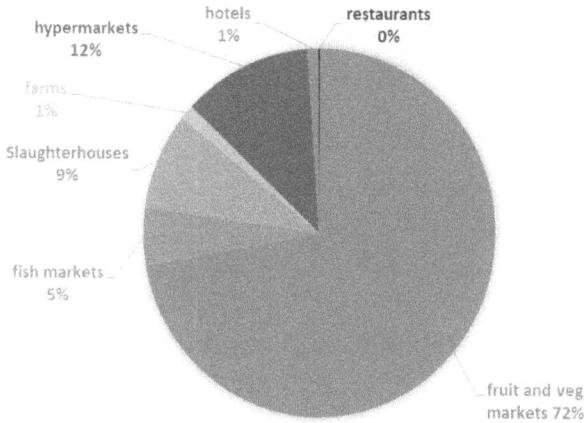

Figure 5   Organic waste generated per year for Muscat governorate

The fruit and vegetable market of Muscat called " Al Mawalih central market of fruits and vegetables" is producing around 50-70 tons of organic waste per day in the normal situation, however, this quantity to be increased to the double or more to reach about 100-150 tons per day of organic waste in different certain occasions including the summer, Ramadhan and Eids.

3.2 Organic waste generation in S. Al Batinah governorate

Figure 5 shows that the slaughterhouses wastes is the common waste generated in Al Batinah South, which considered to be 62% of the total organic waste, followed by fish market waste of 19%. The hypermarkets waste is considered to be the third common one with about 14% of the total waste to be produced. Farms waste is higher in Al Batinah area than Muscat area as Al Batinah area is considered to have a huge areas of green landscaping and more agricultural activities are available. Whereas, fruit and vegetable markets, hotels and restaurants are sharing almost the lowest with about 1% of the total quantities of the organic waste.

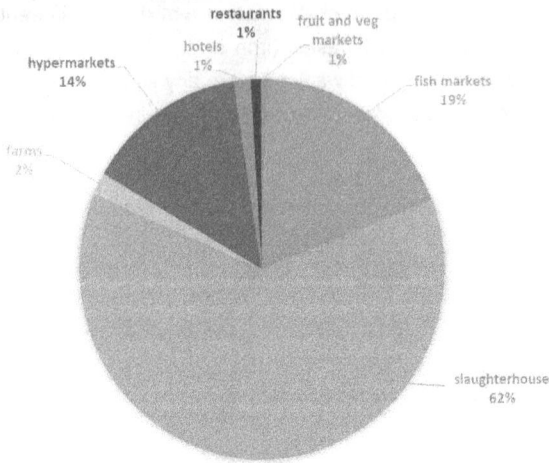

Figure 6   Percentages of total organic waste per year for S. Al Batinah governorate

4   Implementation of biogas plants in Oman for purpose of electricity and thermal energy production

There are three main sources of organic waste generated in Oman includes; Commercial & Industrial (C&I) entities, farms and the municipal solid waste (MSW). Around 600 tons of source segregated organic waste received into the engineered landfills daily which is almost 30% of the total waste received. Food waste alone is considered to be more than 150 tons/day/landfill. The green waste could reach 120 tons/day/landfill, and the remaining 40% is distributed between poultry waste, slaughterhouse waste, fish waste, sludge and dead animals. Implementing ten (10) potential biogas plants for whole of Oman for organic waste diversion is the target for the next five years, till 2023.

4.1 First stage of biogas plants implementaion (2018-2023)

The first stage is for conducting a feasibility study for four different location based on potentiality of feedstock availability and the availability of energy off takers. The table below gives an indicate on these locations and the electricity capacity supposed to be produced:

Table 3:   The first four suggested locations for stage one biogas plants implementation

Sultan Qaboos University, SQU	(medium size plant), to produce around 1 MWel
Oman Agriculture association, OAA	(medium size plant), to produce around 1 MWel + thermal energy for cooling demands
German University of Technology, GUtech	(small size plant), to produce around 0.5 MWel
Barka landfill, BLf	(large size plant), to produce around 2.5 MWel

The feasibility study is already conducted and should be finalized by June 2019, which to cover the following components:

1. Feedstock Assessment and evaluation

2. Technical overview and general project design

3. Feasibility study for development of biogas plant at proposed site provided by German university of technology in Oman (GUtech)

4. Feasibility study for development of biogas plant at proposed site provided by Sultan Qaboos University (SQU)

5. Feasibility study for development of biogas plant at proposed site provided by Omani Agricultural Association

6. Feasibility study for development of biogas at proposed site in Barka landfill

Another feasibility study was conducted for Dhofar governorate as a separate project to handle the sewage treatment plants waste (sludge) beside the other organic waste for purpose of electricity production.

4.2 Second stage of biogas plants implementaion (2018-2023)

Another six (6) projects to be allocated in various locations and governorates, mostly in the areas of the engineered landfills.

In conclusion, be'ah is to divert a total of 900 tons/day of organic waste for whole of Oman and to produce up to 10 MWel.

The following map shows the distribution of 10 biogas plants between 2018 and 2023

Figure 7   Distribution of Biogas plants implementation (2018-2023)

5   Conclusion

In conclusion, it's part of Oman's vision for 2020 and as per the national strategy for the upcoming 10 years to reduce the dependency on oil and gas for power supplement and to produce 10% of the electricity from an alternative source of energy includes biogas implementation. One of be'ah's main objective is to contribute in Oman vision by applying the strategies of implementing 10 biogas plants for the next 5 years to produce 10 MWel, and thus by diverting around 900 t/d of organic waste from the engineer landfills. This plan to be expanded for the next stage to divert more than 2000 t/d organic waste and to produce around 20 MWel. Moreover, biogas to be used for the thermal energy production and to be used as a vehicles fuel or to be upgraded to a biofuel for further uses.

6   References

1. Al Hosni, Suad, 2016-2017, organic waste assessment for Barka engineered landfill, be'ah.

2. Agricultural census, 2014, Ministry of Agriculture and Fisheries, Oman

3. Al-Hooti, S.N., S. Himmo, H. Al-Amiri, and T. Al-Ati, "Food consumption pattern for the population of the State of Kuwait based on food balance sheets", Ecology of Food and Nutrition, vol. 41, 2002, pp. 501-514.

4. Iqbal, S., Khan, N., Bashar, S., "Food Consumption Patterns of Male and Female University Students in Oman", Transactions on Engineering and Sciences ISSN: 2347-1964, Vol. 4, Issue 4, 2016.

5. Al Hosni, Suad, 2017, organic waste strategy for composting and biogas production, be'ah.

Author's address:

Dr. Suad Al Hosni
Oman Environmental services holding co. "be'ah".
Oman, Muscat, Al Khoudh 132. P.O. 1230
Mobile: 00968-96495997
E-Mail: s.s.hosni@gmail.com, suad.alhosni@beah.om

Mechanical wet pre-treatment efficiency of OFMSW for biogas production

Alice do Carmo Precci Lopes, Wolfgang Müller, Anke Bockreis

Universität Innsbruck, Innsbruck, Austria

Abstract
The efficiency of impurities removal by means of a wet pre-treatment system was evaluated in a waste treatment plant located in Scotland. The pre-treatment system consists of pulpers and grit removal systems, which remove impurities from the MSW. The remaining organics are fed into the anaerobic digestion (AD) fermenter for biogas production. Samples from each output of the system were analyzed for total solids, volatile solids, biogas production potential and composition. The results showed that 96% of the incoming impurities were removed by the wet pre-treatment system and that almost 90% of the biogas production potential was transferred to the organics fed to the AD, delivering 545 Nm^3/t VS.

Keywords
Biogas, organic fraction of municipal solid waste, mechanical pre-treatment, grit removal system.

1 Introduction

Municipal solid waste (MSW) management still is a big concern for the municipalities. In countries with large land availability, landfilling has been the most commonly used technology for handling MSW. However, the increase of landfill taxes, stricter regulations, and environmental awareness have been pushing the governments to change the traditional waste treatment systems to more sustainable ones. In Scotland, for example, the Glasgow Recycling and Renewable Energy Centre was designed to reduce by 80% the amount of landfilled waste in the city. The facility, which has been in operation since 2017, generates energy and recovers material for the recycling industry (Glasgow City Council, 2019).

The Glasgow Recycling and Renewable Energy Centre is designed to process up to 200,000 t/a of MSW by means of:

1. Dry pre-treatment of the incoming MSW by screening, to separate the fine fraction (< 80 mm), rich in organic material (denominated organic fraction of municipal solid waste (OFMSW) < 80 mm), from the coarse fraction, rich in recyclable material.

2. Wet pre-treatment of the OFMSW < 80 mm (90,000 t/a), to produce a substrate suitable for the wet anaerobic digestion (AD).

3. Solid-liquid-separation of the digestate and treatment of process water (PW) and effluents.

4. Production of refuse-derived fuel from the solid phase of the digestate and from the residues of the dry and wet pre-treatments.

The wet pre-treatment (referred as BTA® process) constitutes the subject of this study. Its design consists of four parallel pulpers. Each pulper is equipped with an impeller, which mixes the incoming waste with process water (pulping). The pulpers work as a wet screening unit and as an organics defibering machine. The organic material is gently dissolved by means of shear forces introduced by the impeller. During the dissolving process, heavy particles settle down and are removed by means of two heavy fraction traps. After pulping and removal of the settled heavy particles, the mixture is screened through a 10 mm sieve. The fraction < 10 mm (suspension) is sent to the storage tanks prior to further processing. The fraction > 10 mm is flushed out with process water and dewatered, leaving the light fraction. The light fraction is mixed with the outputs of other separation processes of the facility for the production of refuse-derived fuel.

The remaining suspension is sent to two grit removal systems (GRS) working in parallel. Each GRS line comprises four serial grit removal units (GRU 1, GRU 2, GRU 3 and GRU 4). Each grit removal unit consists of one hydrocyclone connected to a grit classifying pipe. In these units, further inert material is removed. The remaining suspension is sieved to reduce the water content and it is ready to be fed into the AD digester. A simplified scheme of the process is presented in Figure 1, where only one pulper and one GRS line is represented.

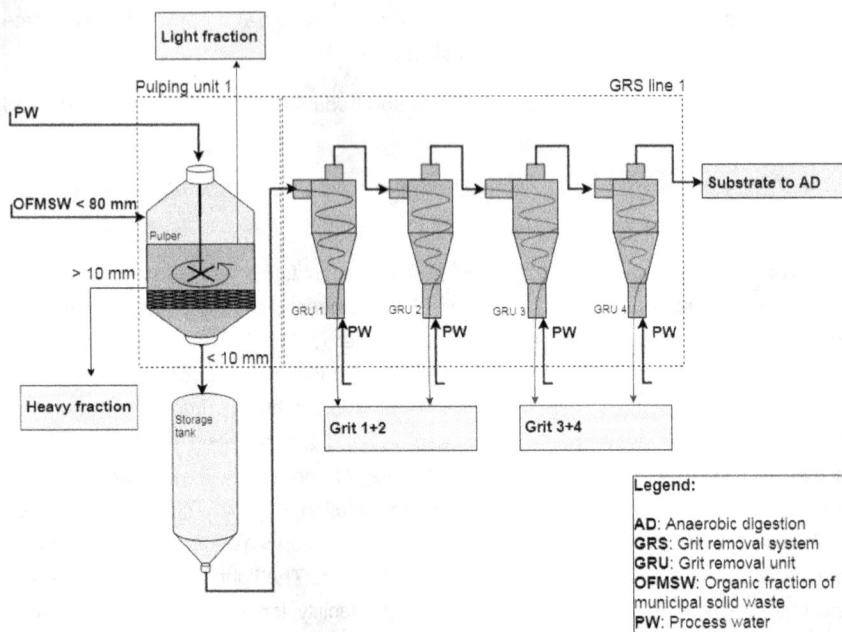

Figure 1 Scheme of the wet pre-treatment system for one line.

The objective of this study was to evaluate the efficiency of the BTA® wet separation process for the preparation of OFMSW < 80 mm for biogas production at a full scale plant.

2   Material and methods

2.1   Material

Samples from each output of the BTA® wet mechanical process were collected on different days during November and December of 2017. The amount of collected samples is presented in Table 1. Due to the plant design, the outputs of the GRU 1 and GRU 2 could only be collected together. The same applies to GRU 3 and GRU 4. The collected samples were dried at 105°C at the laboratory of the waste treatment plant and transported to the University of Innsbruck, where they were further analyzed.

Table 1 Amount of collected samples

Sample	Number of samples	Total amount of collected sample
Light fraction	3	17 kg
Heavy fraction	3	37 kg
Grit 1+2*	3	30 kg
Grit 3+4**	3	30 kg
Substrate to AD	15	27 L

*Outputs from GRU 1 and GRU 2. **Outputs from GRU 3 and GRU 4.

2.2   Methods

The samples were characterized as described by Lopes et al. (2017). The characterization analysis comprised total solids (TS), volatile solids (VS), composition (inert material – glass, stones and ceramics; plastics; textiles; organics and others) and biogas production potential. The biogas tests followed VDI 4630 (2014). The hand-sorted organics (degradable and non-biodegradable) were used as substrate. The results were expressed in Nm^3/t VS_{added} and in Nm^3/t TS. The VS_{added} refers to the VS from the hand-sorted organic material, which was used to feed the biogas batch reactors. The results expressed in TS comprise not only the fed organics, but also include the impurities present in each analyzed output.

The plant operator (Viridor) provided data regarding the weights of each input and output to the wet pre-treatment system, as well as the number of batches per day for the year of 2017. Since the plant operation has started in 2017, there is still a high variation of the data. Thus, the mass balance was calculated based on the median of the values from the period October 2017 to December 2017. Since the outputs of the GRUs were weighed together, the output mass of each GRU was considered equally distributed. Since it was not possible to collect samples from the OFMSW < 80 mm, its composition was calculated based on the composition of the output fractions and on the mass balance of the system. Samples of PW in the pulpers and GRSs could not be collected, thus their TS concentration were assumed to be 0.7% and 1.5% FM (fresh matter), respectively. These values are in the range of the data provided by Viridor. It was also assumed the total solids are constituted of mainly dissolved matter, such as salts or non-bioderadable material, thus, the impurities content in the PW was disregarded.

3   Results

Figure 2 presents the mass and impurities (> 0.063 mm) flows through the entire wet pre-treatment system per batch. From the incoming impurities to the system (2.8 t/batch), 2.7 t/batch could be removed by the combination of pulpers and grit removal system units, i.e. 96% of the incoming impurities were removed before reaching the AD process.

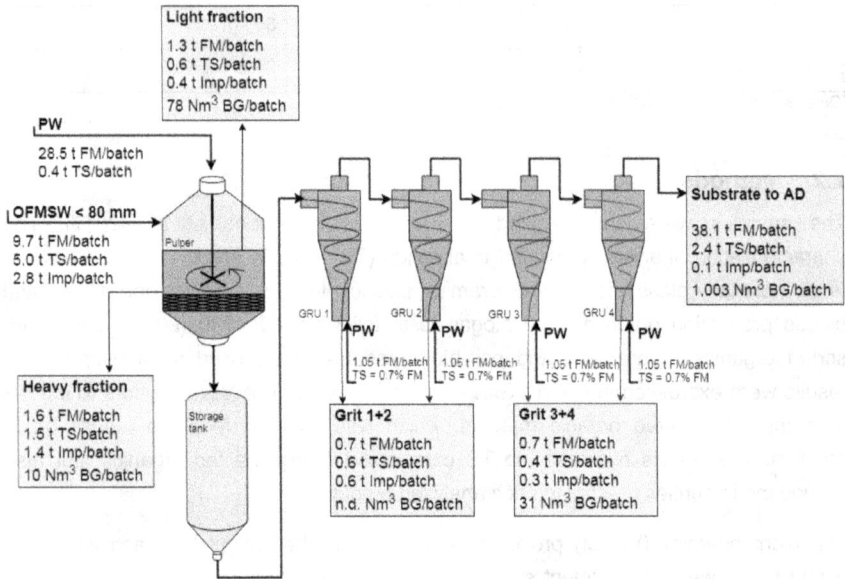

Figure 2   Mass, in fresh matter (FM), and in dry matter basis (TS), impurities (Imp), and biogas (BG) flows in the wet pre-treatment system.

For comparison, in the study of Jank et al. (2017), from 80% to over 90% of impurities > 0.5 mm could be removed from biowaste by means of a hydrocyclone. The removal of such particles prior to the AD is important, since they cause abrasion of pumps and might sediment in the fermenter, reducing its available volume. This is technically challenging and raises costs, since less substrate can be fed to the digester and a higher frequency of cleaning might be needed. Coarse material, such as wood sticks and plastics, might block pipes or form swimming layers in the digesters. Furthermore, plastics can even reduce the dewatering efficiency of the digestate.

Figure 3 presents the cumulative distribution of inert particles (minerals originating from the incineration of the hand-sorted organics, as well as hand-sorted glass, stones and ceramic) in each output of the wet pre-treatment system.

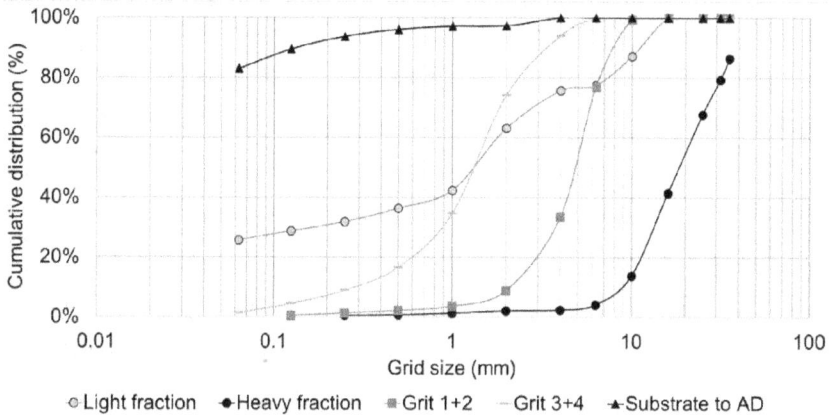

Figure 3   Cumulative distribution of inert particles in each output fraction.

Most of the inert material bigger than 10 mm was removed by the pulpers. Grit 1+2 could take out particles mainly ranging from 1 to 10 mm, while Grit 3+4 removed smaller inert particles, ranging from 0.25 to 4 mm. Over 80% of the inert particles in the substrate to the AD were < 0.063 mm. According to Jank et al. (2016), the particles < 0.063 mm are mainly carbonates and minerals originated from the combustion of organics, thus they are not necessarily impurities.

Figure 4 shows the composition of each output fraction in relation to the total input material to the wet pre-treatment system in dry mass basis. The total input material comprises of not only the OFMSW < 80 mm, but also the process water in the pulpers and GRSs, which still contains some dissolved organics. Summing up the incoming organics, 85% are recovered as anaerobic digestion substrate, 9% remain in the light fraction, being recovered as refuse-derived fuel, and 6% are lost in the heavy fraction, grit 1+2, and grit 3+4.

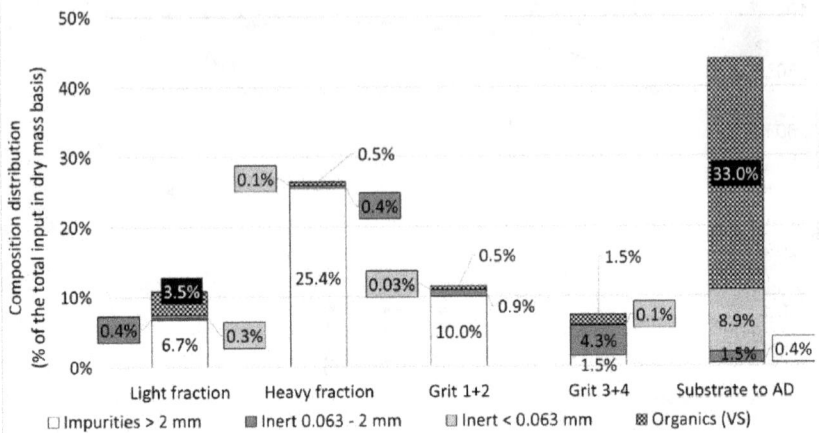

Figure 4   Composition of each output fraction based on the total input material to the wet pre-treatment system, in dry mass basis.

Table 2 shows the biogas production potential in each output fraction based on the TS and VS_{added}. The TS include not only the added organics, but also the impurities in each fraction. On the other hand, the VS_{added} represent only the hand-sorted organics added to the batch reactors. Among the rejected fractions (light fraction, heavy fraction and grit 3+4), the light fraction presented the highest biogas production potential. In fact, during the hand-sorting, it was observed that the organics of the light fraction were rich in peels and vegetables, and the organics of the heavy fraction were rich in bones. The substrate presented the highest biogas production potential (546 Nm^3/t VS). This figure is comparable to source-sorted biowaste (497 to 955 Nm^3/t), considering a methane concentration in the biogas of 60%Vol. (Davidsson et al., 2007).

Table 2 Biogas production potential in each fraction.

Sample	Nm^3/t TS	Nm^3/t VS_{added}
Light fraction*	130	412
Heavy fraction	7	305
Grit 1+2	n.a	n.a
Grit 3+4	76	305
Substrate to AD	415	545

n.a.: not analyzed, since the organic matter content was very low.

Figure 5 shows the biogas transfer rate in the wet pre-treatment system. Almost 90% of the biogas volume is found in the substrate to AD. Although the light fraction presented the highest share of biogas among the rejects, its energy potential is not lost, since this fraction is further processed to produce refuse-derived fuel.

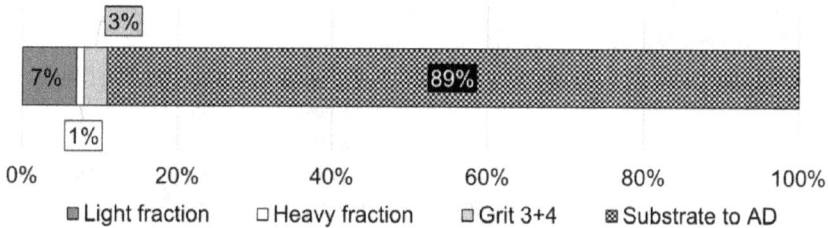

Figure 5 Biogas distribution (in % volume) in the wet pre-treatment system.

4 Conclusions

Although the organic fraction of MSW consists of a complex mixture of organics and impurities (glass, plastics, textile, stones, ceramic etc.), the BTA® wet pre-treatment showed to be an efficient process to remove impurities from OFMSW with the purpose of using the cleaned organics for biogas production. From the incoming impurities, 96% could be removed by means of pulpers and grit removal units, leaving the substrate to the AD almost free of impurities and with a biogas production potential of 545 Nm^3/t VS.

While the pulpers mainly removed inert material > 10 mm and combustible material, the GRUs removed smaller particles: grit 1+2 concentrated particles ranging from 1 to 10 mm and grit 3+4 retained particles ranging mainly from 0.25 to 4 mm. The presented results were derived from a newly built waste treatment facility. Thus, further sampling and analysis are required when the plant is functioning at full capacity.

Acknowledgements

The authors would like to thank BTA® and Viridor for their valuable contributions.

5 Literature

Davidsson, Å., Gruvberger, C., Christensen, T.H., Hansen, T.L., Jansen, J.I.C.	2007	Methane yield in source-sorted organic fraction of municipal solid waste. Waste Management 27, 406-414.
Glasgow City Council	2019	General Waste (Green or Metal) Bin. Related Documents. Glasgow Recycling and Renewable Energy Centre (GRREC). Available at: https://www.glasgow.gov.uk/index.aspx?articleid=16564. Accessed on 19th March, 2019.
Jank, A., Müller, W., Waldhuber, S., Gerke, F., Ebner, C, Bockreis, A.	2017	Hydrocyclones for the separation of impurities in pretreated biowaste. Waste Management 64, 12-19.
Jank, A., Müller, W., Waldhuber, S., Gerke, F., Ebner, C, Bockreis, A.	2016	Impurities in pretreated biowaste for co-digestion: A determination approach. Waste Management 52, 96-103.
Lopes, A.C.P., Müller, W., Bockreis, A.	2017	Mechanical pre-treatment of municipal waste for biogas production, 15th International Conference of Young Scientists on Energy Issues, Kaunas, Lithuania.
VDI 4630	2014	Verein Deutscher Ingenieure. Vergärung organischer Stoffe.

M.Sc. Alice do Carmo Precci Lopes
Ph.D. Student
Universität Innsbruck
Technikerstraße 13
6020 Innsbruck, Austria
Telefon +43 512 507-62128
E-Mail alice.lopes@uibk.ac.at

Biological desulphurisation filter for biogas

Claus Bogenrieder and Andreas Maile

Züblin Umwelttechnik Gmbh, Stuttgart; STRABAG Umwelttechnik Gmbh, Düsseldorf

Biological desulphurisation filter for biogas

Abstract

The new biological desulphurisation filter developed by ZÜBLIN BioBF provides an effective and economical method for external desulphurisation of biogas outside the digester. Especially when faced with high initial hydrogen sulphide loading, pre-purification of biogas by ZÜBLIN BioBF can bring about a significant reduction in gas treatment costs. The life time of activated carbon customarily used for final gas purification can be increased significantly. The addition of excipients to the digestion process, such as iron hydroxide, to lower hydrogen sulphide concentrations in biogas, can be reduced or even avoided altogether. The microbiological oxidation of sulphide to form elemental sulphur, sulphate or thiosulfate allows up to 32 g H_2S per m^3 filter material (FM) and hour to be eliminated. A specific elimination capacity (EC) of 18 g H_2S / (m^3 FM x h) was mostly exceeded during long-term operation of a ZÜBLIN BioBF prototype since December 2015 at a digestion plant run by MKW Großefehn. In times of low hydrogen sulphide loading of biogas (< 250 ppm), complete biological elimination of H_2S was often achieved. The obtained biological oxidation products either form as on the surface of the filter material elemental sulphur or leave the system as soluble components such as sulphate or thiosulfate within the condensate runoff. Based on experiences to date with the ZÜBLIN BioBF prototype, the biofilter material has life times of between 7 and up to 17 months.

Keywords

Biological desulphurisation, hydrogen sulphide, biogas cleaning, biofilter, activated carbon, operating costs

1 Need for biogas pre-desulphurisation

1.1 Biogas desulphurisation methods

More stringent legislative emission control requirements make it essential to largely eliminate hydrogen sulphide from biogas when using biogas as a source of energy in co-generation plants or when upgrading biogas to feed into the natural gas grid. The use of oxidation catalysts to ensure low formaldehyde emissions in exhaust gas from co-generation plants requires biogas to be virtually free of hydrogen sulphide.

In many cases, physical-chemical adsorption filters with activated carbon or iron hydroxide pellets are used for external biogas desulphurisation (outside the digester). Adsorption methods require dried biogas. Prior to treatment in adsorption filters, the biogas's absolute moisture level must be reduced using tube and shell heat exchangers or gas scrub cooling for this reason. Biogas is then reheated in order to achieve optimal relative humidity (approximately 50% rH) for the adsorption method in question. Adsorptive methods allow hydrogen sulphide to be completely removed from biogas as long as the adsorbent is not or only partially saturated. The hydrogen sulphide loading can no longer be fully adsorbed when the filter material reaches a certain material-specific loading. Hydrogen sulphide "breaches" the filter material and the adsorbent has to be replaced due to the laws of physical-chemical equilibrium reactions. The filter material's life time decreases as hydrogen sulphide loading in biogas increases. The required filter changes and consumption of corresponding resources are relevant cost items when operating digestion plants.

Pre-desulphurisation can be an economically viable and necessary addition to the aforementioned adsorptive methods when biogas has high hydrogen sulphide concentrations of > 200 ppm. One proven method is chemical-biological gas scrubbing, such as the ZÜBLIN BioSulfidEx using the biotrickling filter process. In this method, hydrogen sulphide contained in biogas is absorbed into scrubbing liquid. The scrubbing liquid is circulated using suitable pumps and sprayed through fillers using automated irrigation technology. Biofilm on the fillers largely converts hydrogen sulphide into sulphate and thiosulphate. These substances are discharged from the scrubber with the scrubbing liquid at regular intervals. Scrubbing water and nutrients have to be replaced based on the discharge cycles; the pH of the scrubbing liquid also needs to be regulated. This process is relatively sophisticated since it requires a PLC in addition to plant technology (circulation pump, dosing technology and monitoring sensors).

High hydrogen sulphide concentrations in biogas and high hydrogen sulphide loading can also be reduced by adding ferrous preparations to the digestion substrate or by adding air to the digester. The discharge of hydrogen sulphide through biogas is already lowered in the digestion container itself in the process. Consequently, this method is also referred to as internal desulphurisation. Mixing in ferrous oxide/ferrous hydroxide preparations or directly introducing ferrous chloride into the digester can bring about a chemical reaction that produces hardly soluble ferrous sulphide – yielding a relevant reduction in hydrogen sulphide levels in biogas. Completely eliminating hydrogen sulphide is generally not cost-efficient with this method because of the disproportionate increase in expenditure. Reducing hydrogen sulphide levels in raw biogas to 50–100 ppm also involves significant additional input costs.

Introducing air into the digester to eliminate hydrogen sulphide is a fairly low-cost version of internal desulphurisation, but may result in significant corrosion to the digester's structure or parts involved in the process. The use of this technology is principally limited to agricultural wet digestion processes. Many providers of anaerobic process technology rule out an internal desulphurisation method of this kind for structural reasons or because of corrosion concerns.

1.2 External biogas pre-desulphurisation using ZÜBLIN BioBF

ZÜBLIN BioBF shifts the principle of biological biogas pre-desulphurisation from internal desulphurisation in the digester, which creates technical issues, to areas outside the actual digestion process that are accessible, easy to manipulate and not critical. This new technology from Züblin Umwelttechnik GmbH removes and breaks down hydrogen sulphide from biogas with higher levels of contamination. The desulphurisation filter is filled with coarsely shredded and specially processed organic filter material (see Figure 1). Raw biogas is discharged directly from the digester to the BioBF without undergoing additional pre-treatment. The inevitable cooling of water-saturated biogas as it goes through the BioBF lowers the temperature below the dew point. Condensate is constantly generated and discharged from the BioBF through a condensate separator. Biogas flows through the filter material's relatively coarse pore system, its hydrogen sulphide is absorbed into the pore water film and oxidised into elemental sulphur, thiosulphate and sulphate through microorganisms in the biofilm. Elemental sulphur that is formed clings to the surface of the filter material particle (see Figure 2). The soluble oxidation products of sulphate and thiosulphate are desludged with the condensate and removed from the system (Wessel, 2016; Lenhart 2019).

Figure 1   First fill of the BioBF on 8 August 2015, commissioning on 21 December 2015

The filter material's life time essentially depends on the raw biogas's hydrogen sulphide loading and the gas pore volume, which decreases over time through settlement of the filter material and accumulation of insoluble elemental sulphur in the pore system and on the surface of the filter material. In experiences to date with operating the BioBF prototype at the site of the MKW Großefehn digestion plant since December 2015, service times ranging from 7 to 17 months have been achieved to date with different filter material grain sizes. The filter material's cleaning function was still largely preserved at each of the three exchange dates so far. The lowest elimination capacity prior to the last filter material exchange on 19 September 2018 still stood at 15 g H_2S / (m^3 FM x h). The filter material's pressure resistance also increased only marginally compared with its condition at installation and was well below 100 Pa. Apart from the filter material, the BioBF needs no materials, no energy and is a maintenance-free and cost-effective biogas pre-desulphurisation system, apart from the need to change filter material.

Figure 2   Filter material exchange on 11 May 2017 after 507 days in operation

2 Operating experience and characteristics of the BioBF prototype

2.1 MKW Großefehn digestion plant

Materialkreislauf- und Kompostwirtschaft GmbH & Co. KG (MKW) has operated a composting plant to recover biogenic waste at the site of the waste management centre Entsorgungszentrum Großefehn (EZG) since the 1980s. This plant was originally run as a composting plant for waste and sewage sludge and was converted into a plant to compost and undertake biological treatment of residual waste in 1995. In 2010, a digestion stage for separately collected biodegradable waste from the Aurich region was added to the existing set-up. Since December 2010, the EZG's digestion stage has used a portion of the biodegradable waste delivered to generate biogas using biological conversion excluding air in a plug-flow digester. This horizontal digester has an effective filling volume of about 1,300 m³.

Figure 3   Input side of the digestion reactor (plug-flow digester) and co-generation plant

The digestion stage treats up to 20,000 t of biodegradable waste each year. When process control is optimal and the digestate substrate is suitable, the thermophile anaerobic digestion process can generate approximately 2-2.5 million standard cubic metres of biogas per year (Nm³/a). The average biogas flow volume stands at about 250 Nm³/h. This biogas consists of approximately 55-65% methane and approximately 35-45% carbon dioxide. It also contains small amounts of trace gases, such as ammonia and hydrogen sulphide. Hydrogen sulphide concentrations fluctuate between 200 and 800 ppm for seasonal reasons (see Figure 5). Following a multi-stage biogas upgrading process entailing biological hydrogen sulphide elimination, gas scrubbing cooling to remove ammonia as well as activated carbon adsorption to remove all remaining pollutants, the purified biogas is recovered in two co-generation plants with a total electrical capacity of 590 kW and in the thermal exhaust air treatment system for the mechanical-biological treatment (MBT) plant for residual waste at the site. Digestate substrate is removed from the digester after spending an average of about 20 days there and is pressed using screen-conveyer presses. Liquid digestate is recovered in agricultural applications as a high-quality organic multi-nutrient fertiliser. The remaining solid digestate is mixed with surplus fresh biowaste, undergoes biological drying in the existing tunnel composting plant and is turned into quality-assured finished compost. In December 2015, MKW retrofitted a prototype of a new process developed by ZÜBLIN Umwelttechnik for biological hydrogen sulphide reduction in biogas (BioBF). With the help of the BioBF, the company can now forgo relatively costly excipients to reduce hydrogen sulphide (e.g. iron hydroxide) and can significantly extend the service life of activated carbon used to completely remove hydrogen sulphide. Installing the BioBF thus enabled the company to increase plant availability and reduce operating costs considerably.

2.2   BioBF design

The ZÜBLIN BioBF consists of a corrosion-resistant and thermally insulated HDPE cylinder with an HDPE flanged lid.

Figure 4   ZÜBLIN BioBF prototype for the biological reduction of hydrogen sulphide

An excavator fills the cylinder with specially conditioned filter material and removes used filter material from above with the lid detached (see Figures 1 and 2). The filter material is placed on an elevated grate made out of glass fibre-reinforced plastic, which is mounted over the gas distribution and condensate collection area. Raw biogas is discharged into the collection chamber and flows through the filter material from the bottom to the top. Condensate generated in the process is collected on the container floor and discharged from the BioBF using a condensate remover uncoupled from gas pressure and enters the digestion plant's process water collection system. The collection chamber can be accessed to carry out inspection and cleaning work using a 500mm manhole. A flushing nozzle for cleaning purposes and inlet for flushing gas to make the filter's interior inert are located at the base of the cylinder shell. A stainless-steel connecting piece in the lid allows a nutrient or inoculation solution to be added to the filter material, if needed. The lid also has a 400 mm inspection opening and flushing gas outlet connection to make the interior of the filter inert prior to opening the container lid. Additional connection pieces in the cylinder shell and lid allow probes to be installed (temperature, pressure) and gas samples to be taken. The cylinder is designed to have a biogas volume flow of 250 Nm³/h, a maximum gas overpressure of 200 mbar and gas temperature of 50 °C. The filter material has a volume of approximately 5 m³. The driving gradient for gas flow is system pressure in the digester (ca. 35-55 mbar), while the loss of pressure through the bed of filter material is negligible. Even a short time before filter material is changed, filter resistance is still well below 100 Pa. Used filter material is mixed with the digestate and biowaste mix and composted.

2.3   Operating data, cleaning performance, life times

The BioBF was first filled with specially conditioned filter material on 8 December 2015 and started operating on 21 December 2015 (see Figure 1). Filter material has been changed three times during operations to date. Table 1 summarises the change dates, length of use and parameters for the filter material used.

Table 1   ZÜBLIN BioBF filter material: Installation dates, service life, parameters

Installation date	Length of use	Grain size [mm]	Fill height [m] / Fill volume [m³]	Bulk density [Mg solids/m³]	Gas pore volume [m³] / [%]
21 December 2015	507 d	60 / 250	1.90 / 6.0	0.27	4.5 / 76%
11 May 2017	209 d	15 / 60	1.85 / 5.8	0.37	4.1 / 70%
6 December 2017	287 d	15 / 60	1.85 / 5.8	0.56	3.2 / 55%
9 September 2018		15 / 60	n.a.	n.a.	n.a.

The BioBF was first filled with comparatively coarse filter material in a grain size of 60 to roughly 250 mm. Medium-grain filter material with a grain size of 15-60 mm was installed during subsequent fillings. A gas distribution layer around 20 cm thick made out of coarse material (60/250) was placed directly on the grate floor below this filter material. As expected, the coarse filter material has a much lower bulk density and higher gas pore volume than medium-grain filter material immediately after the BioBF is filled.

Table 2   Parameters for filter material for the second filling of the BioBF when the material was installed and removed

Parameter	Installation 11 May 2017	Removal 6 December 2017

Length of operation [d]	209	
Grain size [mm]	15 / 60 (gas distribution layer 60 / 250)	
Water content [% solids]	50.9%	39.6%
Ignition loss (organic dry matter) [% dry matter]	59.1%	54.5%
Sulphur content [% dry matter]	0.14	34.0
Bulky density damp [Mg/m³]	0.37	0.55
Bulky density dry [Mg dry matter/m³]	0.18	0.33
Total fill mass damp [Mg]	2.14	2.84
Total fill mass dry [Mg]	1.05	1.70
Fill height [m]	1.85	1.65
Fill volume [m³]	5.8	5.2
Total porosity [m³] / [%]	5.2 / 89%	4.2 / 81%
Solid content [m³] / [%]	0.6 / 11%	1.0 / 19%
Water porosity [m³] / [%]	1.1 / 19%	1.1 / 22%
Gas porosity [m³] / [%]	4.1 / 70%	3.1 / 59%
Pore water saturation [%]	21%	27%
Sulphur mass in the BioBF [kg]	1.5	583.2
Total sulphur removal [kg]	581.7	

Filter material from the second fill was completely weighed and analysed in depth during installation and removal. Its bulk density, water content, organic substance content and total sulphur level were determined. Porosity and the mass of accumulated sulphur in the filter material resulting from biological hydrogen sulphide oxidation were derived from these figures. The BioBF's elimination capacity (EC) for hydrogen sulphide stood at around 100-125 g H2S/h when the filter material was changed on 6 December 2017 (see Figure 7). Based on the originally installed filter material, this represents a specific EC of approximately 20 g H2S / (m³ FM x h). The BioBF was estimated to have a sulphur accumulation of about 582 kg during the 209 days of operation. The filter material's bulk density increased from 0.37 to 0.55 Mg/m³. Absolute dry mass rose from 1.05 to 1.70 Mg, corresponding to a mass difference of 650 kg and thus deviating about 68 kg or 12% from the calculated sulphur accumulation. Absolute water mass in the BioBF showed no change and also stood at around 1.1 Mg at removal. A significant decrease in the total porosity and gas porosity was also found. Pore water saturation increased from 21% to approximately 27%. Sulphur accumulation is reflected in a growth in the solid content from 0.6 to 1.0 m³. This means that about 0.4 m³ of the decrease in total porosity from 5.2 to 4.2 m³ can be attributed to sulphur accumulation. The other 0.6 m³ is due to material settling by around 0.2 m. Average H_2S removal stood at 104 g H_2S / h or 98 g S / h during operations from 11 May 2017 to 6 December 2017 (see Figure 7). An extrapolation of average sulphur removal over the 209 days of operation results in total removal of 492 kg sulphur. This estimate is less than the identified sulphur accumulation of about 90 kg or 15%. The relationship between estimated sulphur elimination from biogas and sulphur accumulation in the BioBF suggests that elemental sulphur is generated largely as an end product of microbial hydrogen sulphide oxidation. Conversion into the soluble oxidation products of sulphate and thiosulphate appears to play a lesser role in existing microbial reaction processes. This suspicion was confirmed by further examination of the sulphur balance and sulphur dynamics during a master's thesis at Bauhaus University in Weimar (LENHART, 2019). At average condensation of about 110 L/d in the BioBF and average concentration of about 1.3 g/L of thiosulphate and 1.7 g/L sulphate in the condensate, about 4 kg of sulphur was removed from the BioBF as condensate as thiosulphate and about 3 kg of sulphur as sulphate - altogether 7 kg of sulphur as a soluble oxidation product - from 27 September 2018 to 23 November 2018. About 149 kg of sulphur was added to the BioBF with raw gas in the same 57-day period. After biological cleaning, a total of 5 kg sulphur was detected in the clean gas, meaning that 144 kg had been removed during 57 days in the BioBF. This translates into an average eliminated load of about 105 g S / h or 112 g H_2S /h. Just about 5% (7 kg) of the total microbial oxidised sulphur loading of 144 kg leaves the BioBF as soluble sulphur compounds via the condensate. Therefore, 95% (137 kg) is left as elemental sulphur in the BioBF's pore system (LENHART, 2019). Up until 2016, large quantities of iron hydroxide powder (FeOH) were mixed with the digestion substrate in the Großefehn digestion plant to ensure internal reduction of the hydrogen sulphide content in biogas, prior to it entering the digester. Approximately 21,000 kg FeOH was added to the digestion substrate in 2014 and about 22,800 kg FeOH in 2015. Iron hydroxide preparations were gradually lowered from about 90 kg/d to 0 kg/d in 2016 during a first scientific assessment of the effectiveness of the new biological cleaning method (WESSEL, 2016). The Großefehn plant stopped mixing FeOH into the digestion process in September 2016. Up until that time in 2016, some 11,900 kg had been used. The costs of using different iron hydroxide substrate from 2014 to 2016 totalled some €48,500. The Großefehn plant has exclusively used the ZÜBLIN BioBF system to carry out biogas pre-desulphurisation since September 2016, prior to final adsorptive removal of activated carbon. Figures 5 and 6 show the raw biogas and clean biogas' hydrogen sulphide content and the corresponding raw biogas and clean biogas loadings before and after the BioBF. These results are based on manual control measurements of hydrogen sulphide levels using a Multitec 560 instrument made by Sewerin or using Dräger measuring tubes. The Großefehnplant's process control system automatically detects the biogas volume flow to calculate load.

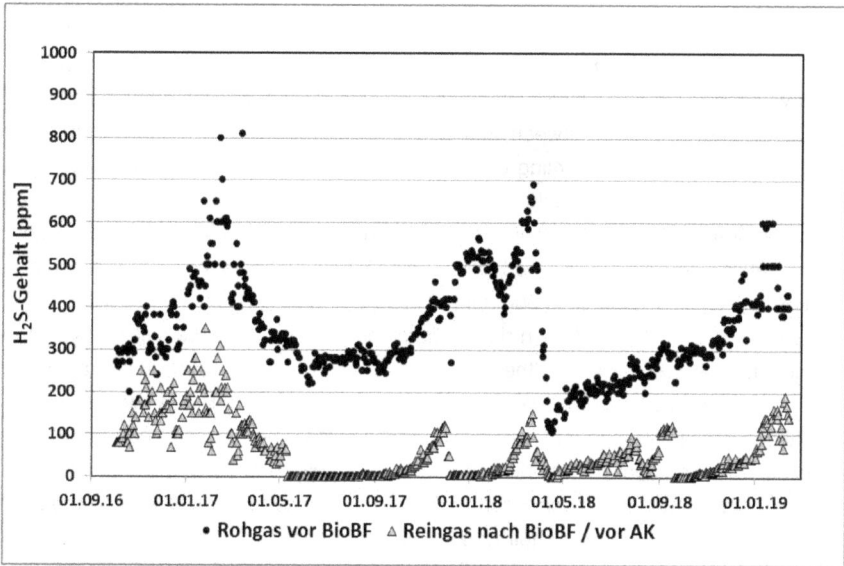

Figure 5 H_2S levels in raw biogas before the BioBF and clean gas after the BioBF

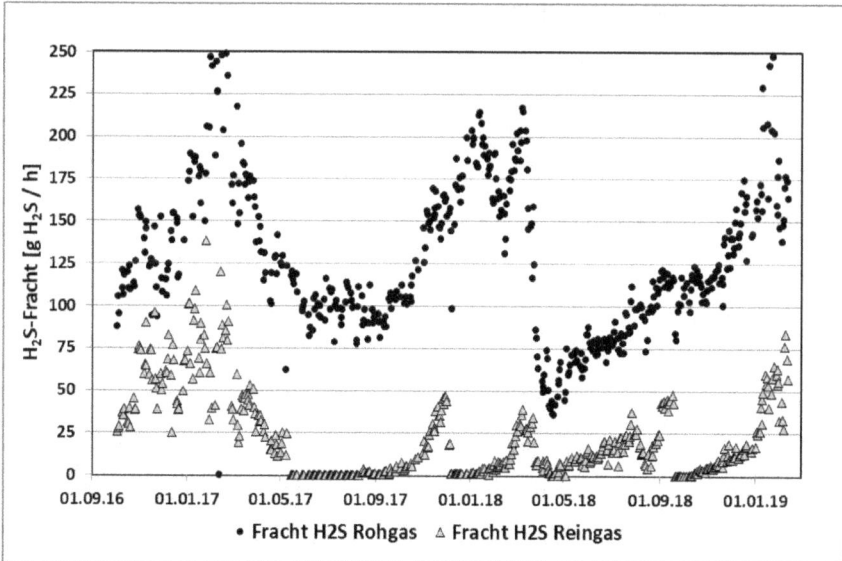

Figure 6 H_2S loadings in raw gas before the BioBF and clean gas after the BioBF

Raw biogas's hydrogen sulphide content fluctuates between about 200 ppm and a maximum of about 800 ppm during the course of the year. Conspicuously low levels of just about 100 ppm from March to April 2018 are not representative since the digestion process encountered significant disruptions during this period. Biological hydrogen sulphide oxidation in the BioBF significantly lowered its levels in biogas to 0-300 ppm. With the exception of the aforementioned period of disrupted gas production, the biogas volume flow averaged roughly 250 Nm³/h during the review period. Some 90% of the measured biogas volume flow was between 220 and 275 Nm³/h. Hydrogen sulphide contents measured manually each working day were used to calculate the corresponding loadings based on the average daily biogas volume flow. Figure 6 contrasts raw gas and clean gas loads. Hydrogen sulphide concentration and loadings show a clear pattern during the course of the year. The highest hydrogen sulphide release was in the winter months with levels in excess of 500 ppm or more than 200 g H_2S/h. Levels fell to about 200-300 ppm or 75-100 g H_2S / h during the summer months. The remaining hydrogen sulphide load in clean gas after the BioBF, which underwent subsequent activated carbon adsorption, was typically lowered to levels well below 100 g H_2S/h because of biological conversion, even in the case of high raw gas loadings. In more than 50% of the review period, clean gas levels after the BioBF were well below 25 g H_2S/h. Immediately after the filter material was changed in May 2017, December 2017 and September 2018, clean gas levels always fell to 0 g H_2S/h and stayed at 'zero level' for several weeks. Figure 7 summarises the BioBF's performance between October 2016 and January 2019. It illustrates absolute separation performance in g H_2S/h and the filter material's specific elimination capacity (EC) in g H_2S/ (m³ FM x h).

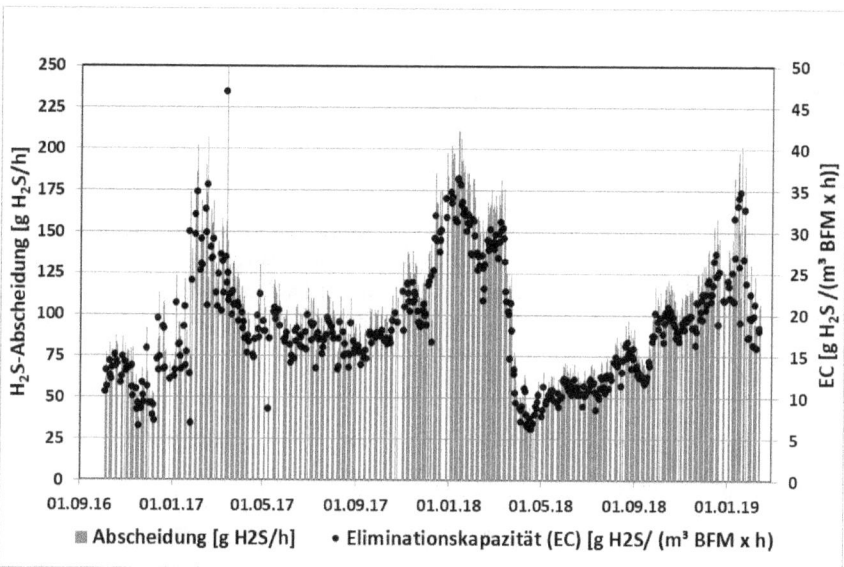

Figure 7 H_2S separation in the BioBF and specific elimination capacity (EC)

Table 3   Descriptive statistical analysis of H_2S separation and EC (513 readings)

Statistical parameters	Separation performance [g H_2S/h]	Elimination capacity (EC) [g H_2S/ (m^3 FM x h)]
5% percentile	52	9
25% percentile	77	13
Mean	107	18
Median	103	18
75% percentile	127	22
95% percentile	183	32

On average, the BioBF had a separation performance that was marginally higher than 100 g H_2S/h. Separation of more than 77 g H_2S/h was found in 75% of readings. Average specific elimination capacity stood at roughly 18 g H_2S / (m^3 FM x h) at a filter material volume of about 5.8 m^3. A statistical analysis of readings from the ZÜBLIN BioBF prototype in the review period to date (see Tab. 3) provides the basis for designing these kinds of systems for additional applications in the future. STRABAG Umwelttechnik GmbH is currently conducting research into process-related ways of increasing the share of the water-soluble oxidation products of sulphate and thiosulphate, exploring the potential to optimise the choice and customisation of the filter substrate as well as analysing the impact of temperature on cleaning performance and the ability to transfer the findings to biogas treatment with mesophilic digestion methods.

3   Literature

Wessel, Imke;   2016   Verfahren zur biologischen Entschwefelung von Biogas am Beispiel eines neuartigen Biogas Biofilters bei der MKW Großefehn. Masterarbeit, Universität Stuttgart, Institut für Siedlungswasserbau, Wassergüte- und Abfallwirtschaft.

Lenhart, Markus;   2019   Schwefelbilanz und -dynamik in Vergärungsprozessen und einer zugehörigen biologischen Biogasreinigungsanlage. Masterarbeit, Bauhaus-Universität Weimar, Professur Biotechnologie in der Ressourcenwirtschaft.

Authors' contact information

Claus Bogenrieder
Züblin Umwelttechnik GmbH
Otto-Dürr-Str. 13
D-70435 Stuttgart
Germany
Tel. +49 711 8202-209
E-mail: claus.bogenrieder@zueblin.de

Dr Andreas Maile
STRABAG Umwelttechnik GmbH
Vogelsanger Weg 111
D-40470 Düsseldorf
Germany
Tel. +49 211 6104-570
E-mail: andreas.maile@strabag.com

Regional concepts for a direct valorization of biogas - Case studies in Austria and Canada

Kübler, Hans; Pellegrini, Roland; Rahn, Torsten; Schulte, Stephan

BTA International GmbH, Pfaffenhofen an der Ilm, Germany

Abstract

„For the substitution of fossil fuels for the production of electricity different renewable energy sources are available. Far more difficult is the substitution of the fossil fuels for the production of heat. This will be a key for any successful transition to renewable energies.

Biogas industry can not only contribute by production of biogas which can be utilized in combined heat and power plants (CHP) or upgraded to natural gas quality and injected into the gas grit, but wherever possible by smart concepts that allow for the regional valorisation of biogas for the direct production of heat. This not only allows reaching a higher energy conversion efficiency, but also contributes to a sustainable development of the region.

In this paper we present two case studies for such a regional valorisation of the biogas for heat production. In Zell am See, a region with a strong touristic impact, the heat is used to partially cover the heat demands of a thermal bath, while in Varennes it is intended to reduce the carbon intensity of an ethanol facility.

Keywords

Source Segregated Organics, biowaste, foodwaste, wet pre-treatment, anaerobic digestion, biological desulfurization, biogas valorization,

1   Introduction

For the substitution of fossil fuels for the production of electricity different renewable energy sources are available. Far more difficult is the substitution of the fossil fuels for the production of heat. This will be a key for any successful transition to renewable energies.

Biogas industry can not only contribute by production of biogas which can be upgraded to natural gas quality and injected into the gas grit, but wherever possible by smart concepts that allow for the regional valorisation of biogas for the direct production of heat. This not only allows reaching a higher energy conversion efficiency (see figure 1), but also contributes to a sustainable development of the region.

In this paper we present two case studies for such a regional valorisation of the biogas for heat production. In Zell am See, a region with a strong touristic impact, the heat is used to partially cover the heat demands of a thermal bath, while in Varennes it is intended to reduce the carbon intensity of an ethanol facility.

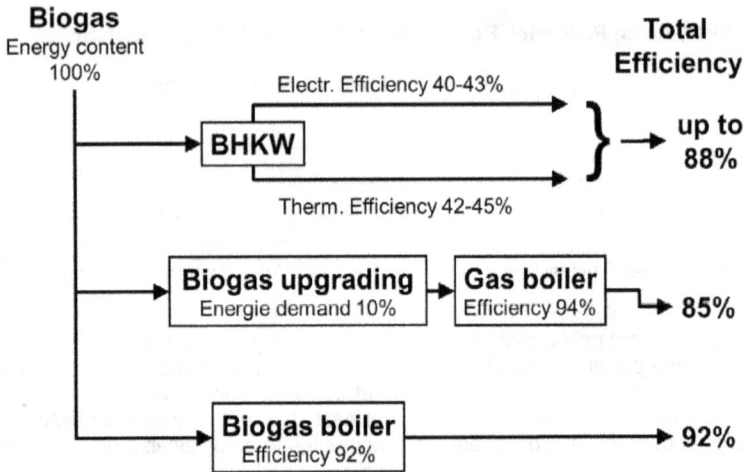

Figure 1   Efficiency degree of different valorization paths from biogas

2   Biogas Plant ZEMKA

2.1   Background and plant description

The Biogas Plant ZEMKA was built with the aim to allow for an energetic valorisation of the Source Segregated Organics (SSO), previously composted at ZEMKA´s facilities, and to be able to process other waste streams produced locally which previously were transported long distances for their treatment in other biogas plants or even thermal valorisation plants.

For this reason, emphasis was made on the possibility to receive and treat different types of organic residues: SSO, food waste, sewage sludge, fat separators' content, and liquid residues. The design capacity of the plant is 18.000 ton per year.

To ensure the required high substrate flexibility, different reception lines were foreseen for the various waste streams. While those residues without impurities are directly fed to the suspension buffer, eventually after a TS-adjustment in the mixing tank, the SSO and food waste first are directed to the BTA® Hydromechanical Pre-treatment.

This wet pre-treatment facilitates efficient removal of impurities as well as nearly complete transfer of digestible organic components into an organic suspension for further anaerobic digestion. This happens in two steps. In the BTA® Waste Pulper the digestible organics are defibred and dissolved, while removing coarse impurities like bones, glass, plastics or woody material. The obtained organic suspension is cleaned from the fine inerts in the next step, the BTA® Grit Removal System.

Thus a clean, easy to manage suspension is won and downstream plant components are reliably protected from wear, silting up, sediments, and obstruction.

The organic suspension is further processed in a one-step wet anaerobic digestion process in the mesophilic range within a fully mixed reactor with a gas mixing system. A suspension buffer allows bridging the weekend, balancing potential peaks and ensuring a continuous feeding of the digester. The discharge of the digestate occurs by a further buffer tank to bridge the different operation time of the digestion and the dewatering step.

Figure 2   Biogas Plant ZEMKA in Zell am See, Austria

The digestate is dewatered in a centrifuge by addition of flocculants. The digestate solids are stored for further processing. The centrate is widely recirculated without any treatment for mixing with the waste or sludge in the pulper and the mixing tank. The remaining centrate passes a 150 µm bow sieve to reduce suspended solids content and recirculated as rinsing water. The surplus amounts are stored in a basin from where they are pumped, in function of the actual load, to the neighboring WWTP "Zeller Becken".

To cover the own heat demand part of the produced biogas is used in a dual boiler (biogas / natural gas). Furthermore, two valorization paths are foreseen. The main consumer of the biogas is thermal bath Tauern SPA in Kaprun, about 2 kilometers away, where the biogas is converted to heat in a boiler. Furthermore, a pilot plant has been implemented by the energy supply company Salzburg AG, for biogas upgrading to natural gas quality.

This leads to special requirements for the cleaning and drying of the gas. The H_2S content is reduced to below 100 ppm by a biological desulphurization unit. The installed facility allows the dosing of air or, optionally oxygen produced on-site. This is necessary in case the biogas upgrading path is active, as this only allows for very low amounts of inert gases (N_2).

Furthermore, the biogas must pass a cooling unit with condensate removal that cools down the gas to -5°C to avoid the formation of condensate in the nearly 3 km long underground gas pipe to the Tauern spa. This is achieved in a three-step condensation drying process (with ambient air, cold water and cold brine).

2.2 Operation experience

The hot commissioning of the Biogas Plant ZEMKA started in August 2013 and acceptance test was already five months later, in February 2014 at nominal capacity.

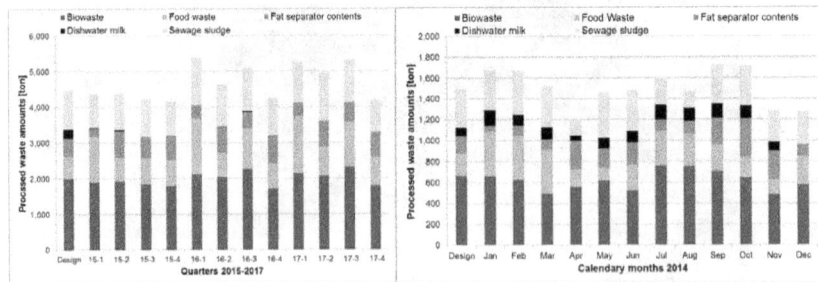

Figure 3   Input streams into the Biogas Plant ZEMKA

The waste streams processed quarterly for the years 2015 to 2017 are given in Figure 3 (left), showing the seasonal fluctuations. Especially the amount of food waste strongly increases in the first quarter of the year as Zell am See is a renowned winter ski region. The monthly variations are even larger, as can be seen on the right side of Figure 3 (right), which shows the monthly distribution of residues processed in the facility in 2014.

Figure 4 shows the methane content in the biogas for the years 2014 to 2017. A widely stable methane content is an indication for a stable anaerobic digestion step. Would it come to an increase of fatty acids in the reactor, CO_2 would be released from the am-monium carbonate buffer leading to a decrease in the methane content.

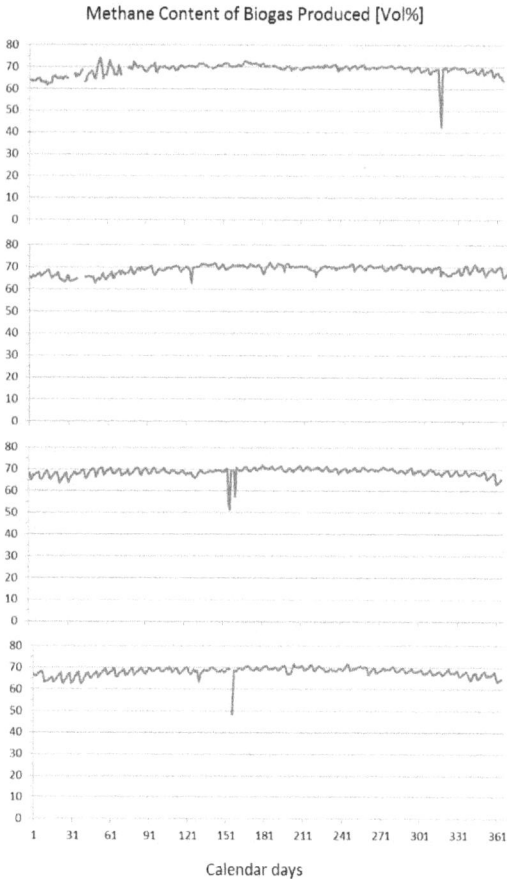

Methane Content of Biogas Produced [Vol%]

Figure 3   Methane content in the produced biogas 2014 – 2017

The storage volume for the different waste streams is limited, therefore fluctuations in the input amounts and in the mix of the waste streams leads to strong fluctuations in the biogas production. Especially the share from food waste in the overall input amounts has a strong impact on the biogas production (please refer to Figure 5).

The methane yield is shown in Figure 6. Due to the co-digestion, the reactor was designed for long hydraulic times of about 34 days referring to the volume of reactor effluent to allow a good degradation rate even for material with worse degradability.

Over the years 2016 and 2017 an average biogas production of nearly 3.9 million Nm³/year are produced. From this, about 5% are needed to cover the own process heat demand, while 95% were provided for external use, mainly in the Tauern Spa.

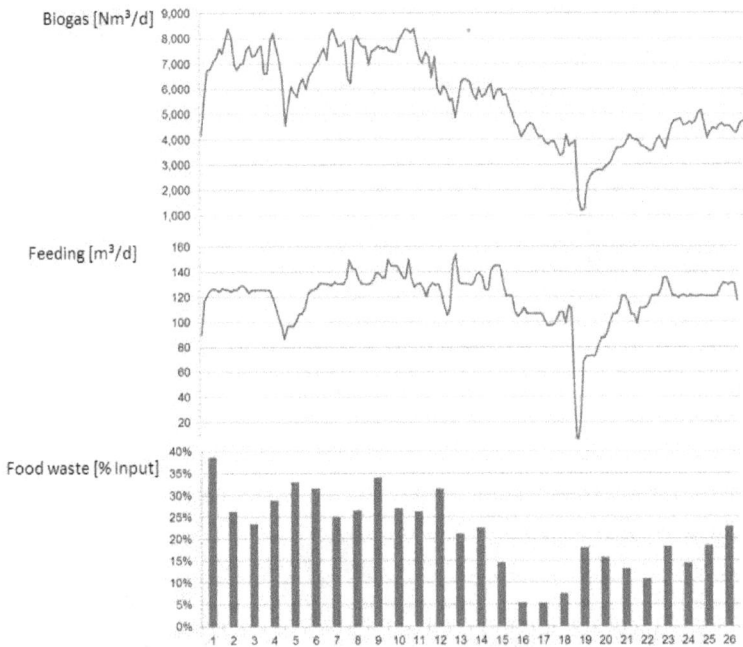

Figure 5   Influence of food waste share on biogas production

Methane yield [Nm³/ton$_{VS}$]

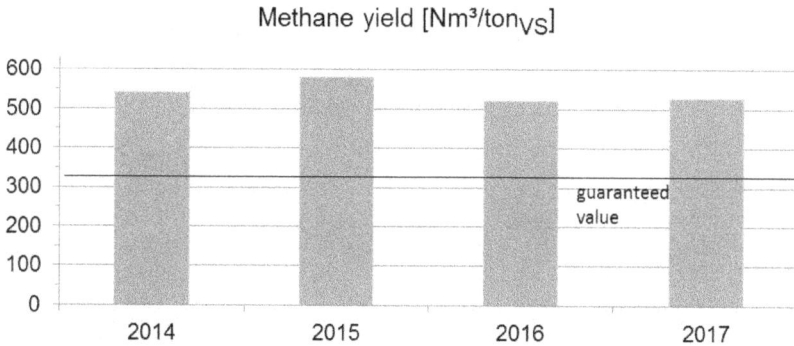

Figure 6   Impact of hydraulic retention time on methane yield

3   SEMECS Organic Waste Treatment Facility Varennes

3.1   Background and Plant Description

The Regional County Municipalities (RCM's) of La Vallee-du-Richelieu, Marguerite-D'Youville and Rouville, representing 27 municipalities totalling more than 235,000 people, are implementing new household SSO diversion programs by providing curbside collection.

A public-private partnership between the consortium of these three RCM's and the private partner, BiogazEG was established under the name "Société d'Économie mixte de l'est de la Couronne Sud Inc." (SEMECS) with the goal to build and operate an anaerobic digestion plant to process these residues: the SEMECS Organic Waste Treatment Facility Varennes in Quebec, Canada. This facility will provide more than 60% of the organics diversion needs to achieve the Quebec government's mandated policy objective in banning organics from landfills as of 2022.

The facility is designed for an annual processing capacity of 40,000 tons (maximally 51.000 ton/year) of wet residential SSO, green waste (grass clippings), septage (septic tank sludge) and liquid industrial and commercial organic wastes. The technological platform for this plant is the BTA® Process, provided by BTA International and its Licensee for North America - CCI BioEnergy Inc.

The SSO, grass clippings, commercial organic wastes and septage are pre-treated via the BTA® Hydro-mechanical Pre-treatment, while the septage and liquid industrial and commercial organic wastes are directly to the anaerobic digestion step.

Figure 7   SEMECS Organic Waste Treatment Facility Varennes in Quebec, Canada

Digestion is carried out in two digesters in series. The first digester, with a volume of 5,400 m³, is executed as vertical reactor, fully mixed by the injection of compressed gas with gas lances. The second digester (3,400 m³) is executed as concrete reactor with horizontal agitators and a gas dome of 1,800 m³ on top.

Part of the produced biogas is used to fuel the boilers to produce the heat required to sustain the mesophilic operating temperature (37°C). The additional biogas will be cleaned-up for use at the on-site Green Field Global Facility to offset up to 20% its natural gas use and thus significantly lowering the carbon intensity of the ethanol production.

The digestate will be dewatered with the reclaim and reuse of the liquid fraction. Excess liquid will be treated on-site via a wastewater treatment plant for effluent discharge to the St. Lawrence River.

The nutrient rich stabilized and dewatered (cake) solids will be hauled to more than 400 farms which grow the corn that supplies the GreenField Global Facility. The high quality digester solids allows for direct land application in accordance with Canadian Food Inspection Agency (CFIA) requirements as a registered organic soil amendment for agricultural uses.

3.2 Operation experience

The plant has been started up in early March 2018 and has been ramped up to normal production capacity by end of July 2018. The acceptance tests have been successfully completed by July 31st 2018.

During the months May to July 2018 the average methane content in the produced biogas was around 63%. The production in July 2018 was about 260.000 Nm³ and specific biogas production lay over 100 Nm³/ton input to the plant.

4 Outlook

The traditional approach for long time has been the conversion in CHP units for the generation of electricity and heat. The efficiency degree strongly depends whether the excess heat can be used or not. In positive case, overall efficiency could climb up to 85%. Similar to the overall efficiency of the biogas upgrading (please refer to fig. 1). The advantage of this solution is that the heat consumer does not need to be in the immediate neighbourhood of the biogas plant, the gas grid acts as a huge gas holder. Yet, the additional upgrading step limits the overall efficiency. Therefore, the direct valorization of the heat may be, from efficiency point of view, the preferred alternative as long as local conditions allow for it. This allows also for a stronger regional rooting of the facilities and supports the sustainable development of the region, as illustrated with the two case studies.

5 Literature

References should be listed like this:

Kübler H., Schulte S., Winter L., Balke H., Pilz G.   2015   Müll und Abfall 6 15. Die Biogasanlage ZEMKA Eine Vergärungsanlage mit hohen Ansprüchen an die Abfallannahme und Aufbereitung sowie an die Biogasbehandlung, pp. 3016 - 320

Author's address(es)

Stephan Schulte, Roland Pellegrini, Hans Kübler and Torsten Rahn
Färberstr. 7
D-85276 Pfaffenhofen an der Ilm

Telefon +49 8441 8086 611
E-Mail s.schulte@bta-international.de

Treatment of Waste Derived Liquids

Bernd Fitzke, Frank Natau

WEHRLE Umwelt GmbH, Emmendingen, Germany

Treatment of Waste Derived Liquids

Abstract

Sanitary landfilling is globally the first choice to start professional waste management. Proven technologies control, eliminate or re-use related emissions like leachate or methane. These are available worldwide with high efficiencies and acceptable costs. To increase recycling rates, Mechanical-Biological Treatment technologies (MBT), which are able to produce "goods from waste", have been installed in Europe. Far higher recycling rates from waste can be achieved – savings of landfill capacity, "waste to energy" and "circular economy in waste" have become reality. Based on examples from installed treatment plants, this presentation shows the origin and quality of waste derived liquids and what is needed to treat them – taking into consideration not only the "classical leachate" from landfilling.

Summary

This publication describes the differences between landfill leachate and effluents from Mechanical-Biological Treatment (MBT) of waste or other waste handling processes. It highlights the challenges in treating each effluent type alone or even in a mixture, i.e. leachate from an old landfill and an MBT plant at the same site. As similar projects over several years have shown, the most cost-effective solution is a high-performance Membrane Bioreactor (MBR) with external cross-flow membranes, depending on site-specific conditions and discharge limits also with additional polishing technologies, such as nanofiltration, reverse osmosis or activated carbon.

Keywords

Sanitary landfill, mechanical-biological treatment of waste, MBT, waste derived effluents, MBT wastewater, RDF, Membrane Bioreactor, MBR, Reverse Osmosis

1 Waste Derived Liquids

1.1 What are waste derived liquids?

The handling of municipal solid waste (MSW) takes place in many ways. It is collected, loaded, sorted, treated, recycled and dumped in a variety of procedures using lots of different methods and processes. This handling of the waste results in the generation of a liquid residual, either through unwanted contact with water and through the release of the liquid components contained in the waste or through the targeted use of water in waste handling and treatment processes. This liquid residual is commonly called leachate or – reflecting the different ways and processes of its generation – waste derived liquid.

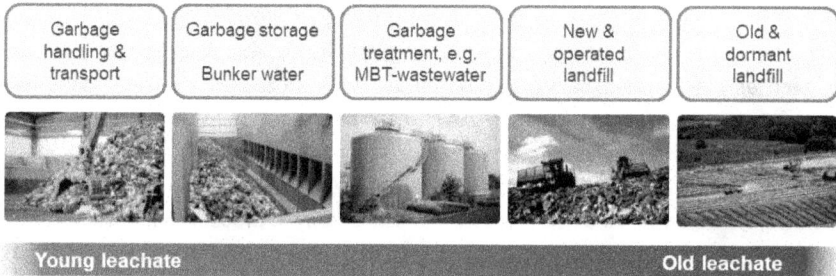

| Garbage handling & transport | Garbage storage Bunker water | Garbage treatment, e.g. MBT-wastewater | New & operated landfill | Old & dormant landfill |

Young leachate   Old leachate

Figure 1   Typical sources of waste derived liquids

1.2   What are the compositions of waste derived liquids and how do they differ?

Waste derived liquids originate from water contained in MSW, from infiltration water that seeps through the waste and from water that is used for or generated during waste processing, treatment and disposal.

The so generated leachates are highly polluted due to a high content of ammonium ions and organic compounds. In most cases they are toxic, acidic and rich in halogenated hydrocarbons. They have a high buffer capacity, possess high amounts of inorganics (salts) and usually show an unbalanced COD to NH4-N ratio. Furthermore, the COD is often hardly or even non-biodegradable, the so-called "hard COD". The leachates can contain elevated values of chloride and sulphate ions as well as high concentrations of common metal ions, especially iron or even heavy metals. In some European countries, new pollutants like pharmaceuticals, pesticides or fluoro-surfactants and even nanoparticles have recently come into public focus and are discussed on expert levels and among authorities.

1.2.1 MSW handling and transport

Leachates generated during the collection, handling and transport of waste (e.g. at transfer stations) can be considered as fresh leachates. They are characterized by a high organic load and comparatively little ammonium nitrogen. Their biodegradability can be considered as good.

1.2.2 MSW storage

The same applies to the effluents generated in collection bunkers for waste incineration plants. Yet due to elevated retention times in bunkers, the nitrogen conversion towards ammonia could be initiated and the ammonia levels might rise.

1.2.3 MSW treatment

Wastewaters from waste treatment plants using anaerobic biological process technology, such as pure fermentation plants or various MBT processes, generally have less
dissolved organic pollution. However, because of the biological transformation processes and the reductive milieu, they contain large amounts of ammonium, which is generated from the previously organically bound nitrogen. As those systems usually shred the
waste before treatment and add more mechanical energy to processes for mixing, many
ingredients are broken down into very fine particles. However, this mode of operation,
which is required for the hydrolysis and mass transfer in the reactors, is responsible for
the fact that both organically largely inert and biodegradable organic matters leave the
processes with the process effluents in large quantities as a very fine fraction, which
leads to considerably increased solids contents in such wastewaters and finally these
solids significantly contribute to the total COD load of the effluents.

Since many of these processes provide post-composting, the wastewater produced during composting also contributes to the high particulate loading of the process effluents.
Other auxiliary process steps like biogas drying and desulphurisation, biofilters, cleaning, etc. also generate effluents in different amounts (see Table 1).

Table 1   Example for different effluents on an MBT site (Mataró, Spain)

Source	m³/a
Fermentation	28,780
Pre-humidification	7,300
Biofilter	3,650
Bunker	2,850
Incineration	2,080
Composting	2,000
Desulphurisation	2,000
Cleaning	2,000
Total	**50,660**

1.2.4 MSW disposal

The quality and pollution of the landfill leachate strongly depends on the type and composition of the waste disposed of. Another factor is the climate (quantity of precipitation
on and evaporation from the landfill). In addition to that, landfill leachate changes ac-

cording to the age of the waste disposed of. Whereas leachate with high but easily degradable organic substances (BOD/COD = 0.3 – 0.5) is produced in younger landfill parts, the organic loads decrease on older landfills, yet including lower degradability of the substances (BOD/COD ≤ 0.1) and higher ammonium concentrations.

Table 2 shows some basic pollution parameters of MSW derived liquids.

Table 2   Basic pollution parameters of MSW derived liquids

	COD [g/l]	BOD5 [g/l]	TKN [g/l]	TSS [g/l]
Waste transfer stations	65 – 80	40 – 50	1 – 2	2 – 5
Waste bunkers	50 – 70	25 – 30	2 – 3	3 – 6
MBT effluents	12 – 50	5 - 15	2 – 6	5 – 30
"Young" landfill leachates	10 – 40	3 – 18	1 – 3	0.2 – 1.5
"Old" landfill leachates	0.5 – 5	0.1 – 1	1.5 – 5	0.1 – 0.8

2   Treatment of waste derived liquids

As described above, the pollution of waste derived liquids is very much determined by the handling / treatment / discharge process they originate from. Yet also the types of waste as well as climatic and social constraints are crucial (FITZKE, 2013).

In order to select a suited treatment process, those factors as well as the required discharge limit values for certain parameters must be considered.

Should there be only limit values for COD and BOD, a biological treatment process is certainly the best solution. For nitrogen elimination, a biological process (with nitrification / denitrification) is also the most suitable process. Should there be obligatory limit values also for inorganic substances (salts), further treatment steps must be added to the biological treatment. For this purpose, filtration systems such as nanofiltration (NF) and reverse osmosis (RO) are often used. NF and RO can also be used as sole treatment processes for smaller effluent quantities.

Whereas the dimensioning of a treatment plant mainly depends on the actual load and quantity of the effluent, the determination of the appropriate process or process combination is above all a matter of observing the respective limit values. The processes available may hence be classified according to the discharge limits fixed.

2.1   Low discharge requirements

If the only requirements are BOD5 reduction and partial oxidation of ammonium (NH4-N), so-called Sequencing Batch Reactors (SBR) are often used. In this activated sludge process, the different process steps (charging, mixing, aerating, sedimenting, discharging) are combined in one reactor (see Figure 2).

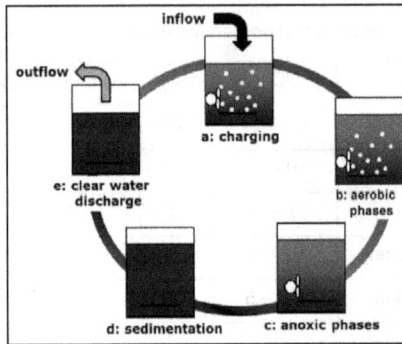

Figure 2   SBR process

Due to the simple assembly (only one reactor), the SBR process is cost-effective and quickly to install. Discharge qualities of NH4-N < 100 mg/l and BOD5 < 50 mg/l are easily achieved for average leachate compositions. Difficulties may appear in case of higher pollution loads or installation in colder regions.

2.2   Medium discharge requirements

In case of higher requirements with regard to the discharge quality of the leachate treatment plant, not only BOD5 elimination and ammonium oxidation but also COD and TN (total nitrogen) reduction are necessary.

For this purpose, Membrane Bioreactors (MBR) have become the standard process. The MBR consists of a bioreactor system and an ultrafiltration stage (see Figure 3), being an activated sludge process at the same time. MBRs achieve a more comprehensive pollutant reduction than other aerobic systems, requiring less space. Due to the high treatment capacity and full automation, this technology can also easily cope with variations of leachate quantity and composition.

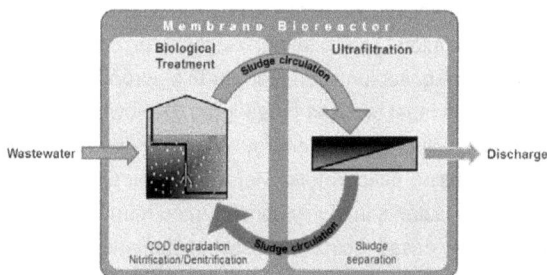

Figure 3   MBR process

Besides, should the discharge requirements increase, a clear outlet free of solids serves as basis for directly adding activated carbon adsorption plants or nanofiltration systems as further treatment steps at reasonable costs (see Figure 4), making it possible to comply with very high discharge requirements without producing concentrates.

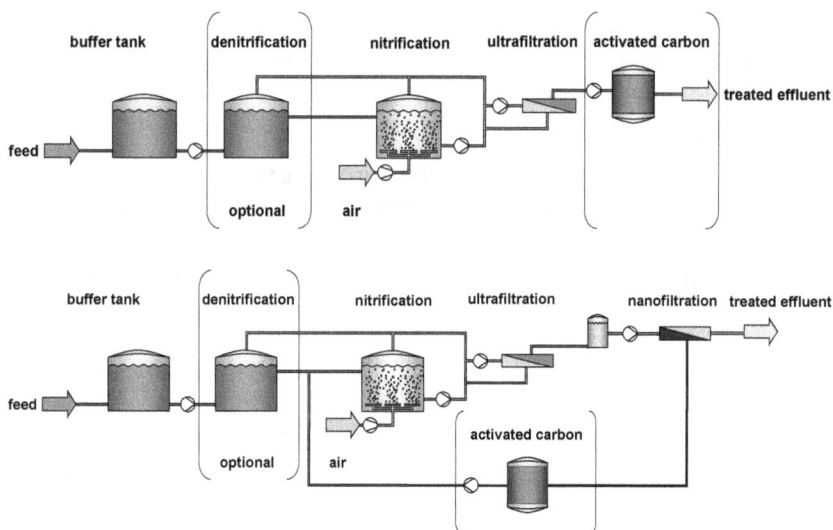

Figure 4   MBR combination process

2.3   High discharge requirements

As mentioned above, certain countries have established legal regulations prescribing not only full elimination of organic pollutants and total nitrogen but also the reduction of salts. Some of those inorganic substances can be reduced by using physical-chemical processes, such as flocculation / precipitation. Yet those processes do not eliminate the salts of monovalent ions, e.g. sodium chloride. For this purpose, either evaporation processes or reverse osmosis must be used (see Figure 5). However, both processes have several disadvantages, including high energy consumption and significant emission output, especially when using evaporators. Moreover, their final product is concentrate containing the leachate pollutants in highly concentrated form. This concentrate must be disposed of, being in many cases recirculated back to the landfill.

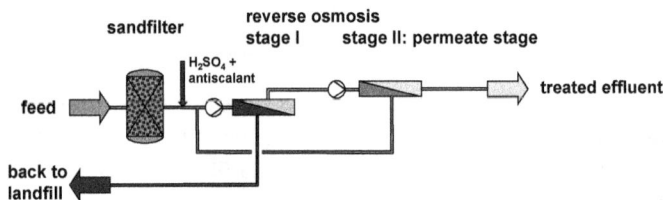

Figure 5   Reverse osmosis process for leachate treatment

Compared to evaporation processes, reverse osmosis plants are considerably less expensive and also involve lower operating costs. Their biggest disadvantages are the relatively high concentrate quantities and additional salt loads caused by chemicals used for operation, having to be recirculated back to the landfills.

A solution to reduce the concentrate quantities ensuring at the same time the very good outlet quality of the RO is a combination of MBR and RO (see Figure 6). As mentioned before, this is unproblematic due to the particle-free MBR outlet. Another advantage of this combination is the fact that the nitrogen has already been eliminated in the MBR, making the RO less complex and, consequently, less expensive.

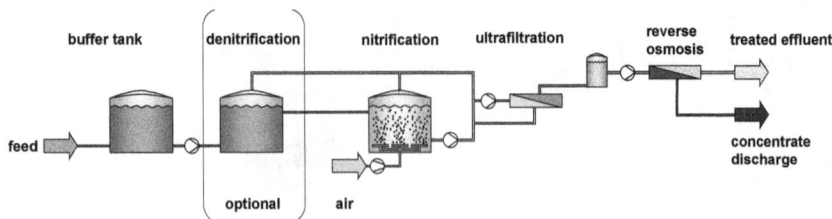

Figure 6   Reverse osmosis process for leachate treatment

3 Synergies through the joint treatment of different waste derived liquids

In many projects, MSW landfills and waste sorting and treatment facilities are located in the immediate vicinity. Landfilling is often supported by appropriate separation and MBT pre-treatment, which significantly reduces the amounts of waste to be landfilled and the emission potential of the remaining landfilled fraction due to the biological treatment and substantial reduction of biogenic waste components.

Due to the fact that different waste treatment and disposal processes are operated simultaneously at such sites, the effluents described above also occur simultaneously. There are often already existing facilities treating the leachate of the existing landfill. Yet are such plants also able to treat the effluents from sorting and treatment processes? And how does the changeover from landfilling alone to waste treatment affect the amount and quality of the landfill leachate?

3.1 Landfill leachate

The characteristics of leachate alter with time. Leachate obtained at acetogenic, methanogenic and stabilization phases of landfills are classified as young, intermediate and stabilized leachate. COD and biodegradability decrease but pH and TKN usually increase from young leachate to stabilized leachate. If the discharged wastes furthermore result from MBT treatment with minimized biogenic content, the organic pollution in the leachates will remain low but the biodegradability will become even lower while the TKN remains high or even rises.

3.2 MBT effluent

Within MBT treatment of MSW, the fermentation units produce highly contaminated (COD, N) wastewater with very specific morphology, inhibitory effects and a general high level of TSS. This wastewater cannot be recycled to the process and has to be treated externally according to the local requirements for discharge.

In the last 10 years, various projects in Europe have shown that this kind of MBT wastewater cannot be treated alone with different single technologies (CAS, SBR, membrane processes UF/NF/RO, evaporation, AOP processes, etc.) on a commercially acceptable level with long-term process reliability.

Due to unbalanced C/N levels and the requirement to establish a biocoenosis comprising either slow growing heterotrophic specialists or sufficient autotrophic nitrificants, extremely high sludge ages are required to ensure the conditions for an adequate growth of the required microorganisms. Stable temperatures within the optimum of the mesophilic range and non-fluctuating but always low reactor concentrations of the inhibitory pollutants (steady and non-phased operation) are mandatory for a successful operation. To establish the required sludge ages and to also guarantee the full retention of biomass and proper hydrolysis of the inflow TSS, mainly membrane bioreactors (MBR) are applied for the treatment of these effluents, which are characterized by an extreme proportion of suspended organic solids (see Table 2)

Table 2   Basic pollution parameters of MSW derived liquids

	MBT Process Location	Volumetric flow [m³/d]	COD [mg/l]	TSS [mg/l]
ECOPARC 1	LINDE / BTA Barcelona, ES	204	15,000	13,800
ECOPARC 2	VALORGA Barcelona, ES	142	50,000	30,000
MATARO	BTA Mataró, ES	180	10,500	4,500
LAS DEHESAS	VALORGA Madrid, ES	192	40,000	30,000
LA RIOJA	KOMPOGAS Logroño, ES	120	23,000	9,800

3.3   Joined treatment of landfill leachate and MBT effluent

Current state-of-the-art technology is the mixture of MBT wastewater with other suitable streams in a combined treatment to overcome shortages of water accrual and to assure an economic operation at acceptable throughputs with sufficient nutrients, micronutrients and trace elements. To maximize the activity of even slow growing biomass and to intensify mass transfer, higher shear is required as well as good mixing in combination with an advanced and maintenance-free aeration system. For this reason, a jet aeration system in combination with tall reactors is the best solution. It is a matter of known knowledge that biological systems with these properties tend to be composed of very homogenous biomass without flocs or aggregates. Their properties (viscosity, particle size, particle concentration...) can be considered as extreme for every available solid-liquid separation system. Therefore a flexible but reliable biomass retention system with minimum negative impact on the degradation processes in the bioreactor and maximum particle retention and slight retention capabilities of even dissolved high molecular COD is indispensable. These properties are covered by MBR technologies with side-stream UF membranes (BARMI, 2016).

The dimensioning and design of such plants requires not only know-how for the treatment of such wastewater but also knowledge of the operation of MBT processes and their internal water management and operation regime. This is due to the fact that, for a specific location, the quantities and qualities of leachate predominantly affected by climatic events are to be controlled in line with the flow characteristics and quantities of MBT sites that change, especially as a result of the operating regime of the MBT plants.

As an example for the successful treatment of landfill leachate and MBT process water, see Figures 7 and 8. The example shows a compact treatment plant equipped with MBR technology using external ultrafiltration (containerised installation). At this site, the MBT plant was built in the immediate vicinity of a landfill. So it was appropriate to combine the treatment.

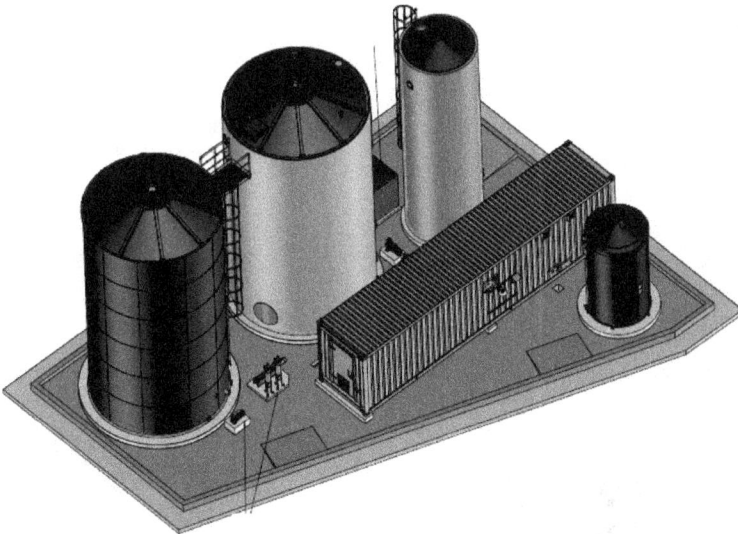

Figure 7   Layout of a combined leachate and MBT effluent treatment plant, Poland
(WEHRLE 2018)

Figure 7   Combined leachate and MBT effluent treatment plant, Poland (WEHRLE 2018)

Yet besides the organisational and commercial advantages which such a project offers (one project, one plant, lower specific investment, less operating expenses, less personnel…), there are also many synergies in the operation of the plant. The highly polluted and partially problematic process water of the MBT plant is diluted down by the leachate to enable biological treatment without inhibition (see Figure 8).

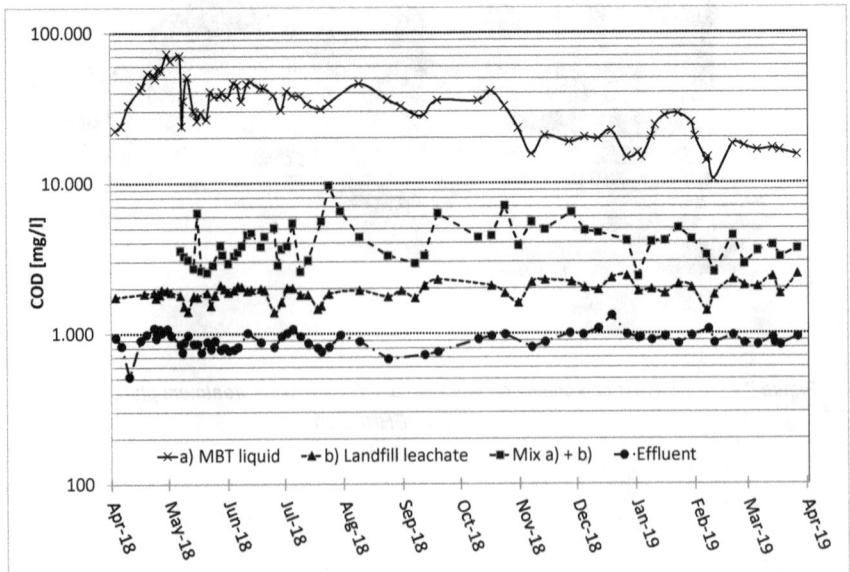

Figure 8   COD elimination in a combined treatment process

The sometimes very high solids contents of the MBT effluents undergo dilution. As MBR technology with full solids retention is used and the leachate contains comparatively little biogenic COD, these solids undergo a much greater degradation due to complete retention, hydrolysis and increased mineralization. Thereby the effluents are less polluted and no longer accumulate in quantities formally known as surplus sludge that needs to be disposed of at expensive costs.

Due to the mixture with the MBT effluent, the unbalanced leachate with its very low BOD to N ratio receives enough BOD to allow for complete elimination of total nitrogen without having to add a carbon source, caustic soda or other chemicals (see Figure 9).

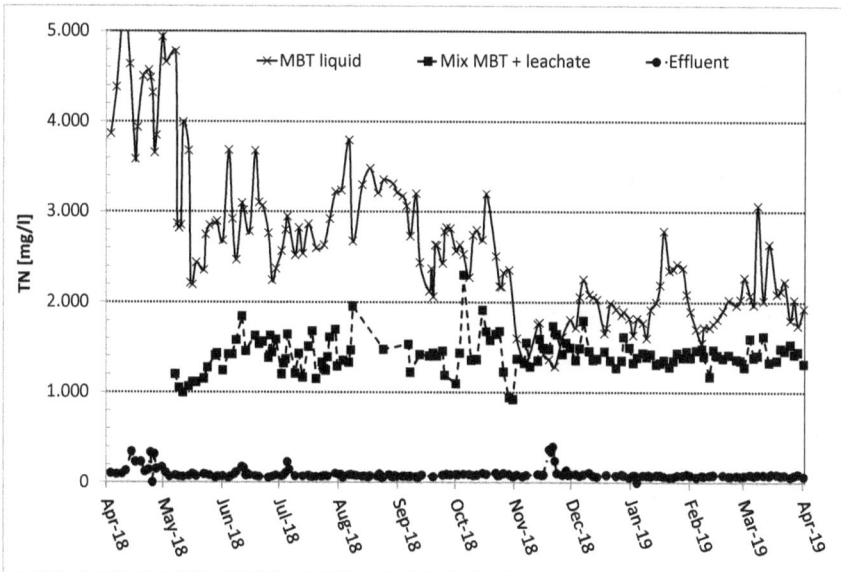

Figure 9   TN elimination in a combined treatment process

In general, the example mentioned above shows very well that a joint treatment of different effluents from waste handling and treatment processes is possible and that, with an appropriate design and an adapted operation of the treatment plant, it offers many synergies to significantly reduce the treatment costs. The single effluents, which are problematic themselves, lose some of their disadvantages in the mixture. The plant operation becomes more stable with increasing effluent qualities.

4   Literature

Barmi, A.; Bennett, A.   2016   The hidden problem of landfill leachate. Filtration +
                                        Separation, ISSN: 0015-1882, Vol: 53, Issue: 2,
                                        Page: 30-35

Fitzke, B.; Blume, T.:   2013   Hybrid Processes for the Treatment of Leachate from
Wienands H.; Cambiella, A.   Landfills in Economic Sustainability and Environmen-
                                        tal Protection in Mediterranean Countries through
                                        Clean Manufacturing Methods (2013), Springer Dor-
                                        drecht, ISBN 978-94-007-5078-4, S. Page 107-126

Authors' address:

Dr.-Ing. Bernd Fitzke
WEHRLE Umwelt GmbH
Bismarckstr. 1-11
79312 Emmendingen, Germany
Phone: +49 7641 585 355
E-mail: fitzke@wehrle-umwelt.com

Dipl.-Ing. Frank Natau
WEHRLE Umwelt GmbH
Bismarckstr. 1-11
79312 Emmendingen, Germany
Phone: +49 7641 585 225
E-mail: f.natau@wehrle-umwelt.com

First Investigation of the Occurring Stresses During the Collection, Transport and Unloading of Waste Electrical and Electronic Equipment

Ralf Brüning and Julia Wolf

Dr. Brüning Engineering UG

Abstract

Waste electrical and electronic equipment (WEEE) can contain a variety of hazardous substances under certain circumstances. Monitors are of particular concern, as they can contain mercury (in flat screens) and other substances like luminous coatings that contain heavy metals (in analogue screens). In practice, monitors are frequently destroyed or damaged during logistics and tipping processes, and pollutants, such as mercury from flat screens, are released. To optimise logistics processes and prevent pollutants from escaping from WEEE in practice, scientific readings were taken for the first time that can help to determine which stresses occur during logistics processes. The findings should now be used in the development and refinement of logistics processes and collection and transport containers.

Keywords

Stresses, collection, transport, unloading, waste electrical and electronic equipment (WEEE)

1 Introduction

More than 700,000 tonnes of waste electrical and electronic equipment (WEEE) are collected in Germany each year. The majority of this equipment is collected and transported in roller containers about 40m³ in size. WEEE frequently contains substances that can pose a significant risk to the environment and human health if not handled properly. They include, for instance:

- mercury

- heavy metals

- toner dust

- batteries.

In this vein, it is important to note that logistics processes are very different in the recycling business than when handling new products. For example, careful trans-shipment processes, weather protection and packaging goods are not the rule in the recycling business. While new electrical and electronic appliances are generally transported carefully as piece goods, bulk cargo transportation is the norm in the recycling business. For instance, WEEE being tipped from a roller container on to a concrete floor from a height of one or two metres is everyday practice. Unlike new appliances, WEEE is generally not packaged. Weather protection is required under German WEEE legislation, but is rare in practice.

Pollutants, such as mercury, may be released from flat screens if WEEE is damaged during logistics processes. This can cause considerable damage to the environment, especially when pollutants like mercury and heavy metals in luminous coatings are emitted. It should be noted that, in addition to having a negative environmental impact, pollutants can also put employees at risk - something that must be avoided at all costs. Nonetheless, WEEE is still repeatedly destroyed by logistics processes in practice.

2   Analysing WEEE logistics processes

2.1   Identifying typical transport and transhipment processes

The research project presented here aimed to use of data loggers to help to determine which stresses generally occur during different logistics processes since pollutants are released from WEEE time and again in practice because of very high mechanical stress.

Therefore, transportation and transhipment processes were documented at several public waste management authorities and recyclers in a first phase. Typical processes were identified, which were to be examined in greater detail in this research project. The use of data loggers was chosen for documentation purposes.

One typical process when unloading WEEE is tipping, which repeatedly occurs in practice and is shown in the picture below.

Figure 1: WEEE being tipped

Alternatives, such as loading by hand or mechanical loading, were taken into account during the assessment so that data could be compared during subsequent analysis of readings, if necessary.

2.2   Suitable measuring instruments

As the next step after identifying the processes to be analysed, it was important to select data loggers[1] that can register very high accelerations because WEEE falling during tipping was also supposed to be mapped.

Since loading and tipping processes are not directed movements, data loggers were chosen that can record accelerations in all three axes.

Data loggers should also be capable of being used in the harsh environment of the recycling business and allow readings to be recorded for a period of at least 24 hours. This step is necessary to be able to document real transportation processes in practice.

2.3   Testing

Data loggers were screwed tightly into power packs for practical testing in order to guarantee an additional layer of mechanical protection.

---

1 The MSR 165 data logger was used in the project, for more information, visit: https://www.msr.ch/de/.

Figure 2: Attaching the data loggers

Moreover, power packs containing data loggers were also packed in plastic bags to guarantee weather protection.

Since a WEEE container can contain several thousand small appliances, power packs were marked with about 2 m of barrier tape, which made them easier to find in piles (see image below).

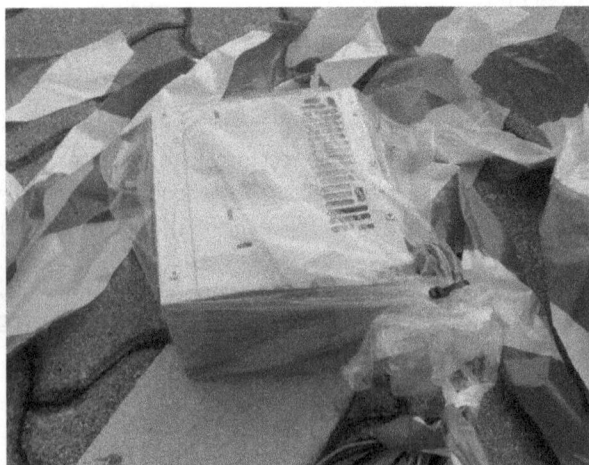

Figure 3: Protection and marking

Several data loggers were placed at different locations in roller containers so that readings could be taken.

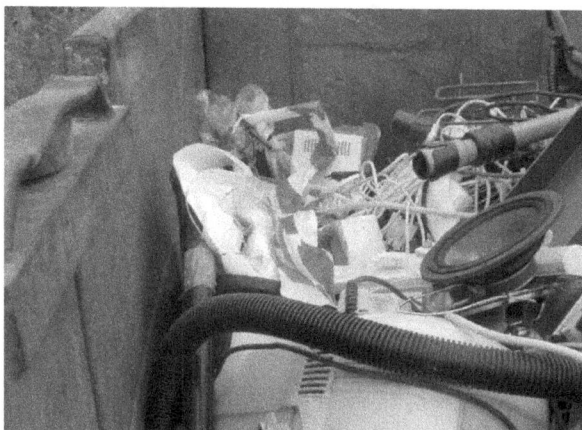

Figure 4: A power supply pack with a data logger in use

This step ensured that data loggers undergo the same typical transportation, transhipment and storage processes as other WEEE in the containers. Both transport and transhipment processes, such as container tipping, were then illustrated, with peak contamination expected during transhipment processes. The following processes were mapped in particular:

- Loading containers by hand

- Loading containers mechanically (e.g. using wheel loaders)

- Raising the container on to the transport vehicle

- Transporting by road across different distances (about 100 – 300km)

- Tipping WEEE from the roller container

- Tipping WEEE from the set-down container

3   Initial findings

All of the tests carried out aimed to document accelerations at all three axes and the respective exposure time. This approach allowed the stresses to which WEEE is exposed during the aforementioned logistics processes to be documented and analysed for the first time.

As expected, the readings sometimes show very high acceleration levels, with peak levels of up to 200g, depending on the scenario.

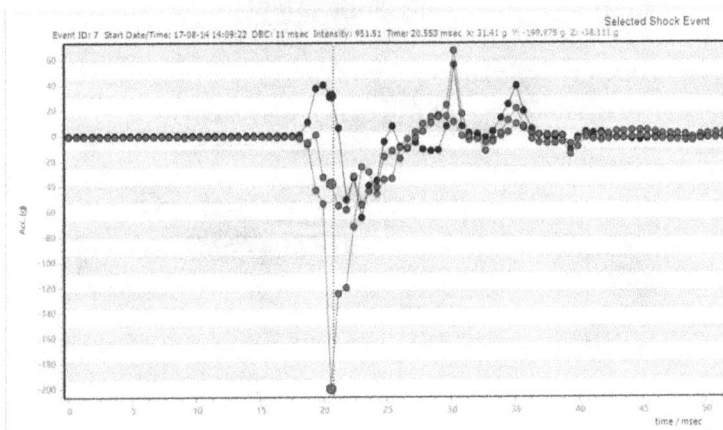

Figure 5: Sample readings

These first readings clearly show that WEEE is subject to the greatest stresses when being thrown into or unloaded from a receptacle into a container. On the other hand, stress value readings are much lower during transportation.

The plan is to ascertain during further tests which appliances are damaged from what stresses and from what point pollutants are emitted. The findings should be used to develop and improve logistics processes and collection and transport containers.

Figure 6: Alternative containers

The measurements now carried out and documented have for the first time laid the foundation for further research into this environmental issue. The goal must be to be able to better protect the environment and employees against pollutant emissions in the future.

Author's contact information

Dr Ralf Brüning
Julia Wolf
Dr. Brüning Engineering UG
Kirchenstr. 26
D-26919 Brake
Germany
Tel. +49 4401 7049760
E-mail info@dr-bruening.de

Use of air-separation tables for density sorting of waste materials

A.P. Kindler, J.J. Cebrian de la Torre and M. Trojosky

Allgaier Process Technology GmbH, Forschung und Entwicklung, Uhingen, Germany

Abstract
A new separation table to conduct density sorting of heaps of material has been developed on the basis of air-separation tables for gravimetric sorting of solid streams. This table is particularly suitable for sorting large streams of solids in the mineral and recycling industry. This presentation describes the development stages and results of laboratory and field testing to optimise this technology and details ex-amples of applications in industrial practice. The new separation table offers a raft of benefits over the machinery available to date, including the ability to adjust the angle of the work surface, easy adjustment of the table lifting frequency and the ability to change the inflow plate quickly and easily.

1 Introduction

The separation of raw and waste materials is playing an increasingly important role in the most varied of sectors, for instance when separating solid waste materials, such as cullet, scrap electrical and electronic goods, rubble, compost, plastic and mineral slag to extract recyclable materials in an environmentally sound and sus-tainable process.

Density sorting (more accurately known as gravimetric sorting) of waste materials separates substances and particles based on their specific weight. For instance, pre-crushed glass can be separated from its concomitant substances and metal slag generated from metallurgical processes can be separated from components with a lower metal content. A wide variety of different separation methods exist for density sorting, which can also be operated with air or water (Bunge 2012, Wills & Finch 2016).

Water-based processes have played a role for decades now, especially when it comes to sorting heavy mineral substances. Its drawback is that recycling the proc-ess water that is used is a complex undertaking and does not always happen consis-tently. The environmental ramifications are huge. A scarcity of water is also intensify-ing in many countries (Harder 2014).

Therefore, for a few years now, there has been a trend towards dry converting proc-ess which are more environmentally sound, can operate at lower costs and can also help to bring about a significant reduction in the overall plant's size in the recycling process (Harder 2013).

One disadvantage of dry separation processes is that sorting quality is worse than with water-based processes for physical reasons. Optimising existing air-based methods is thus urgently needed, especially given rising demands placed on separa-tion processes (e.g. higher purity, poorer ore quality and more stringent environ-mental legislative re-quirements).

In efforts to meet these requirements, Allgaier Process Technology GmbH has opti-mised the processes and mechanics of a technology that has been proven for dec-ades. In the process, the company paid attention to user friendliness, plant costs, maintenance costs and safety and security in order to meet the latest safety and se-curity requirements. The result is the newest product in its portfolio, the GSort air-separation table.

2

2.1   Operating principle

Density separation using the GSort is based on an inclined vibrating base. A flow of ris-ing air crosses this base (Figure 1):

•   Less dense products float with little or no contact with the base and slide to the lower portion of the inclined table because of the incline.

•   More dense products frequently contact the table and are pushed towards the top due to the vibration.

The greater the difference in density of the products, the larger the grain size bands can be. Variables can be set individually, quickly and easily with the GSort – allow-ing ma-chinery to be perfectly tailored to the materials. The best precision is achieved by set-ting the following variables:

•   air speed, which can be adjusted by zone, across the entire base,

•   overflow flap height,

•   the table incline and

•   vibration frequency.

When starting up the GSort, air flow is adjusted and air flow distribution is calibrated using an efficient mechanism over the table's cross-section.

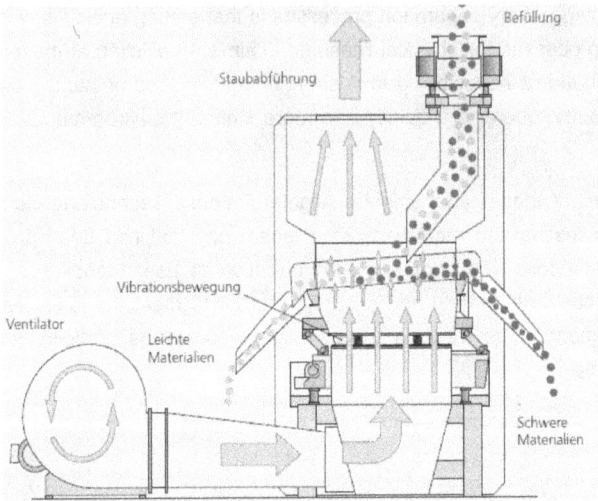

Figure 1: Operating principle of the new GSort air-separation table.

2.2   Analysing the influence of new configuration options on separation behaviour

Experimental testing with new configuration options were systematically carried out based on an air-separation table (GOSAG) that has been on sale for many decades now at a pilot plant at Allgaier Mogensen S.L.U. in Spain. To this end, an existing pi-lot plant (Figure 2) that remixes and recycles material after separation was gradually re-built. The plant's air and bulk material cycle is shown below in diagram form.

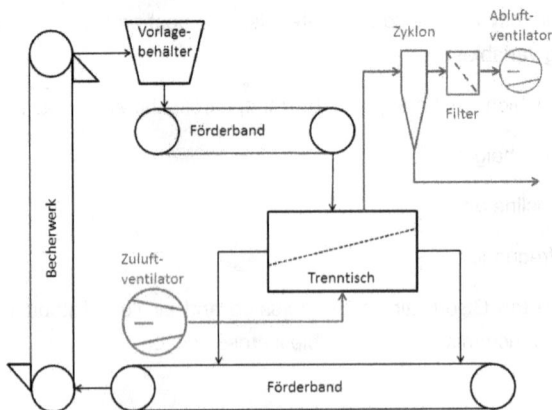

Figure 2: Flow chart for the continuous pilot plant.

During this testing, the influence of several process parameters on the separation be-haviour of different binary bulk material mixtures of varying particle density and grain size (mixtures of pebbles (2.6 g/cm³, magnetite (4.9 g/cm³), shot peening pel-lets (7.0 g/cm³), plastic (~1.0 g/cm³) and glass (~2.6 g/cm³)) was analysed. The pa-rameters include the frequency of table vibration, the table incline, air speed that can be regulated by zone, the overflow flap height (on the light and heavy material side) and different ba-ses.

Besides just reviewing the impact of process parameters on separation material, at-tention was paid to making these parameters are sturdy, and easy and simple to ad-just when used in practice. As outlined in the introduction, the goal was to optimise the effi-ciency, reliability, maintenance effort, transportability, space needs, user friendliness and design of the existing technology.

The findings are illustrated below based on the example of a blend of magnetite (4.9 g/cm³) and pebbles (2.6 g/cm³). The mixture has a particle size range of 0-25 mm. While it is not a waste material, the results can be replicated for other free-flowing bulk solids with a good level of quality. A decision was made to forgo the customary use of separation curves (Bunge 2012) when presenting the results for reasons of practicality and comprehension. Instead, the measured process parameters are illus-trated directly.

2.2.1 Influence of air flow

Inflow velocity has to be optimised for bulky solids that need to be separated, re-gardless of vibration frequency, in order to achieve good fluidisation of bulky solids. Fig-ure 3 illustrates partial mass flows for the separated lightweight and heavy mate-rial as well as their absolute density after the separation process. It is evident that a narrow, optimal speed range of 5.5-6.5 m/s yields the best separation results

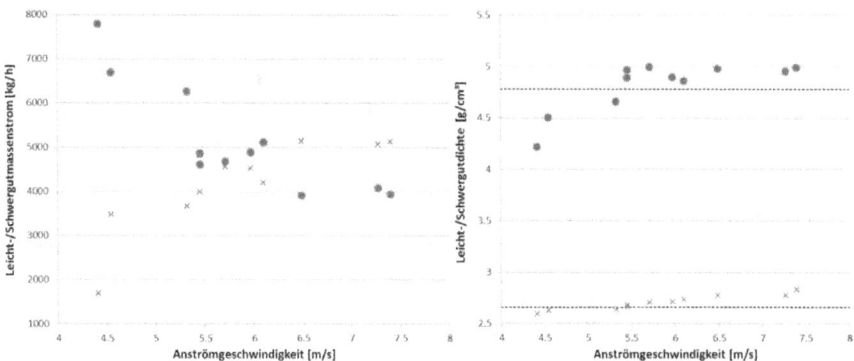

Figure 3: Impact of average air-flow speed on partial mass stream and the absolute density of lightweight and heavy material. Mixture ratio of magnetite to pebbles= 1:1, working surface: 0.45 m². Dots: Heavy material, crosses: lightweight material, dotted line: pure material density. Other process parameters are confidential.

This insight is in principle nothing new. However, it is important that this high through-put, based on the working area, can only be achieved if the base is divided into several zones (in this case, five zones). As shown in Figure 4, separation behaviour leads to a counterflow between the light material (to the left) and heavy material (to the right). The composition thus constantly changes along the base, meaning that different inflow velocities are needed in the zones. An efficient mechanism that can be easily adjusted was developed to this end.

Figure. 4: A look inside the GSort in operation.

2.2.2 Influence of vibration frequency

Besides air flow velocity, the frequency of table vibration also has a decisive impact on separation behaviour. Vibration frequency has an impact on two effects: fluidisation and transportation of the heavy material against gravitation to the upper discharge chute. If vibration frequency is not high enough, the bulky material is not adequately fluidised, lightweight material moves upward and not enough heavy material is transported, more of which slides downward. If vibration frequency is too high, though, too much heavy material is transported upward, bringing lightweight material along with it. Therefore, both of these extreme cases should be avoided and an optimum balance found. In Figure 5, for instance, this optimum is between 300 and 400 1/min. This range is pretty wide in most cases, depending on in-flow velocity.

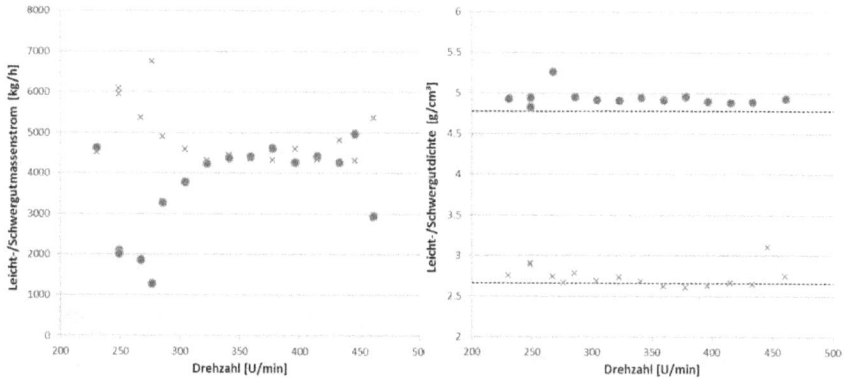

*Figure. 5: Impact on vibration frequency on partial mass stream and the absolute density of the lightweight and heavy material. Mixture ratio of magnetite to pebbles= 1:1, working surface: 0.45 m². **Dots**: Heavy material, **crosses**: lightweight material, **dotted line**: pure material density. Other process parameters are confidential.*

Another key insight of these analyses is that the separation process can be conveniently controlled through vibration frequency in the event of a change in the composition of the feedstock (with the same material to be separated).

2.2.3 Influence of table incline

While vibration frequency and air speed primarily influence fluidisation and heavy material transportation, table incline is primarily a key process parameter for optimal slide of the lightweight material. Since lightweight material floats over the heavy material because of fluidisation, it slides downward because of the plate's incline. Heavy materials, which come into contact with the plate more often statistically, are transported upward. The table incline should thus be optimised for every material so that a high enough driving force occurs in the direction of the bottom discharge chute.

2.3   Typical overall process

When developing the GSort, attention was paid to making sure that it could be integrated into the overall portfolio. As a system provider, Allgaier Group offers system services using drying, screen, sorting, washing and cooling technology.

Figure 6 shows an example of a recycling process in which cullet, for instance, can be separated from concomitant materials and sorted by colour. The process starts with pre-classification of very coarse components using a bar sizer. Material to undergo processing can then by dried in a drum dyer and then sorted into several fractions using a screen machine (MSizer). Classified cullet can then be separated from concomitant materials using the new GSort air-separation table. In the final phase of the process, glass is then separated into components of different colour with a colour sorter (MSort) for further processing.

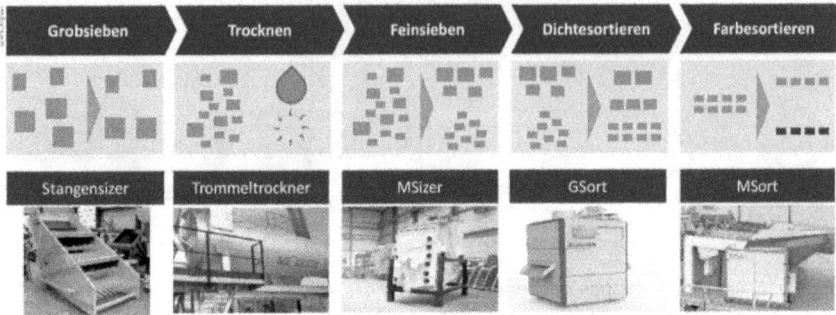

Figure 6: Sample recycling process at Allgaier Process Technology GmbH.

3   CONCLUSION AND SUMMARY

The latest developments on waste recovery and minerals and mining markets are placing increasingly stringent requirements on separation technology. Moreover, there is a trend towards more environmentally friendly techniques and towards investing in and operating lower-priced dry separation processes. With the new GSort air-separation table, Allgaier has developed a particularly efficient method for dry density sorting.

Modifications to a separation table that has proven its worth in-house for decades (GOSAG) were systematically undertaken and their impact on the separation process examined during development work. The new separation table gives customers the ability to make their separation process much more efficient than comparable products on the market through new configuration options (table incline, vibration frequency, air speed (by zone), overflow flap height, work surface).

The impact of these 'new' process parameters was analysed for different material mixtures of varying particle sizes. In particular, the GSort allows customers to sort heavy products (e.g. slag and ore) efficiently based on the criterion of particle density using an air-separation table.

4   Literature

Bunge, R. (2012) Mechanische Aufbereitung – Primär- und Sekundärrohstoffe. Wiley-VCH, Weinheim. ISBN: 978-3-527-33209-0.
Harder J. (2013) Chancen – Entwicklung bei Trockenverfahren. AT MINERAL PROCESSING EUROPE Issue 07-08/2013.
Harder J. (2014) Wassermanagement – Wasserknappheit beim Mining. AT MINERAL PROCESSING EUROPE Issue 09/2014.
Wills, B.A. & Finch J.A. (2016) Wills' Mineral Processing Technology – An Introduction to the Practical Aspects of Ore Treatment and Mineral Recovery. Elsevier, Waltham (USA). ISBN: 978-0-08-097053-0.

Waste glass reuse in geopolymer binder prepared by combining fly ash and metakaolin

Abdelhadi BOUCHIKHI[1,2], Mahfoud BENZERZOUR[1], Nor-Edine ABRIAK[1], Walid MAHERZI[1], Yannick MAMINDY-PAJANY[1]

[1]IMT Lille Douai, LGCgE-GCE, Douai, France

[2]Université de Lille-1, France

Abstract

In this work, waste glass (WG) was activated by alkaline solution to liberate some elements contained in glass and specifically silicium, which is present in more quantity in WG. The formation of geopolymere binder N-A-S-H requests combinations with aluminosilicate source, for example Metakaolin (MK) or Fly Ash (FA). The development of N-A-S-H can be analysed by several chemical and physical methods. The experimental approach of this study has been divided into two parts. The first stage is to characterize the raw materials by helium pycnometer (absolute density), gas adsorption method (specific surface), X-ray crystallography and X-ray Fluorescence (chemical composition of materials) and Optical Emission Spectroscopy (ICP-OES) to quantify the major element present in the liquid phase after the activation of WG. The second step is dedicated to the formulation of mortar 4*4*16 cm and determination of mechanical properties of the matrix. This step allows too to follow the degree of reactivity between the sodium silicate source (waste glass treated) and the aluminosilicate source (MK and FA). Mechanical performances and porosity accessible to water were performed at 1, 7 and 28 days of curing. Study was completed with MEB observation and Fourier Transformation InfraRouge (TFIR) spectra.

Keywords : waste glass packaging, alkaline activation, geopolymerization, mechanical properties, accessible waste porosity

1   Introduction

The reduction of natural resources and the accumulation of solid wastes affect both the current and future generation. The exponential need of raw materials in the industry and the ecological problem caused by wastes involve working on the use of waste in alternative materials. Many studies focus on the valorisation of natural and industrial wastes in several applications, and specifically in cementitious matrix and concrete applications (Soliman 2017)(Tam 2006)(De Larrard 1994). The durability and the technical performances remain critical points for the validation of materials (according to standards and regulations). The use of solid waste for concrete applications leads to interesting results but the nature of waste has an important impact for the route of valuation chosen. Indeed, in some case, specific treatment are required before use of waste.

The valorisation of solid wastes by geopolymerisation and alkaline activation is one of the main routes of stabilisation and reuse of waste in the construction matrix due to the advantage in terms of energy and the ecological positive impact compared to Ordinary Portland Cement (Benhelal 2013).

This paper deals with the valorisation of waste glass packaging in cementitious matrix by geopolymerisation reaction. This study is based on the characterisation of materials and activation of the glass structure to liberate silicium and make it more reactive. The attack efficacy has been controlled by ICP analysis. The reactivity of treated waste glass is determined by use in mortars and addition of aluminosilicate sources MK and FA.

2   Experimental program

The experimental program is divided into three parts. The first one is to characterize from a physico-chemical point of view all the materials used. In the second step, WG is activated by alkaline attack and free elements dissolved are quantified by ICP-OES. Finally, the objective of the last step is to study the effect of alkaline activation of WG in the solidification of mortars with MK and FA as aluminosilicate sources.

2.1   X-ray diffraction analysis of WG

Mineralogical characterization has been realised by X-ray diffraction analysis (XDR). The analysis parameters were 40Kv and 40mA, with samples in powder form. The results presented in Fig.1 show that WG has a large part of amorphous phase with the presence of crystallisation peaks compatible with the quartz phases. Other peaks are present but their nature is difficult to identify because many chemical elements have common peaks.

The presence of crystallized sites means that these sites are not available or not reactive. This observation is the subject of comparisons between waste glass with and without impurities (crystallized quartz) (Bouchikhi 2019).

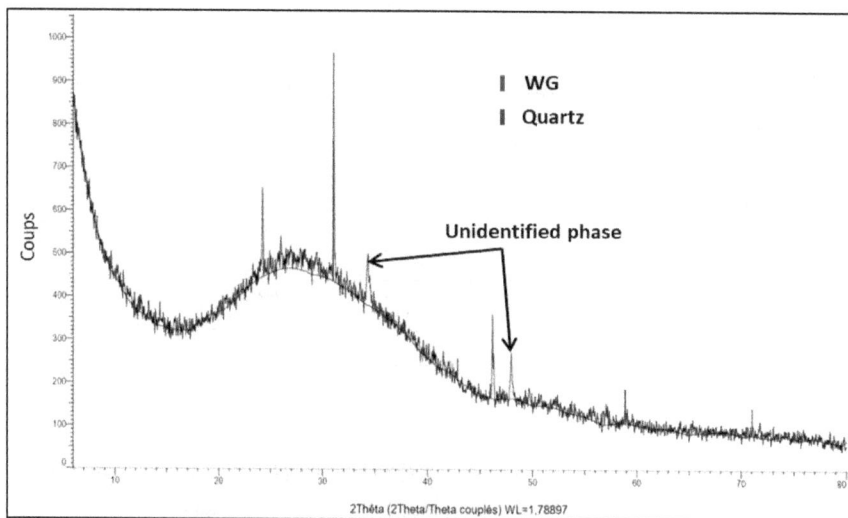

Fig.1 : XRD-Analysis of WG

2.2 Particle-size distribution of Samples

The particle-size-distribution of MK, FA and WG (Fig.2) is determined by dry route on a laser diffraction particle-size analyzer Beckman Coulter's LS13320. In order to give more reactivity to WG, the parameters selected was d_{50} = 6.3µm and d_{90}=16µm. For MK and FA, the size distribution used was not changed (used as received) with d_{50} = 15.7µm) and d_{90} = 53 µm for MK and d_{50} = 34µm and d_{90} = 115µm for FA.

This large difference of particle-size distribution between MK and FA has an impact on the packing density of mixture and directly on the mechanical properties (Sedran 1994) (Thiery 2005). That's why mechanical performances are presented as a function of time of cure. There is no comparison between the two sources of aluminosilicates.

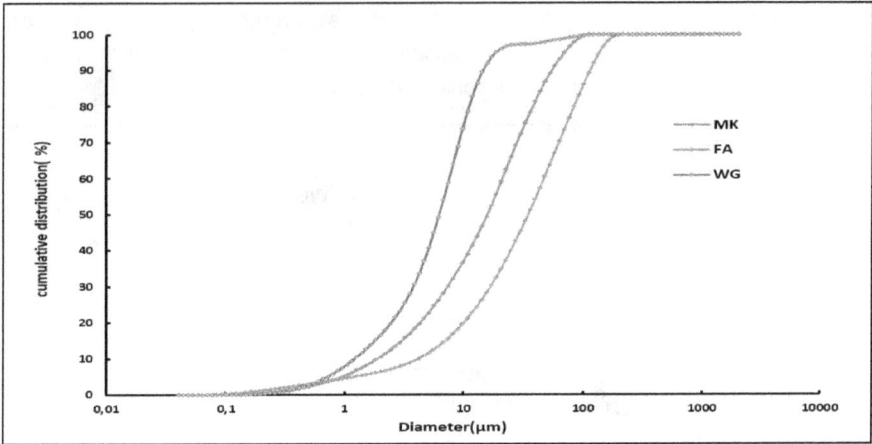

Fig.2 : Particle-size distribution of WG, MK and FA

2.3 Physical and chemical analysis

Chemical composition of MK, FA and WG was determined by x-ray Fluorescence(XRF) analysis. The data given in Table 1 show that WG contains an higher quantity of SiO_2 (71%), CaO (12.4%) and Na_2O. These mineral oxides are the basic elements for the glass structure. Other elements are also present in the glass, but in lower content, in order to strengthen the structure (such as Al_2O_3 and FeO_3) or to give a specific coloration by using the transition elements (TiO_2, Fe_2O_3, CrO…).

Chemical analysis of the aluminosilicates sources (MK and FA) confirms that they are rich in Al and Si. MK contains 74% SiO_2 and 22% Al_2O_3 with the existence of a low percentage of Fe_2O_3, K_2O, CaO and MgO. FA sample have 50% SiO_2, 25.3 Al_2O_3 and a non negligible part of Fe_2O_3, CaO, MgO and K_2O. This different chemical composition between MK and FA can influence their reactivity in geopolymer matrix. This factor is added to the other that will affect the nature of the hydrated gel formed (J. Davidovits 1991). The presence of Ca^{2+} modifies the reaction environment, which leads to the apparition of $Ca(SiO_{5/2})_2$ phases, more stable than $Na(SiO_{5/2})$ according to the steps herein below:

- Intermediate phases between Na^+ and ($SiO_{5/2}$):

$$(SiO_{5/2})^- + Na^+ \longrightarrow Na(SiO_{5/2}) \qquad (1)$$

- Reaction between $Na(SiO_{5/2})$ and Ca^{2+}

$$2SiO_{5/2}Na + Ca^{2+} \longrightarrow Ca(SiO_{5/2})_2 + 2Na^+ \quad (2)$$

Table 1 : Physical and chemical analysis of samples MK, FA and WG

	Elementary chemical analysis									Physical properties		
Sample	Na₂O	MgO	Al₂O₃	SiO₂	K₂O	Fe₂O₃	CaO	P₂O₅	SO₃	Sp (g/cm^3)	d (cm^2/g)	d₅₀ (μm)
WG	13.2	1.8	1	71.1	0.4	0.2	12.4	0.1	-	2.53	851	6.3
MK	-	0.3	22.4	73.7	0.2	1	1.3	0	1.1	2.54	792.6	15.7
FA	0.5	1.3	25.3	50.5	3.7	7.6	2.7	-	-	2.26	1143.7	35.6

3   Activation of the glass structure by alkalinity reaction

Treatment of WG by sodium hydroxide 10M leads to deconstruction of glass structure and liberation of the major elements Si, Ca, Al and Na which reorganize themselves in other phases very soluble in water. This new structure brings more reactivity for these elements and precisely for Si, Al, Ca and Na in the geopolymerisation. The program for heat treatment was chosen according to the study of Torres and al. [Torres 2014], study in which treatment is made at 90°C during 6 hours and d₉₀ of powder is lower than 45μm. In order to reduce the duration of treatment from 6 to 4 hours, d₉₀ was fixed to 16μm. Table 2 presents the results of leaching test realised in 10M of NaOH at 90°C (WG10M90°C) compared to WG leached in 0M of NaOH at 22°C (WG0M22°C). These results show the efficacy of treatment by sodium hydroxide.

The role played by silicates available is indispensable in the geopolymeration of alumi-nosilicate ($Si_2Al_2O_2$ (MK or FA) in presence of water (H_2O) and sources of alkaline (Na^+ or K^+). The following reaction shows the global mechanism of geopolymer reaction:

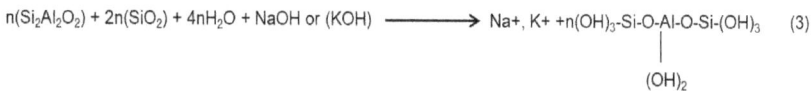

$$n(Si_2Al_2O_2) + 2n(SiO_2) + 4nH_2O + NaOH \text{ or (KOH)} \longrightarrow Na+, K+ +n(OH)_3\text{-Si-O-Al-O-Si-}(OH)_3 \quad (3)$$
$$\mid$$
$$(OH)_2$$

The presence of calcium (Ca^{2+}) available after the dissolution of glass favours the condensation as well as strengthens the gel of geopolymer. Other elements can play the same role like Mg^{2+} (Joseph Davidovits 2015).

Table 2: Leaching test after treatment of WG

(mg/kg)	SiO₂	Al₂O₃	CaO	Na₂O
WG0M22°C	3	2	0	3
WG10M90°C	172270	4513	280	Excess

4   Formulation

In this part, mechanical comparisons were made on prismatic samples 4*4*16 cm. Mortars were realised considering a binder formed by MK or FA with add of WG10M90°C as activator. B = (MK+ WG10M90°C) or B = (FA + WG10M90°C). The sand used is normalized according to CEN 196-1 ISO standard. The particle size distribution is determined by sieving: 0.08-2 mm with D_{50} = 0, 9 mm. Mixes were performed according to standard NF EN 196-1. Table 3 presents the compositions of mortars.

Table 3 : Composition of mortars

Designation	Quantity of materials (g)				
	MK	FA	Water	Sand	WG10M90°C
M-MK	450	0	160	1750	100
M-FA	450	450	160	1750	100

4.1 Mechanical properties

Mortars undergo a cure at 50°C in order to accelerate the reaction and avoid cracking (Khale and Chaudhary 2007). Their mechanical properties were determined after 1, 7 and 28 days of cure. Compressive strength test was realised an INSTRON press of 15 tons with a load speed of 144KN/min. Fig. 3 shows the evolution of compressive strength as a function of time of cure.

The evolution of the compressive strength between 7 and 28 days goes from 29 to 33, and from 31 to 36MPa for M-MK and M-FA respectively. This increase of compressive strength shows that the main part of solidification has been realised after 7 days of cure at 50°C (Khale 2007). The comparison between the two types of alumino-silicate sources MK and FA proves the similar reactivity of the both sources of Al/Si.

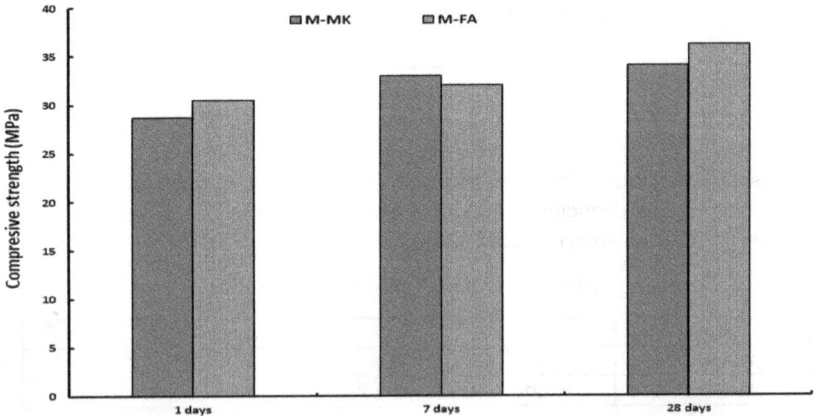

Fig.3 : Compressive strength of M-MK and M-FA mortars at 1, 7 and 28 days of cure

4.2 Porosity accessible to water

The evolution of the porosity accessible to water (Fig. 4) is inversely proportional to the evolution of the development of the mechanical properties of mortars. A large decrease in porosity was noticed between 1 day and 7 days with a drop from 27% to 17%. This observation remains compatible with the mechanical evolution. Between 7 and 28 days, small change has been noticed. The development of the geopolymer network leads to the reduction of the mortar porosity, which has an influence on the absorption of water. It can also be noted a decrease of porosity accessible to mercury and air according to the literature (Torres-Carrasco 2015). These results of porosity accessible to water are comparable to those presented in literature for cementitious mortar formed by use of Ordinary Cement Portland.

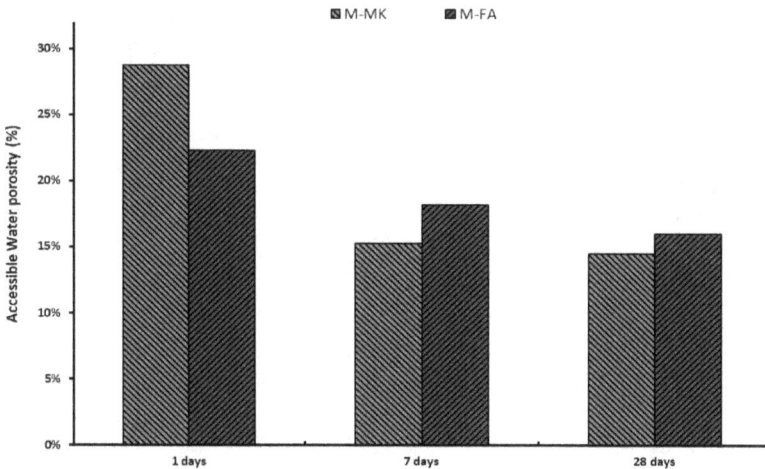

Fig.4: Accessible water porosity (%)

5   Conclusion

The use of residual waste glass as a source of silicate activators is one of the avenues contemplated for the valorisation of WG. This study shows the advantage of a treatment of WG with sodium hydroxide in order to liberate the elements required for the geopolymerisation, precisely Si and Ca. The use of alumino-silicate sources as MK and FA is necessary to obtain molar ratios Si/Al, Si/Na and Na/H2O adapted to geopolymer reaction. Mechanical results show the benefit of WG treated in the development of geopolymers as well as the evolution of porosity accessible to water reduced to 17%. This level of porosity remains comparable with the porosity accessible to water of cement-based mortars which gives a vision on durability of these materials, including their freeze-thaw behaviour.

6   References

Benhelal, Emad, Gholamreza Zahedi, Ezzatollah Shamsaei, and Alireza Bahadori. 2013. "Global Strategies and Potentials to Curb CO2emissions in Cement Industry." *Journal of Cleaner Production* 51. Elsevier Ltd:142–61. https://doi.org/10.1016/j.jclepro.2012.10.049.

Bouchikhi, Abdelhadi, Mahfoud Benzerzour, Nor-Edine Abriak, Walid Maherzi, and Yannick Mamindy-Pajany. 2019. "Study of the Impact of Waste Glasses Types on Pozzolanic Activity of Cementitious Matrix." *Construction and Building Materials* 197. Elsevier Ltd:626–40. https://doi.org/10.1016/j.conbuildmat.2018.11.180.

Davidovits, J. 1991. "Geopolymers - Inorganic Polymeric New Materials." *Journal of Thermal Analysis* 37 (8):1633–56. https://doi.org/10.1007/BF01912193.

Davidovits, Joseph. 2015. *Geopolymer Chemistry and Applications.* Edited by Geopolymer Institute.

Khale, Divya, and Rubina Chaudhary. 2007. "Mechanism of Geopolymerization and Factors Influencing Its Development: A Review." *Journal of Materials Science* 42 (3):729–46. https://doi.org/10.1007/s10853-006-0401-4.

Larrard, F. de, and T. Sedran. 1994. "Optimization of Ultra-High-Performance Concrete by the Use of a Packing Model." *Cement and Concrete Research* 24 (6):997–1009. https://doi.org/10.1016/0008-8846(94)90022-1.

Sedran, T, and F de Larrard. 1994. "RENE - LCPC - Un Logiciel Pour Optimiser La Granularité Des Matériaux de Génie Civil." *Bulletin Des Laboratoires Des Ponts et Chaussées* 194:87–93.

Soliman, Nancy A., and Arezki Tagnit-Hamou. 2017. "Using Glass Sand as an Alternative for Quartz Sand in UHPC." *Construction and Building Materials* 145. Elsevier Ltd:243–52. https://doi.org/10.1016/j.conbuildmat.2017.03.187.

Tam, Vivian W.Y., and C. M. Tam. 2006. "A Review on the Viable Technology for Construction Waste Recycling." *Resources, Conservation and Recycling* 47 (3):209–21. https://doi.org/10.1016/j.resconrec.2005.12.002.

Thiery, M, G Platret, E Massieu, G Villain, and V Baroghel-Bouny. 2005. "Un Modèle d'hydratation Pour Le Calcul de La Teneur En Portlandite Des Matériaux Cimentaires Comme Donnée d'entrée Des Modèles de Carbonatation." *Journées Ouvrages DArt Du Réseau Des LPC.* http://media.lcpc.fr/ext/pdf/sem/2005_joa_er6.pdf.

Torres-Carrasco, M., and F. Puertas. 2015. "Waste Glass in the Geopolymer Preparation. Mechanical and Microstructural Characterisation." *Journal of Cleaner Production* 90. Elsevier Ltd:397–408. https://doi.org/10.1016/j.jclepro.2014.11.074.

www.ingramcontent.com/pod-product-compliance
Lightning Source LLC
Chambersburg PA
CBHW060759220326
41598CB00022B/2488